Introduction to Environmental Science and Technology

Introduction to Environmental Science and Technology

Gilbert M. Masters

University of Santa Clara
Santa Clara, California

Stanford University
Stanford, California

JOHN WILEY & SONS

New York London Sydney Toronto

to Laurie, Val, and Guy

This book is printed on recycled paper.

Library of Congress Cataloging in Publication Data:

Masters, Gilbert M.
Introduction to environmental science and technology.

Includes bibliographies.
1. Pollution. 2. Human ecology. 3. Power resources. I. Title. [DNLM: 1. Air pollution. 2. Environmental health. 3. Sanitary engineering. 4. Water pollution. WA670 M423i 1974]

TD174.M38 301.31 73-23088
ISBN 0-471-57607-7

Printed in the United States of America

10 9 8 7 6 5 4 3 2 1

Preface

It would be difficult to overemphasize the importance of population, pollution, and resource depletion as primary factors which limit the range of probable alternative futures still available to mankind. As we enter the last quarter of the twentieth century, it is becoming increasingly necessary for policy makers to become aware of these environmental constraints and to include them in their decision-making processes.

The first environment books were largely nontechnical in nature and served the purpose of increasing the public's awareness of our worsening situation. However, it is becoming apparent that to convert that awareness into action requires a more quantitative understanding of our environmental problems, and to meet that need universities across the country are introducing environment courses of one sort or another, frequently in their engineering or physics departments.

It has been my experience in teaching such courses at both the University of Santa Clara and Stanford University, that one of the most challenging aspects of the problem is to find a way to present technical material to an audience composed of students from every department of the university. I have found that the best approach is to rely heavily on graphs, tables, and illustrations, and to use equations only infrequently. That has been the approach taken in this book.

The book is divided into four parts. In the first, an introduction to some basic but important principles of ecology is presented, followed by an analysis of population growth and a discussion of some of the problems encountered in trying to provide an adequate food supply for our growing population.

In the second part, water pollution is discussed, with an emphasis on how man's activities can overload the natural ability of a body of water to purify itself. The importance of proper water and sewage treatment to public health is described along with the principles of operation of modern treatment plants.

In Part III, the problems associated with air pollution are presented with an emphasis on the five principal types of emissions. A chapter on air pollution meteorology is included in which the processes which cause the formation of inversion layers are described along with the possible effects of air pollution on climate. The technology of air pollution control, with an emphasis on the automobile, is included.

Part IV discusses energy and raw materials, the essential ingredients of our technological world. While the population and pollution aspects of our environmental problems have been generally accepted, it is only just now that the importance of resource depletion is being acknowledged. The problem is treated as a combination of growing demand, dwindling reserves, and increasing environmental disruption associated with the acquisition and use of these resources. Considerable attention is given to the new and future sources of electric energy: nuclear fission, fast-breeder reactors, and geothermal, solar, and fusion power.

The material has been written so as to be understood by anyone having a fairly simple science background which, ideally, would include a little chemistry, biology, physics, and mathematics (calculus is not necessary), although none of these is essential. At the end of each chapter, questions have been included; some of these are qualitative, but the majority are reasonably straightforward numerical calculations.

There have been a number of people whose assistance in the preparation of this material I gratefully acknowledge. Most especially, I would like to thank John Randolph of Stanford University for his many helpful criticisms and suggestions, as well as Professors Alfred W. Hoadley of the Georgia Institute of Technology, Alonzo W. Lawrence of Cornell, and Dragoslav M. Misic of California Polytechnic State University, San Luis Obispo. I also take pleasure in acknowledging Professor Shu Park Chan of Santa Clara for arranging the financial support for this project through a grant from the Sloan Foundation.

Santa Clara, California *Gilbert M. Masters*

Contents

Part IV
Energy and Raw Materials

Part I

Ecology and Population

Chapter 1

Some Principles of Ecology

Relative to the history of all forms of life on earth, man is a recent phenomenon. Perhaps 2 billion years ago, algae were already releasing oxygen into the atmosphere through photosynthesis, thus helping to create the conditions that were necessary for the higher forms of life which followed. Man appeared much later, somewhere around 2 million years ago, but his numbers and his effect on the biosphere remained relatively small until the invention of agriculture some 8000 years ago. We can use that date as the beginning of man's manipulation of nature, the unwanted side effects of which, are the subject of most of this book.

Before proceeding to the study of man's impact on the earth's environmental systems, we need to consider how those systems normally function; that is, we need to study *ecology*. The word *ecology* was coined about 100 years ago from the Greek *oikos*, which means "house," and has come to mean the study of "the totality or pattern of relations between organisms and their environment" (Webster's Unabridged Dictionary).

1.1 *Ecosystems*

There are several important terms which need to be defined before proceeding. A group of individuals of one kind of organism is called a *population* and all of the populations living in a given area form a *community*. The community and the nonliving environment with which it interacts is an *ecosystem*. And if we consider all of the earth's living organisms interacting with the physical environment as a whole, then we are talking about the *ecosphere* or *biosphere*.

Ponds, streams, meadows, and forests are typical ecosystems that are frequently studied, but for now, let us examine the major components that exist in any ecosystem and see how they operate together.

The initial source of all of the energy used by an ecosystem is the sun. Green plants capture solar energy during photosynthesis and store it in chemical form for subsequent use by the plant itself or by any other organism that consumes the plants. The plants are called *producers* or *autotrophs* (self-feeding) while organisms that feed on the plants or on each other are called *consumers* or *heterotrophs* (other-feeding). The consumers can be further subdivided as shown in Figure 1.1. This linear description of feeding relations is called a *food chain* and each level is called a *trophic* level. We will see later that at each trophic level so much energy is converted to heat and lost that food chains seldom have more than four or five levels. A simple four-level food chain might be

$$\text{plant} \rightarrow \text{insect} \rightarrow \text{frog} \rightarrow \text{man}$$

In addition to the producers and consumers, the biotic (or living) portion of an ecosystem must have *decomposers*. The decomposers, which are mainly bacteria and fungi, derive their energy from the waste products and dead organisms in the ecosystem, and in doing so perform the invaluable service of converting complex organic molecules into simple inorganic forms for reuse by the producers.

Figure 1.2 shows the movement of energy and nutrients (nitrogen, phosphorus, sulfur, etc.) through an ecosystem. There are two fundamental points which should be emphasized here. The first is that nutrients move through the ecosystem in a cyclic fashion. As they do so, they are not degraded in any way and hence can be recycled over and over again. Later in this chapter we will trace the movement of several of the most important nutrients through the biosphere in what are called *biogeochemical cycles*. The second point is that energy flow through the ecosystem is *not* cyclic and is in effect a one-way process. Energy is employed by organisms to fuel their movements and growth and is subsequently lost as heat. Thus energy must be supplied continuously by the sun to keep

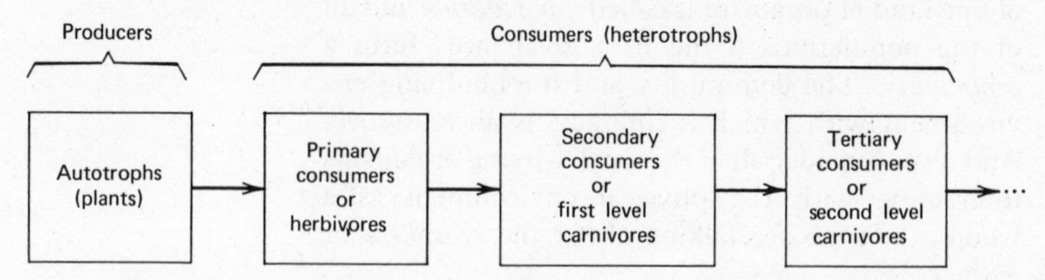

FIGURE 1.1 Various designations for each of the trophic levels in a food chain.

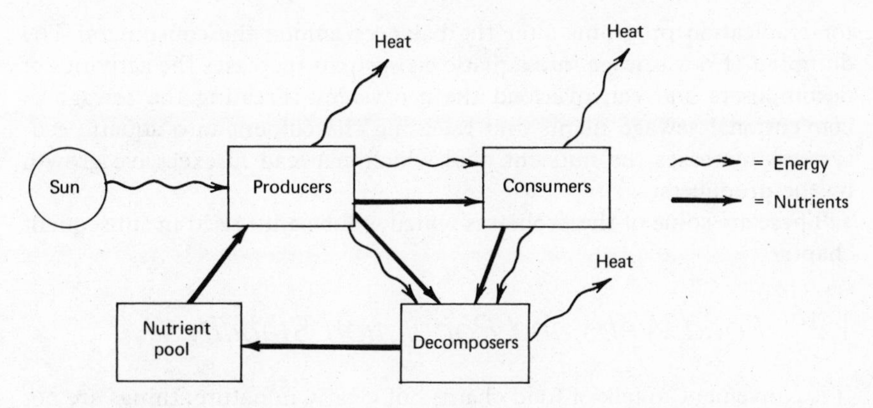

FIGURE 1.2 Nutrients move cyclically through the ecosystem but energy must constantly be renewed by the sun.

the system operating. Both of these important concepts will be discussed in detail in succeeding pages.

Figure 1.3 illustrates some of the many ways in which man's activities can upset the balanced operation of an ecosystem. Air pollution not only damages plants but also reduces the amount of sunlight reaching them. On the other hand, combustion of fossil fuels increases the carbon dioxide in the atmosphere which increases photosynthesis. Pesticides and preda-

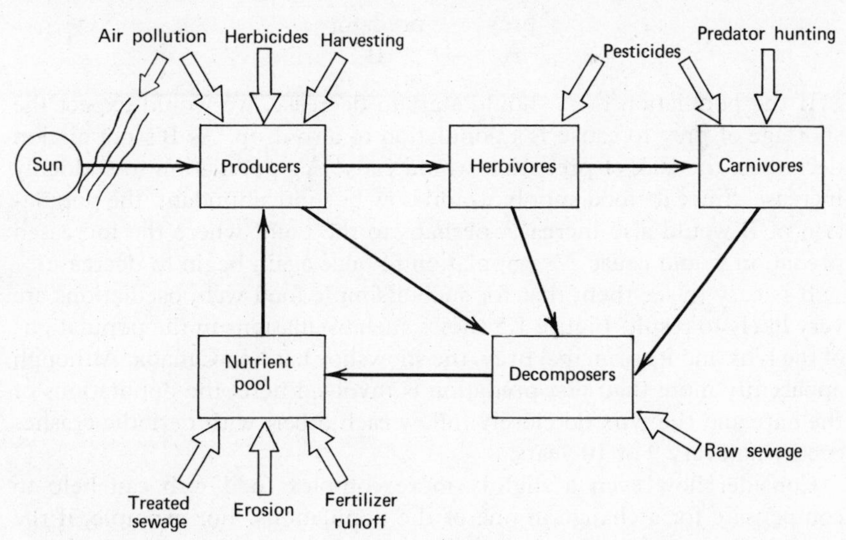

FIGURE 1.3 Some of the ways man's activities can upset ecosystem balance.

tor-eradication programs alter the balances among the consumers. The dumping of raw sewage into aquatic ecosystems increases the activities of decomposers and can overload the ecosystem. Treating the sewage in conventional sewage plants and releasing the effluent into aquatic ecosystems increases the nutrient pool which can lead to excessive growth by the producers.

These are some of the problems which will be addressed in subsequent chapters.

1.2 *Food Webs and Population Stability*

It is convenient to talk of food chains but clearly in nature, things are not so simple. Each link in a food chain interconnects with many other food chains and the complex set of feeding relations that results is called a *food web*, a relatively simple example of which is shown in Figure 1.4.

It is conventional wisdom in ecology to suggest that the stability of a food web is related to its complexity. Complex food webs tend to be stable, that is, the number of individuals in each population tends to remain nearly constant. Conversely, minor perturbations in a very simple food web can cause such major fluctuations in population sizes that species may even be driven to extinction.

Consider, for example, the simple two-species food chain:

$$\begin{array}{ccc} \text{prey} & & \text{predator} \\ A & \rightarrow & B \end{array}$$

If the population of A should start to decrease, we would expect the shortage of prey to cause B's population to also drop. As B's population decreases, the lack of predation would cause A's population to suddenly increase. Since its food supply would now be more abundant, the population of B would also increase, perhaps to the point where the increased predation would cause A's population to once again begin to decrease.

It is easy to see then, that for such a simple food web, oscillations are very likely to result. Figure 1.5 shows such oscillations in the populations of the lynx and its principal prey, the snowshoe hare, in Canada. Although apparently more than just predation is involved here, the populations of the hare and the lynx do closely follow each other, with periodic crashes occurring every 9 or 10 years.

Consider how even a slightly more complex food web can help to compensate for a change in one of the populations. For example, if the

$$\begin{array}{l} A \searrow \\ \quad\quad C \\ B \nearrow \end{array}$$

population of A decreases, then C can switch its eating habits to prey more on B, thus allowing A's population to recover without seriously upsetting the ecosystem balance.

There are many exceptions, but a great deal of evidence exists to substantiate this diversity-leads-to-stability rule, which basically suggests that the more species the food web contains and the greater the number of links connecting the species, the greater will be the stability.

This concept has many practical applications to man. For example, an agricultural ecosystem is extremely simple and hence quite vulnerable. If not carefully managed, it would quickly be overrun by weeds and insects. As man becomes more and more dependent on a few strains of highly productive food crops, the danger of some new pest or plant disease evolving, which could sweep through these monocultures unchecked, increases.

The Irish potato famine in the 1840s, which was caused by a fungus, provides a grim lesson of the consequences of such a dependence. Before the famine hit, the population of Ireland was about 8 million. In the 10 years during and after the famine, more than 1 million died from direct and indirect effects, and 2 million more emigrated (mostly to the United States).

One of the arguments against man's interfering with the ecology of the Arctic, such as would occur in the construction and operation of the Alaska oil pipeline, is that food webs there are rather simple and it would be quite easy to cause major disruptions with rather minor actions.

1.3 *Biogeochemical Cycles*

Although the list of chemical elements required by each organism to maintain life is quite lengthy, the five elements, oxygen, carbon, hydrogen, nitrogen, and phosphorus, account for more than 97% of all protoplasm. It is possible to trace the cyclic movement of these essential chemical elements through the biosphere and by so doing, much can be learned about the interdependencies which exist. Biological, geological, and chemical systems are usually involved and the cycles are called biogeochemical cycles. We shall consider the biogeochemical cycles for nitrogen and phosphorus in some detail as they will be important in our study of water pollution. The cycles for carbon dioxide, water, and oxygen are presented in a somewhat simplified form in Figure 1.6.

The water cycle, which will be discussed further in Chapter 4, involves evaporation from open bodies of water and transpiration (evaporation of the water which has passed through plants), followed by precipitation. The movement of carbon dioxide and oxygen is largely determined by

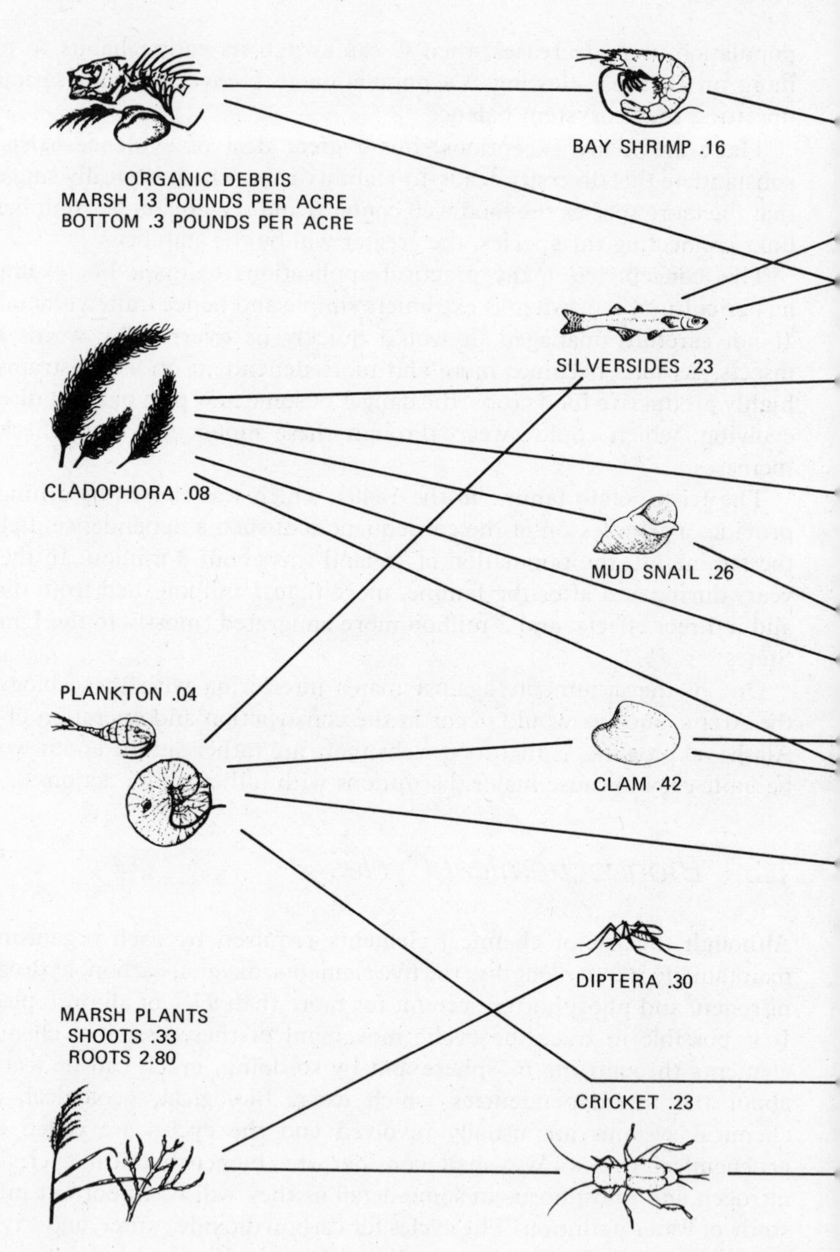

FIGURE 1.4 Portion of a food web in a Long Island estuary. Numbers indicate residues of DDT in parts per million. From Woodwell, G. M., "Toxic Substances and Ecological Cycles."

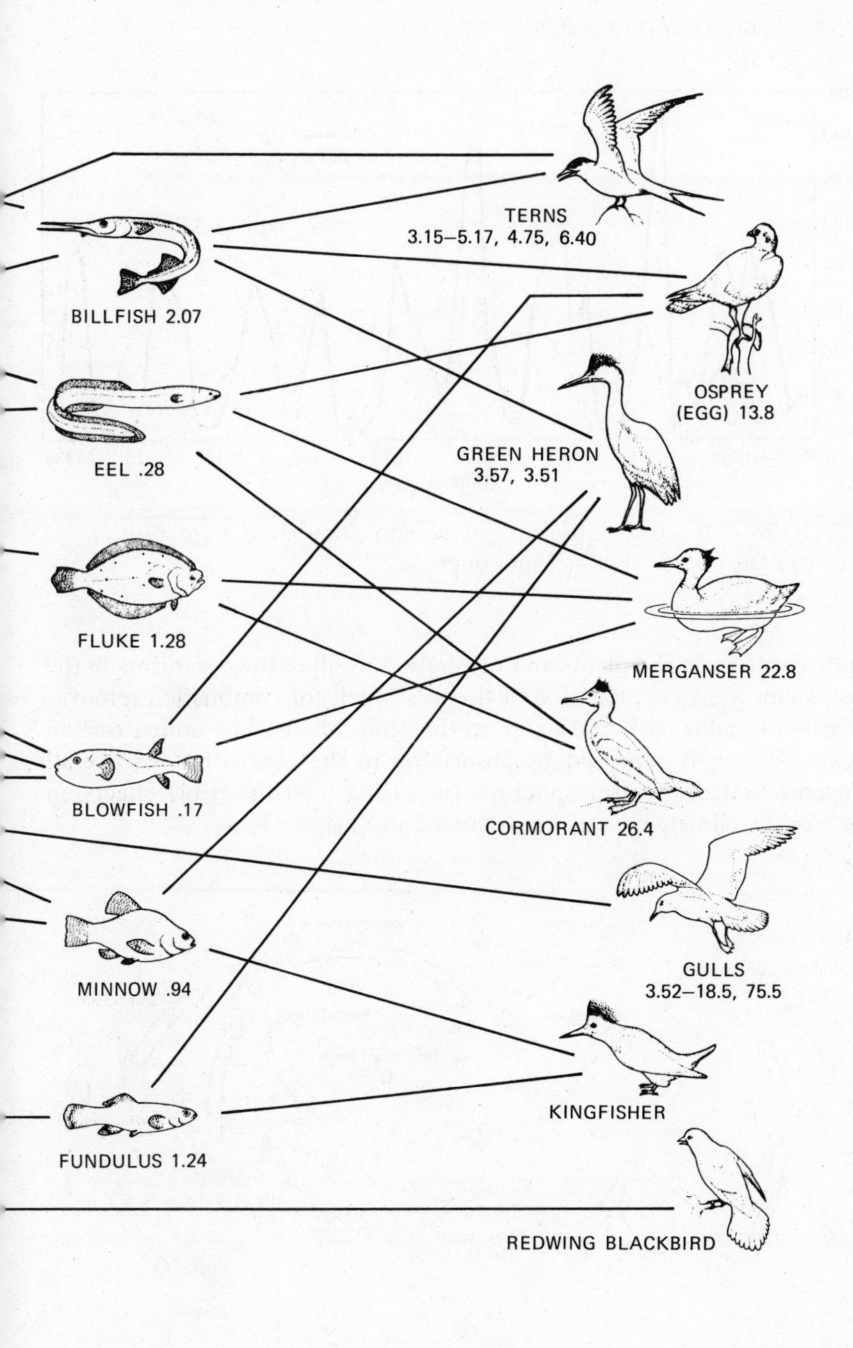
TERNS
3.15–5.17, 4.75, 6.40
BILLFISH 2.07
OSPREY
(EGG) 13.8
EEL .28
GREEN HERON
3.57, 3.51
FLUKE 1.28
MERGANSER 22.8
BLOWFISH .17
CORMORANT 26.4
GULLS
3.52–18.5, 75.5
MINNOW .94
KINGFISHER
FUNDULUS 1.24
REDWING BLACKBIRD

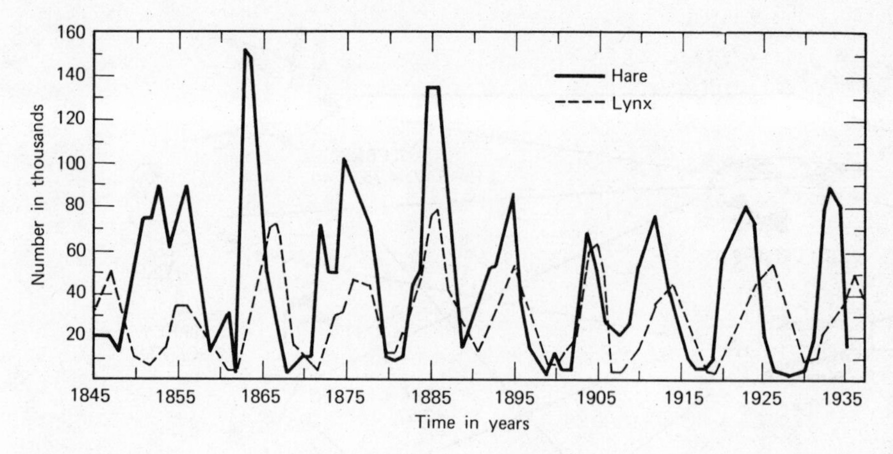

FIGURE 1.5 Oscillations of lynx and snowshoe hare populations (Odum 1971 after MacLulich 1937).

photosynthesis in the plants and respiration by all of the organisms in the ecosystem. Man's exploitation of the fossil fuels for combustion removes oxygen and adds carbon dioxide to the atmosphere. The added carbon dioxide is largely removed by absorption in the oceans but its overall concentration in the atmosphere is increasing—with possible effects on the world's climate as will be discussed in Chapter 9.

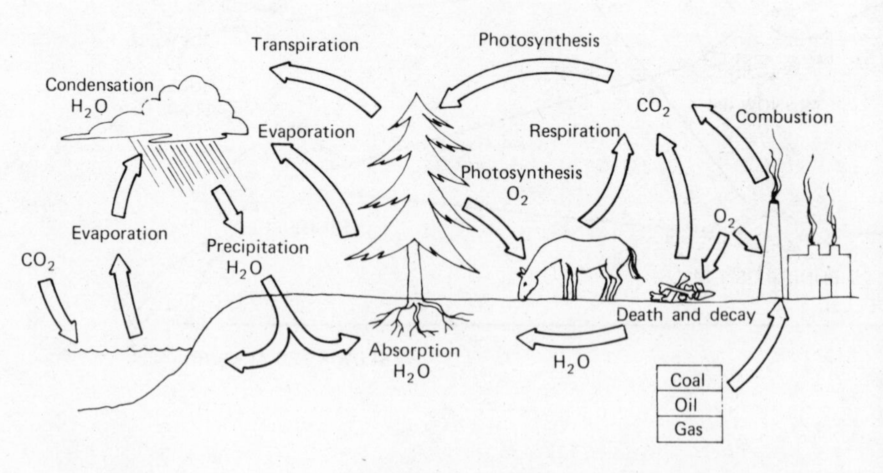

FIGURE 1.6 Water, oxygen, and carbon dioxide cycles in the biosphere.

The Nitrogen Cycle

It is instructive to study the biogeochemical cycle for nitrogen because nitrogen plays such an important role in so many environmental problems. As a nutrient, nitrogen is essential because it is a major component of protein, and lack of protein in the diet is the principal cause of human malnutrition. We shall also discuss the use of nitrogen in artificial fertilizers and see how it contributes to water pollution. Furthermore, when oxidized during combustion, nitrogen forms one of the principal components of photochemical smog.

The atmosphere contains an abundant supply of nitrogen (about 79% of air) but this nitrogen is not in a form which is directly usable by plants. Part of the nitrogen which ends up in plants is "fixed" from atmospheric nitrogen by nitrogen-fixing bacteria infecting the roots of leguminous plants such as clover and alfalfa. These bacteria fix atmospheric nitrogen (N_2) by oxidizing it to nitrates (NO_3) which can be used by the plant. This is an example of *mutualism* wherein the bacteria and plant live together to their mutual advantage. The bacteria receive energy from the plant in the form of carbohydrates and the plants receive a source of nitrogen.

Since most plants do not have this ready source of nitrate, they must obtain it from the soil. However, constant harvesting can deplete the soil unless some mechanism of nitrate replacement is provided, which is the purpose of crop rotation involving legumes. The nitrogen-fixing bacteria associated with legumes provide a natural means of maintaining soil fertility without the use of artificial fertilizers. However, as our population continues to increase, we are less and less likely to allow land to lie fallow and instead are becoming dependent on artificial fertilizers which become principal sources of water pollution.

Notice how normal agricultural practices upset the natural ecological balance of not only the land-based ecosystem where the crops are grown (by removing nutrients at the harvest), but also of the aquatic ecosystem where the food wastes will normally end up (perhaps after passing through a sewage treatment plant). The aquatic ecosystem gets overloaded with nutrients from the land, which can result in undesirable algal "blooms."

Nitrogen can be fixed also by free-living bacteria in the soil or water, or by certain blue-green algae in the water. In the Orient, the naturally occurring blue-green algae in the rice paddies are very important in maintaining soil fertility under intensive cropping.

Figure 1.7 shows the nitrogen cycle. In it we see that nitrates (NO_3), or ammonia (NH_3), are taken up by plants and used in protein synthesis. Decay products and excreta from the members of the food chain get converted to ammonia by bacteria and fungi. Then nitrite bacteria *(Nitro-*

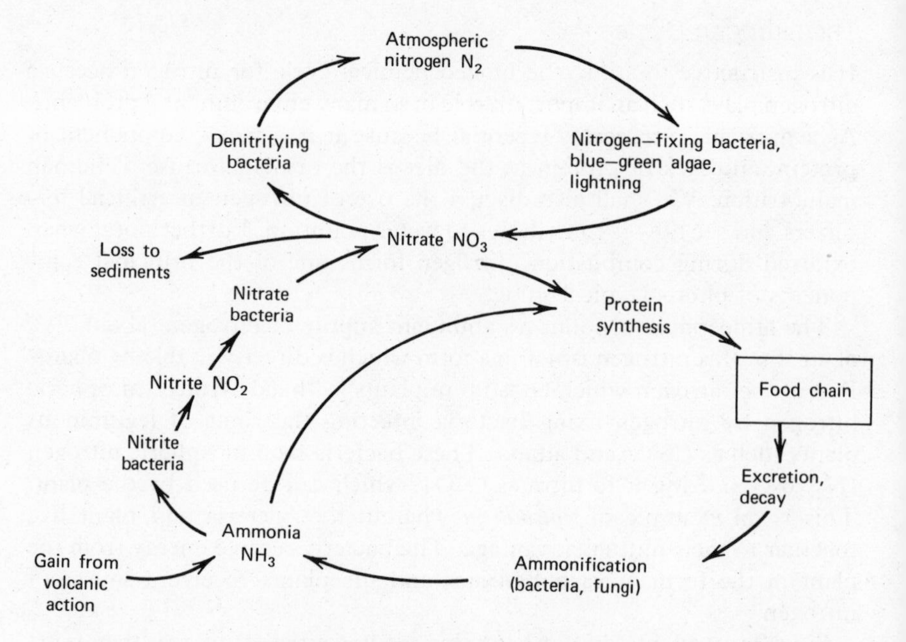

FIGURE 1.7 The nitrogen cycle.

somonas) convert ammonia to nitrite (NO_2), and nitrate bacteria *(Nitrobacter)* convert NO_2 to nitrate (NO_3). Nitrogen returns to the atmosphere when denitrifying bacteria convert nitrate back into molecular nitrogen (N_2). Not shown in Figure 1.7 is the man-made fixation of atmospheric nitrogen into ammonia that occurs in the manufacture of artificial fertilizers. Presently more nitrogen is artificially fixed than is fixed by natural means, but the effects of this new input to the biosphere are unknown.

The Phosphorus Cycle

Phosphorus is another element that is absolutely essential to life. It is required by all life forms in the metabolic processes that provide energy for the cells. Phosphorus will be seen to be a very important nutrient in our discussion of water pollution, as its heavy use by man in detergents and artificial fertilizers can result in excessive plant growth in our waterways.

Whereas the main reservoir for nitrogen was seen to be the atmosphere, the main sources of phosphorus are phosphate rocks and natural phosphate deposits of guano and fossilized animals. The phosphorus cycle is unusual in that it is more of a unidirectional process than a cycle. As shown in Figure 1.8, phosphorus moves from rocks by erosion or mining, into the

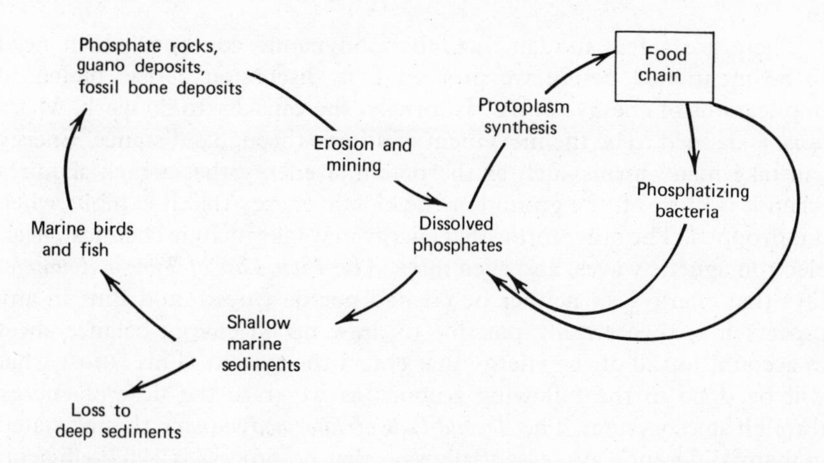

FIGURE 1.8 The phosphorus cycle. Note the loss to deep sediments.

water cycle and food chains where it may cycle for a short period, but eventually it ends up in deep sea sediments where it is lost until geologic activity can uplift it again. Although some of the phosphates in the sea are returned to the land by the excrement of sea birds (guano) or by fish that are caught by man, most of it is lost. Man is accelerating this loss by mining phosphate rock for fertilizers which eventually are washed into the sea.

Since phosphorus is a nonrenewable resource of great importance, it is important to have some idea of how long it may last before reserves are exhausted. If the world continues to consume phosphate rock at the 1970 rate of about 94 million tons per year, world reserves would last about 100 years [U.S. Department of Interior (USDI) 1972]. This must be considered a rather rough estimate in that more reserves may be discovered, but on the other hand our rate of consumption is accelerating which could deplete the reserves much more quickly. If we should ever run out of this vital resource, the effect would be catastrophic.

1.4 *Solar Energy and the Atmosphere*

There were two main points mentioned with regard to Figure 1.2. The first, that of the cyclic nature of the biosphere's use of nutrients, has now been considered. The second point dealt with the unidirectional flow of energy, and that is what will be discussed in the remainder of this chapter.

There are a few fundamental thermodynamic concepts which need to be mentioned before we proceed to a discussion of the biological implications of energy. *Energy* is, briefly, the capacity to do work, where *work* is defined to be the movement of a force through a distance. Energy can take many forms such as the potential energy that a rock acquires when it is lifted off the ground or the kinetic energy that it exhibits when it is dropped. The other forms that energy may take include chemical, heat, electromagnetic waves, and even mass. The *First Law of Thermodynamics* says that energy can neither be created nor destroyed, and thus in any process it is theoretically possible to draw up an energy balance sheet to account for all of the energy that enters the system. This is just what will be done in the following sections as we trace the flow of energy through an ecosystem. The *Second Law of Thermodynamics*, though stated in many different ways, essentially says that no process is 100% efficient, so that more energy must be put into any system than can be extracted as useful work. Thus for example, when a motor is supplied with a certain amount of electric energy, it will always deliver a lesser amount of mechanical energy. The difference is lost as heat from such places as the windings and the bearings. As a final practical point, there are many systems of units used to measure energy, but for now we will be using the *calorie* (cal), which is defined as the amount of energy required to raise 1 gram (g) of water from 14.5 to 15.5°C.

With these preliminaries out of the way, let us begin our consideration of the energy flow through the biosphere with a discussion of solar energy. The sun supplies the energy which is used by essentially all forms of life on earth (the exception being a few kinds of chemosynthetic bacteria which obtain their energy from oxidation of inorganic substances). The sun derives its energy through the process of thermonuclear fusion (the energy source for the hydrogen bomb) wherein mass is converted to energy in accordance with Einstein's relation $E = mc^2$. Hydrogen atoms are fused together in a series of reactions to form helium and the energy released is transmitted through space in the form of electromagnetic waves.

The amount of solar energy arriving at the earth's outer atmosphere per unit area (perpendicular to the rays) and per unit of time, called the *solar flux*, is about 2 cal/cm^2/min. Because of the interaction with the earth's atmosphere not all of this energy reaches the earth's surface, in fact only about half does. About 30% of the incoming energy is reflected back into space, while the remaining 20% is absorbed by the atmosphere (Oort 1970), as is indicated in Figure 1.9.

There are several important comments to be made about Figure 1.9. The first concerns the spectrum shift between the wavelengths of the incoming solar radiation and the wavelengths of the energy that is re-

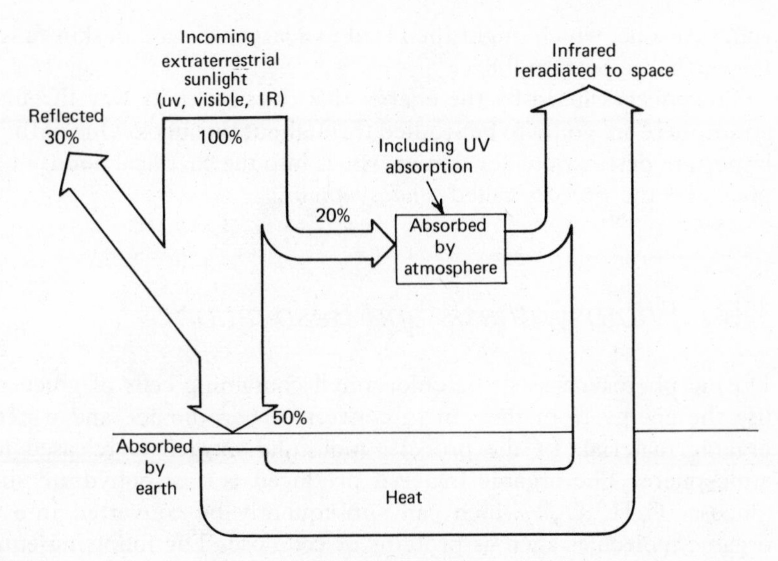

FIGURE 1.9 About half of the solar energy reaches the surface of the earth, is converted to heat, and reradiated back into space.

radiated back into space. About 99% of the incoming solar energy is contained in the portion of the spectrum extending from 0.2 to 4.0 micron (μ) (from ultraviolet, through the visible, and into the infrared). The energy that is absorbed by the earth and the atmosphere is changed into heat (infrared) and reradiated into space with much longer wavelengths. If the earth's average temperature is to remain constant, then the energy arriving at the earth must equal the energy being reradiated into space. Later when we discuss the "greenhouse effect" we will see how the increasing concentrations of carbon dioxide in the atmosphere can upset this balance by absorbing this reradiated infrared radiation, which may result in an overall increase in the temperature of the earth.

The second important point to be made concerns the incoming ultraviolet (UV) radiation. It is a very significant fact that most of the UV energy does not reach the surface of the earth. The UV portion of the spectrum has sufficient energy to break chemical bonds, so living systems must be shielded from an excess of this radiation. Oxygen molecules (O_2) in the upper atmosphere are decomposed by UV into oxygen atoms (O) which can then combine with molecular oxygen to form ozone (O_3). Ozone strongly absorbs UV radiation and so, acts as a protective filter. One of the arguments against the supersonic transport (SST) suggests that a large fleet of SSTs could diminish this protective layer of stratospheric ozone, which might then lead to a large increase in skin cancer on the earth (see Section 9.8).

The effects caused by the energy that does make its way through the atmosphere are going to be studied throughout this book. One of the most important destinations for that energy is into the chemical bonds of green plants by the process called *photosynthesis*.

1.5 *Photosynthesis and Respiration*

During photosynthesis, the chlorophyll-containing cells of green plants use the energy from the sun to convert carbon dioxide and water into organic material. In the process, molecular oxygen is released to the atmosphere. The organic material produced is a carbohydrate such as glucose ($C_6H_{12}O_6$), which can subsequently be converted into other organic molecules such as proteins or cellulose. The following equation summarizes the complex series of reactions that actually take place:

$$6\,CO_2 + 6\,H_2O + \text{light energy} \rightarrow C_6H_{12}O_6 + 6\,O_2$$

As a sidelight, it should be pointed out that the oxygen liberated does not come from the carbon dioxide but rather is taken from the water.

During photosynthesis plants are performing several vital functions in the ecosphere. They are providing the basic source of energy (carbohydrates) for themselves and the consumers, and they are providing the oxygen required for respiration. Moreover, the producers create complex carbon compounds such as glucose out of simple carbon dioxide molecules. These complex molecules are needed by the consumers as basic building blocks for the synthesis of other complex compounds.

Respiration is roughly the opposite of photosynthesis, being the process whereby an organism derives its energy by the combustion of organic material. However, while photosynthesis occurs only in the presence of light, respiration is a continuous process. Besides being used to do muscular work, the energy released is used in such activities as the generation of nerve impulses and the synthesizing of proteins and other molecules for the building of new cells.

The following chemical reaction showing the oxidation of glucose during respiration is simply the reverse of the photosynthetic equation:

$$C_6H_{12}O_6 + 6\,O_2 \rightarrow 6\,CO_2 + 6\,H_2O + \text{energy}$$

Notice that in respiration oxygen is consumed while in photosynthesis oxygen is liberated.

1.6 *Energy Flow through the Biosphere*

We have considered the interaction of the atmosphere with the energy arriving from the sun and have seen how some of the energy that does reach the earth's surface is captured and stored by the producers. It is of considerable importance, especially when dealing with the problem of supplying enough food for man, to understand the movement through the biosphere of this stored energy. Man, of course, must consume sufficient food to meet his energy requirements and as we shall see, the amount of energy available depends directly on the trophic level from which he is deriving his food.

Consider the producers. Producers are synthesizing a certain amount of organic material per unit of time, called the *gross production* (usually expressed in terms of energy rather than biomass). From a large number of studies, it appears that under natural conditions the efficiency of this energy capture, from total incident energy to gross production, is seldom more than 3% and is more typically about 1% (Kormondy 1969). The 1% efficiency figure applies to ecosystems under favorable conditions. E. P. Odum (1971) estimates the gross primary efficiency, averaged out over the entire biosphere and over the full year, to be only about 0.2%. He also estimates the total gross production of the biosphere to be about 43.6×10^{16} kcal/yr from marine organisms and about 57.4×10^{16} kcal/yr from terrestrial sources, for a total of about 10^{18} kcal/yr.

Now, to go on, the producers are not only capturing energy, they are also consuming a certain amount of this gross production for their own respiration. What remains is called *net production*, and is the energy that is potentially available to herbivores and decomposers. The simple relationship is thus, gross production (GP) is equal to net production (NP), plus respiration (R). Some of the net production is consumed by herbivores, some supplies energy to the decomposers, and the rest is left as the standing crop.

$$GP = NP + R$$

It will be helpful now to use a specific example. Figure 1.10 summarizes some of the productivity data for a river ecosystem, Silver Springs, Florida (H. T. Odum 1957). As can be seen from the data, the gross production equals about 1.2% of the incident solar radiation. Since respiration accounts for about 58% of the gross production, the remaining 42% is net production.

Let us continue this example beyond the autotrophs to consider the flow of energy through all of the trophic levels. We see in Figure 1.10 that 3368 kcal/m^2/yr are consumed by the herbivores. This will be their

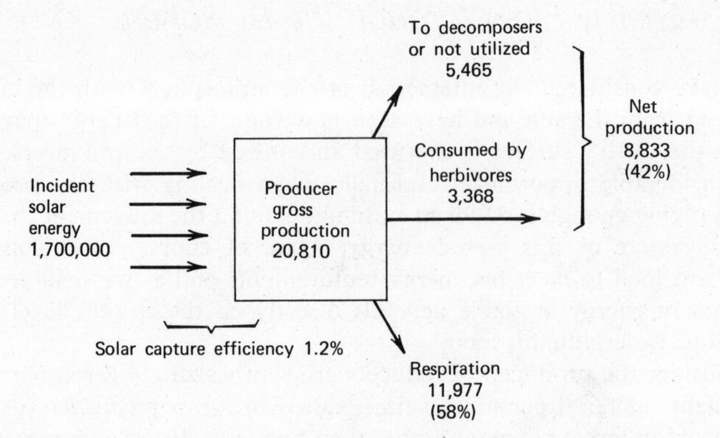

FIGURE 1.10 Annual energy budget for producers, Silver Springs, Florida in kilocalories per square meter per year (H. T. Odum 1957).

gross production; notice it is only about 16% of the producers' gross production. Table 1.1 summarizes the data for all four trophic levels while Figure 1.11 presents the data as a simple energy flow diagram. Notice the drastic decrease in net production from one trophic level to the next.

There is another commonly used diagram for illustrating the decrease in energy available at each succeeding trophic level, called an energy pyramid (Figure 1.12). Each box in the pyramid represents the energy acquired in gross production by all the organisms at that particular trophic level. The gross production is shown divided into respiration and net production components.

What these diagrams so graphically illustrate is that the energy available to carnivores is less than that available to herbivores. Every step up the food chain results in a severe decrease in energy available to the consumers

TABLE 1.1 Summary of the Energy Flow in Silver Springs, Florida in Kilocalories per Square Meter per Year (H. T. Odum 1957)

	Producers	Herbivores	First level carnivores	Top carnivores
Gross production	20,810	3,368	383	21
Respiration	11,977	1,890	316	13
Net production	8,833	1,478	67	6

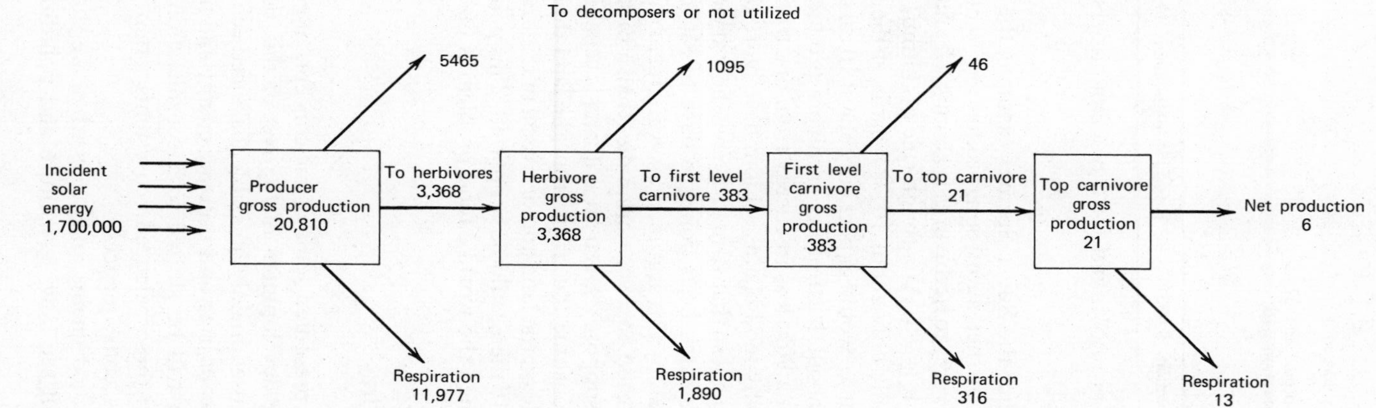

FIGURE 1.11 Energy flow diagram for Silver Springs, Florida, in kilocalories per square meter per year (H. T. Odum 1957).

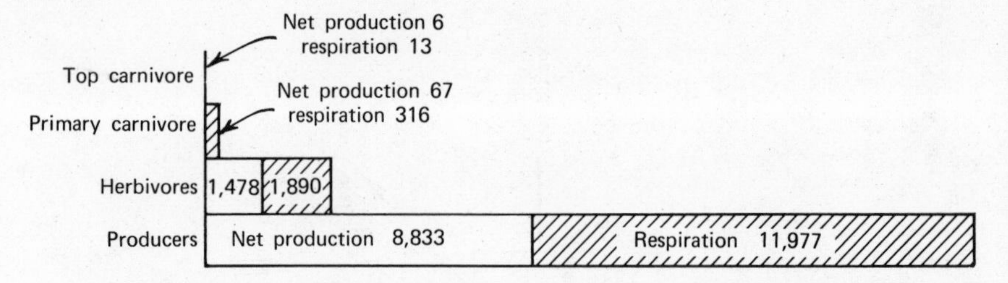

FIGURE 1.12 Energy pyramid for Silver Springs, Florida (H. T. Odum 1957).

feeding at that level. For the Silver Springs example, the ratio of the net herbivore production to net autotroph production is about 17%; this ratio for primary carnivores to herbivores is about 4.5%; for top carnivores to primary carnivores it is 9%. As a rough rule of thumb, this *net production efficiency* is generally only about 10 %. That is, if there exists 100 cal of net plant production we would expect only about 10 cal of net herbivore production and only about 1 cal of net carnivore production. Thus, we would typically expect a 90% loss in energy from one trophic level to the next which explains why food chains seldom have more than four levels.

These arguments suggest that if man is somehow trying to maximize his population on this earth he is going to have to eat low on the food chain. Consuming meat, by this argument, is wasteful of calories and more people could be supported on a vegetarian diet than a diet heavy in meat. As a simple example, suppose a given area of land is capable of supporting 100 men when they consume the crops from the land directly. If instead they were to raise cattle on the land and derive all of their calories from the beef, then only about 10 men could be supported. This point will be raised again in Chapter 3 when the world's food problem is discussed.

1.7 *Conclusions*

This first chapter has presented some of the concepts of ecology that will be most useful in the development of the rest of this book. So far our attention has been focused mainly on stable, balanced ecosystems as they go about their task of capturing and utilizing energy and materials. As we proceed, our concern will be shifted to a consideration of how man's activities can overload these balanced ecosystems, often with unknown, but potentially serious, consequences.

We will return to food chains and webs when we discuss pesticides and radiation; we will see how agriculture and pollution decrease the complexity of ecosystems, causing potential instabilities. Biogeochemical cycles will be important as we look at the greenhouse effect or the eutrophication of lakes due to nutrient increases. And the subject of energy, which is so important, will be treated extensively in Part IV.

Bibliography

Kormondy, E. J. (1969). *Concepts of ecology*. Englewood Cliffs, N.J.: Prentice-Hall.

Lehninger, A. L. (1971). *Bioenergetics*. 2d ed. Menlo Park, Ca.: W. A. Benjamin.

Lindeman, R. L. (1942). The trophic-dynamic aspect of ecology. *Ecology* 23: 399–418.

MacLulich, D. A. (1937). Fluctuations in the numbers of the varying hare (*Lepus americanus*). *Univ. Toronto Studies. Biol. Ser.*, No. 43.

Odum, E. P. (1971). *Fundamentals of ecology*. 3rd ed. Philadelphia: W. B. Saunders.

Odum, H. T. (1957). Trophic structure and productivity of Silver Springs, Florida. *Ecol. Monographs* 27: 55–112.

Oort, A. H. (1970). The energy cycle of the earth. *Sci. Am.*, Sept.

Smith, R. L. (1966). *Ecology and field biology*. New York: Harper and Row.

U.S. Department of Interior (USDI), (1972). *First Annual Report of the Secretary of Interior under the Mining and Minerals Policy Act of 1970 (P.L. 91–631)*. Washington, D.C., March.

Wilson, E. O., and Bossert, W. H. (1971). *A primer of population biology*. Stamford, Conn.: Sinauer.

Woodwell, G. M. (1967). Toxic substances and ecological cycles, *Sci. Am.*, Mar.

Questions

1. Explain how an energy flow diagram conforms to the First and Second Laws of Thermodynamics.
2. Consider the simple food chain

$$P \rightarrow H \rightarrow C$$

If the population of C declines, what would you expect to be the short-term effects on the populations of H and P?

3. One way to estimate the gross production of a field is to simply harvest the crop, dry it, and weigh it. Will this estimate be too low or too high? Explain.

4. Suppose a field is capable of supporting 100 farmers when they eat the crops directly. If the farmers decide to eat the crops from half of the field and feed the other half to cattle which they subsequently consume, how many men could the field support? (Assume a 10% production efficiency.)
ans. 55

5. Arrange the various components of the nitrogen cycle (Figure 1.7) into the generalized ecosystem model of Figure 1.2.

6. Do a stoichiometric analysis of the photosynthesis equation to determine the weight of oxygen released for each gram of carbohydrate synthesized.
ans. 1.07

7. Which of the following food chains would you expect to be the most stable? Why?

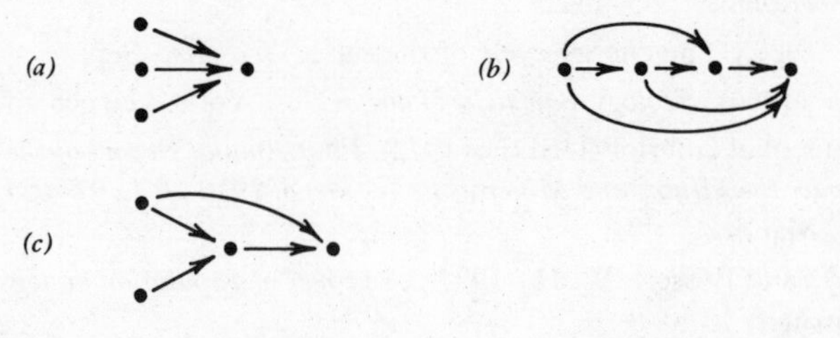

ans. (b)

8. The following table summarizes the energy data, in kilocalories per square meter per year, for Cedar Bog Lake (Lindeman 1942).

	Producers	Herbivores	Carnivores
Gross production	1,113	148	31
Respiration	234	44	18
Net production	879	104	13

(a) Draw an energy pyramid for this ecosystem.
(b) Calculate the net production efficiencies for each energy transfer.
(c) Calculate the ratio of respiration to gross production for each trophic level. Do you see a pattern? Can you think of a physical explanation for the pattern?

Chapter 2

Population Dynamics

There are many factors that contribute to man's impact on the environment, and one of the most important of these is population. As long as our numbers were few and our technology simple, it was quite reasonable to rely on the earth's ecosystems to dilute, decompose, and recycle our waste products. In the last few centuries, however, technology has reduced the death rate and our numbers have since been rising explosively. In succeeding chapters, we shall look into some of the effects of this rapid growth, but in this chapter the emphasis will be on the growth itself.

2.1 *Some Biological Growth Curves*

Man's population is growing exponentially, which is typical of any biological population given conditions which allow unlimited growth. But obviously continuous growth is impossible in any finite system and eventually environmental constraints will restrict any further increases.

To gain insight into the transition that must occur in the human population, it is useful to consider what happens to other biological populations as their growth becomes subjected to environmental limitations. We have already seen some of the ways food webs respond to growing populations. Species either run out of food or increased predation by their natural enemies may limit the growth. Since man is at the top of the food chain, it is the former effect that is of most interest.

There are several different population curves that are observable in nature. Oftentimes the growth curve will make a smooth transition to some constant value called the *carrying capacity of the environment*. This curve is called a *sigmoidal* or *logistic*

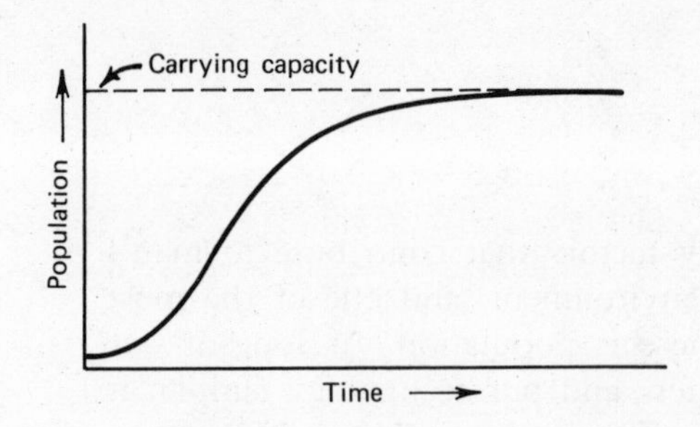

FIGURE 2.1 Logistic growth curve. The population smoothly approaches the carrying capacity of the environment.

curve and is shown in Figure 2.1. When the population size is small, the growth is exponential, but as the carrying capacity is approached, the growth rate goes to zero, and the population size remains constant. It should be noted that the carrying capacity is not necessarily constant throughout the year. Weather, for example, can be strongly influential, especially in locations where the winter is severe.

Figure 2.2 shows two other ways that equilibrium can be reached. In 2.2(a) the population overshoots the carrying capacity and is subsequently cut back, gradually approaching equilibrium with damped oscillations. Figure 2.2(b) shows a more serious possibility wherein the population overshoots the carrying capacity and in the process damages the environment to such an extent that the carrying capacity itself is reduced. The population is cut back to some value below the original carrying capacity. This can happen, for example, if herbivores overgraze an area and so damage the plants that the vegetation cannot recover.

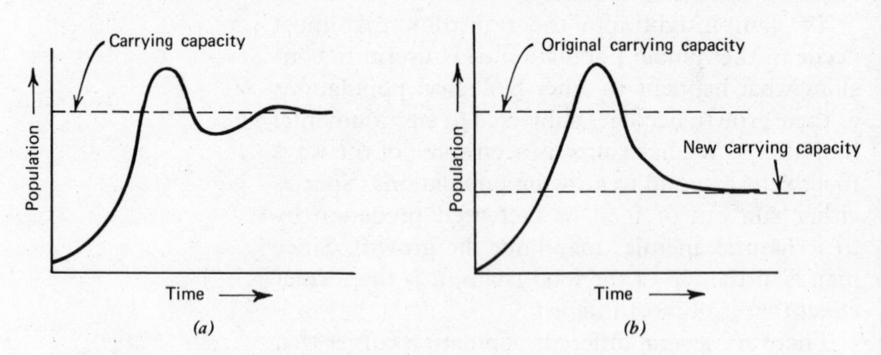

FIGURE 2.2 Two possibilities for a population that overshoots the carrying capacity of the environment: (*a*) damped oscillatory return; (*b*) overshoot damages the environment resulting in lower carrying capacity.

The most extreme version of Figure 2.2 results in what is known as an "outbreak-crash" or J-shaped curve wherein a population sustains rapid growth, then precipitously drops to such a low value that extinction is a likely possibility. The drop can be caused by outside factors such as an extremely severe winter, or by such density-dependent factors as the exhaustion of the available nutrient supply or by such severe pollution of the environment with the population's own wastes that life cannot be sustained. Figure 2.3 shows such a curve for a reindeer herd on St. Matthew Island off the coast of Alaska (Klein 1968). In 1944, 29 reindeer were put ashore on the 128 square mile island by the U.S. Coast Guard. Lacking natural enemies and finding an adequate food supply (lichen) the population grew exponentially to 1350 members in 1957 and finally to 6000 in 1963. But by 1966 the herd had suddenly dropped to only 42 members as a combination of overgrazing and a bad winter led to a massive die-off.

Of course, it is not necessary for equilibrium to ever be achieved as was demonstrated in Figure 1.5 for the lynx and hare example, a system which is characterized by severe oscillations.

2.2 *Exponential Growth*

Since exponential growth is such an important subject—not only in a discussion of population but in many other areas—it is important to

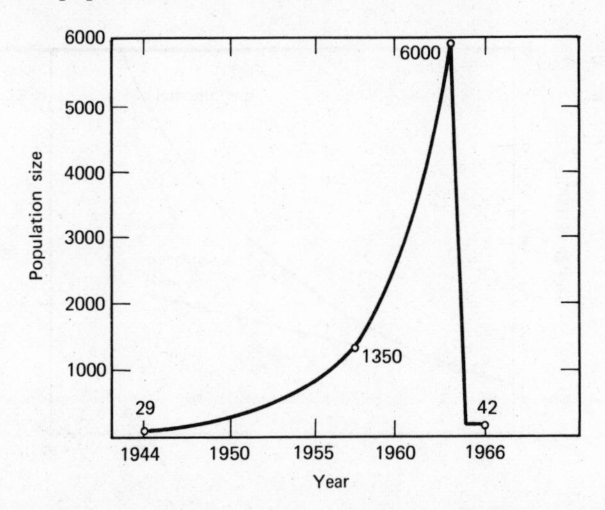

FIGURE 2.3 Assumed population growth of the St. Matthew Island reindeer herd. Actual counts are indicated on the population curve (Klein 1968).

explore it in some detail. Any time something grows by a fixed *percentage* in each time period, then the growth is *exponential.* Contrast this with *linear* growth where the quantity in question grows by a fixed *amount* in each time period. For example, if $1000 is invested in something that earns 10% per year, the total capital will grow exponentially. However, if the $1000 is not invested but each year $100 is added, then the total will only grow linearly. In both cases the capital will be $1100 after 1 year, but after 21 years the capital in the exponential example will be worth about $8000 while in the linear case it will be worth only $3100, as shown in Figure 2.4.

Exponential growth is deceivingly fast. For example, if you were to make an arrangement to earn 1 cent on the first day of the month, 2 cents on the second, 4 cents on the third, etc., thereby doubling your earnings each day, you would be a millionaire before the end of the month—even if you were unlucky and it happened to be February.

The most important parameter for characterizing exponential growth is the *growth rate, r,* usually expressed as percent per year. For a population where migration is not a factor, the growth rate is equal to the birth rate minus the death rate. Thus, very simply, if

r = growth rate

b = birth rate

d = death rate

then

$$r = b - d \tag{2-1}$$

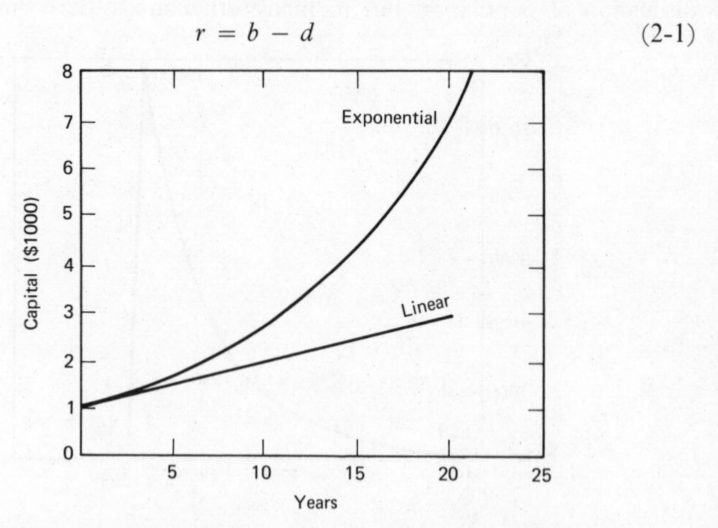

FIGURE 2.4 Exponential growth at 10% per year compared to linear growth at $100 per year.

For example, in the United States in 1973, the birth rate was 15.6 per thousand (usually referred to as the crude birth rate) and the death rate was 9.4 per thousand, which would give a growth rate of 6.2 per thousand, or 0.62 %. To be more precise, however, when speaking of a country's population growth, migration must be included. If m is the net migration rate into the country, then the growth rate is given by

$$r = b - d + m \tag{2-2}$$

In the United States, net migration is about 400,000 per year into the country, or about 2 per 1000 population, so that the growth rate in 1973 was actually

$$r = \frac{15.6 - 9.4 + 2}{1000} = 0.0082 = 0.82\%$$

If a population starts with N_0 people and grows at the constant rate r, then it is possible to predict the population at any time, t, in the future using the following equation

$$N = N_0 e^{rt} \tag{2-3}$$

where N is the future population size. Equation (2-3) is quite simple to derive if you know a little calculus (see Appendix). To actually use the equation it is necessary to be able to raise e (the base of natural logarithms and approximately equal to 2.7183) to whatever power is required. This can be done with a slide rule or from tables or from a series expansion (see the Appendix and Question 8 at the end of the chapter). Since this requires a little bit of mathematical sophistication we should look for an easier way to handle exponential growth.

A very convenient, simple way to compare various rates of exponential growth is by means of the *doubling time*. The doubling time is simply the length of time required for the quantity in question (population here) to double in size, when growing at the constant rate r. As shown in Figure 2.5, if the population starts with N_0 people, then at time T_d the population will have 2 N_0 people and at time 2 T_d it will have 4 N_0 people. The following simple relation for doubling time is derived in the Appendix and is so handy for any quantity that is growing exponentially, that it is well worth committing to memory:

$$T_d \cong \frac{70}{r} \text{ years} \tag{2-4}$$

where r is the growth rate expressed as a percent per year.

For example, knowing that the growth rate in the United States is 0.82 % per year says that if that rate were to continue for 86 years our

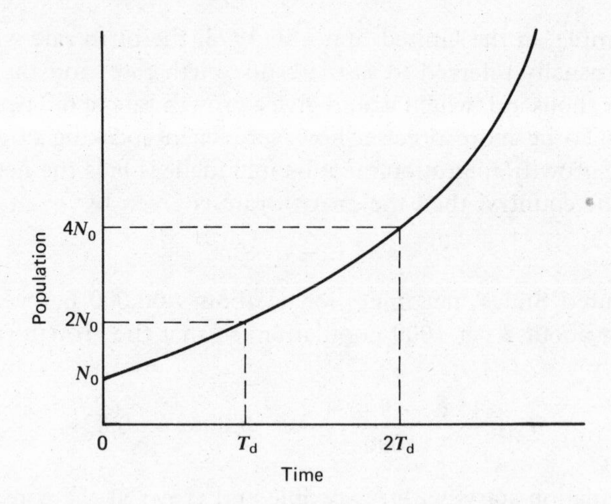

FIGURE 2.5 The doubling time for an exponential.

population would double. The world's population in 1973 was 3.86 billion and growing at 2 % per year. If that rate continues, then in 2008 (35 years later), the population will have doubled to about 7.72 billion.

Another way to make the calculation of population growth is by generations. For example, suppose one female on the average leaves two females in the next generation. Then the number of females in succeeding generations would follow the series 2, 4, 8, 16, ... This is also exponential growth but the calculation is done somewhat differently. This calculation is especially applicable to annual plants and many kinds of insects which breed in one season of the year and whose generations do not overlap.

The equation which describes this growth pattern is

$$N = N_0 R_0^{t/T} \tag{2-5}$$

where N is the number of people in the generation being calculated; t is the time into the future; T is the time per generation (actually t is taken to be some multiple of T, for example, the fourth generation would have $t = 4T$); R_0 is called the *net reproductive rate* and is equal to the average number of surviving female children per female in the population. In 1970 this parameter was equal to 1.9 for the world. If the number of male children is equal to the number of female children, then the average number of children surviving to their reproductive years per couple is two times the net reproductive rate.

Notice that reference is made to numbers of females and not males since it is the number of females which control the number of births. Also note that for $R_0 = 1$ (two surviving children per couple) the new generation size is the same as the old—that is, there is exact replacement.

The interesting thing about equation (2-5) is that it indicates that population growth can be slowed not only by reducing the number of children per person (R_0) but also by increasing the average time between generations (T). In fact, it can be shown from this equation that having three children per couple with a time of 18 years between generations would cause the same growth rate as having four children per couple when the time between generations is extended to 30 years.

2.3 *Human Population Growth*

What is the present situation in the world? The 1973 world growth rate was 2 % per year which corresponds to a doubling time of 35 years. The United States, at 0.8 % per year, would double in 86 years. The Latin American growth rate of 2.8 % implies a doubling time of 25 years! If that Latin American rate holds constant, then, in order to simply maintain the present low standard of living, within 25 years every school, hospital, house, road, gas station, market, water treatment facility, food supply, and power plant must at least be doubled (not even counting replacing those which will wear out). Table 2.1 summarizes some important population statistics for 1973, including crude birth rates, death rates,

TABLE 2.1 Summary of 1973 World Population Data (Population Reference Bureau 1973)

	Population mid-1972, millions	Annual births per 1000 population	Annual deaths per 1000 population	Growth rate r, percent per year	Doubling time, years	Population[a] estimate year 2000, millions
World	3,860	33	13	2.0	35	6,494
Africa	374	46	21	2.5	28	818
Asia	2,204	37	14	2.3	30	3,777
North America	233	16	9	0.8	87	333
Latin America	308	38	10	2.8	25	652
Europe	472	16	10	0.7	99	568
U.S.S.R.	250	17.8	8.2	1.0	70	330
Oceania	21	25	10	2.0	35	35

[a]U.N. medium estimate.

and annual growth rates. Notice that birth rates for what would normally be considered to be the most developed areas in the world seem to be less than 20 per thousand (e.g., North America, Europe, U.S.S.R.) while the less developed areas have birth rates which are typically twice as high, around 40 per thousand (e.g., Asia, Africa, Latin America). The difference is so clear that it is often used as the most convenient criteria for categorizing a country as developed (DC) or underdeveloped (UDC).

What do these numbers mean? The number 2% is a deceptively small quantity but it means a net addition to the world's population of about 77 million people per year. This is an addition of about 1 million people (equivalent to a city the size of San Francisco) every 5 days or about 2.5 people per second. Every 3 years a population equal to the present United States population is being added to the world. We will see in the next chapter that most of the world's population is not even being adequately fed today and that problem will be compounded as growth continues. Obviously growth cannot continue at this high rate for very long before we exceed the carrying capacity of the planet, if we have not already (in Chapter 15 it will be argued that it would be impossible for the earth to support the present world population if everyone lived at the same standard of living that present day Americans do; so from that point of view, the carrying capacity has already been exceeded).

Historically, what has been the picture? Figure 2.6 shows a synthesis of estimates of world population through the last 1000 years. In this figure the world's population in the year 1000 is estimated to have been about 275 million. The half billion mark was reached in about 1650, 650 years later, which corresponds to an annual growth rate of about 0.1%. Just after 1800, only 150 years later, the population doubled to 1 billion. It took about 120 years to double to 2 billion and the next doubling, to 4 billion, will have required only about 45 years. Since the doubling time has not been constant, we know the growth has not been exponential —it has been worse than that! The rate of growth r has itself been growing, increasing from about 0.1% per year during the first half of this millennium to the present value of about 2% per year. Figure 2.7 shows the increase in r for the developed and underdeveloped regions of the world, and for the whole world. The most dramatic increase, one that surely warrants the term *explosion*, has taken place in the underdeveloped regions of the world (which includes about 70% of the world's population). For these regions the average annual population increase has risen from 0.3% during the period 1850–1900, to 0.8% during 1900–1950, to a huge 2.3% expected from 1950 to 2000.

What is causing this sudden rise in population growth rate and why does it seem to be so much greater in the underdeveloped regions of the world?

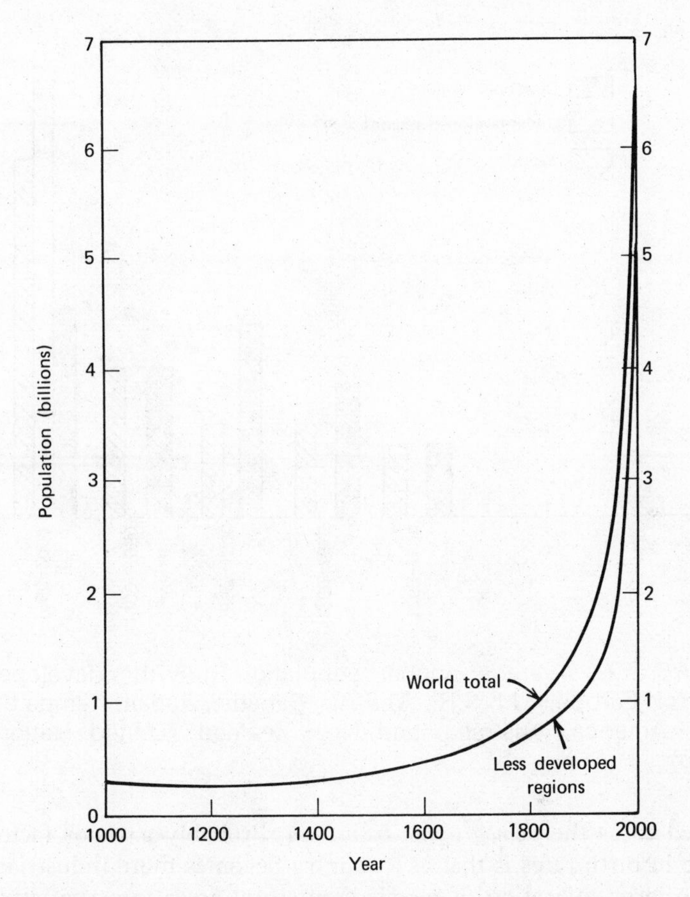

FIGURE 2.6 History of world population growth and projection to year 2000. Less developed regions include everywhere but Europe, U.S.S.R., U.S.A., Canada, Japan, temperate South America, Australia, and New Zealand. Historical data from Bennett (1954), projection is U.N. median estimate; less developed region figures from United Nations (1971b).

Recall that the growth rate is the difference between the birth rate and the death rate and, as Figure 2.8 indicates, it is decreasing death rates, not increasing birth rates, that is causing the sudden growth. As shown in that figure, there is a distinct difference between the demographic histories of the (presently) developed and underdeveloped countries.

Countries in the modern world have generally had their death rates drop during industrialization and shortly thereafter the birth rates have also fallen. This drop in birth rates that follows the drop in death rates

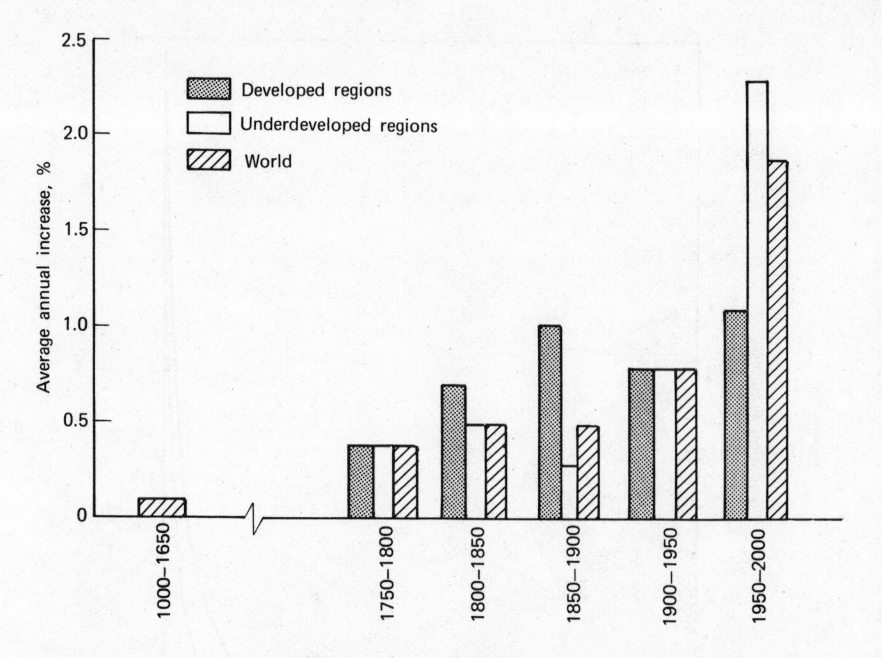

FIGURE 2.7 Average annual population growth; developed regions: Europe, U.S.S.R., U.S.A., Canada, Japan, temperate South America, Australia, and New Zealand (United Nations 1971b).

is referred to as the *demographic transition*. Probably a major factor in this drop in birth rates is that as a country becomes more industrialized there is a large migration of people from rural areas into the city (see Figure 2.9). In the city children change from being an economic asset (free farm labor, etc.) to an economic liability (e.g., university students).

As shown in Figure 2.8, the UDCs maintained high death rates until the beginning of the twentieth century. Since then the death rates have dropped precipitously largely due to the importation of modern public health measures, but they have not yet had the corresponding industrialization which typically leads to a drop in birth rate. In addition to the free-labor aspect of children in a rural society, there are many factors which contribute to the maintenance of high birth rates. For example, the high infant mortality rates in UDCs makes it prudent for a couple to continue to have children until at least two sons are born to increase the likelihood that at least one will survive to help support the parents. This would mean that typically at least four children would be born before there would be any desire to practice birth control techniques.

In fact, in India, on the average, a man must have 6.3 children to be 95 % sure that one son will survive to the father's 65th birthday. Paradoxically then, it seems that a reduction in infant mortality should eventually reduce rather than raise the population growth. Other factors which would tend to maintain high growth rates would include social pressures to have large families and various religious and government restrictions on birth control and abortion.

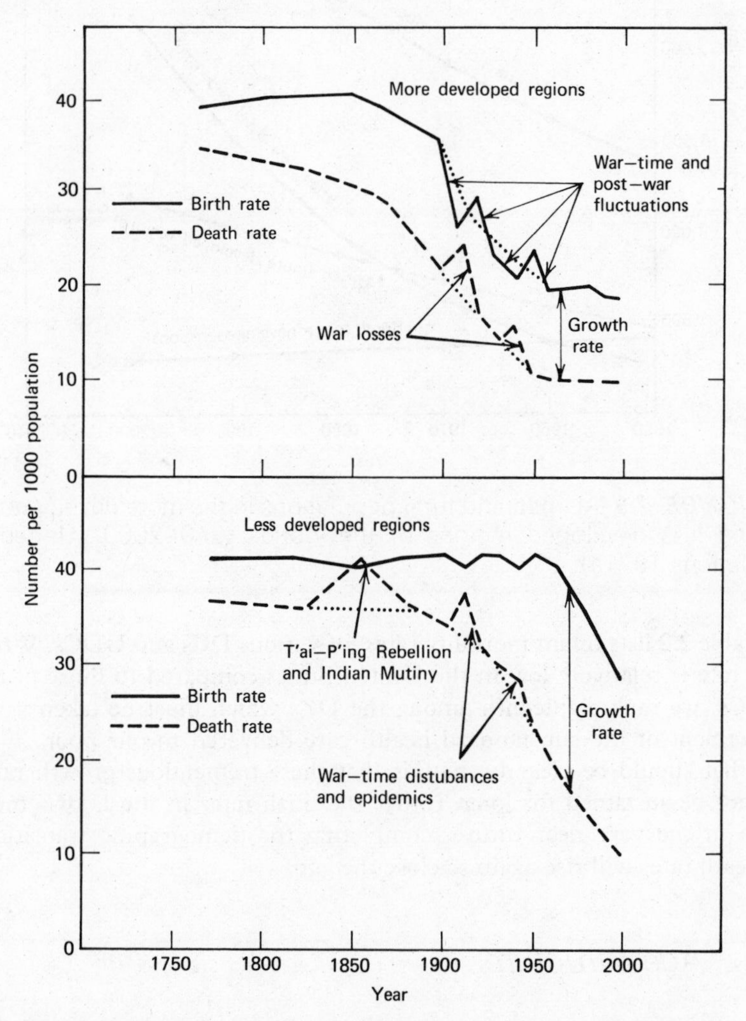

FIGURE 2.8 The difference between birth rate and death rate is the growth rate, which has suddenly increased in the less developed regions of the world (United Nations 1971a).

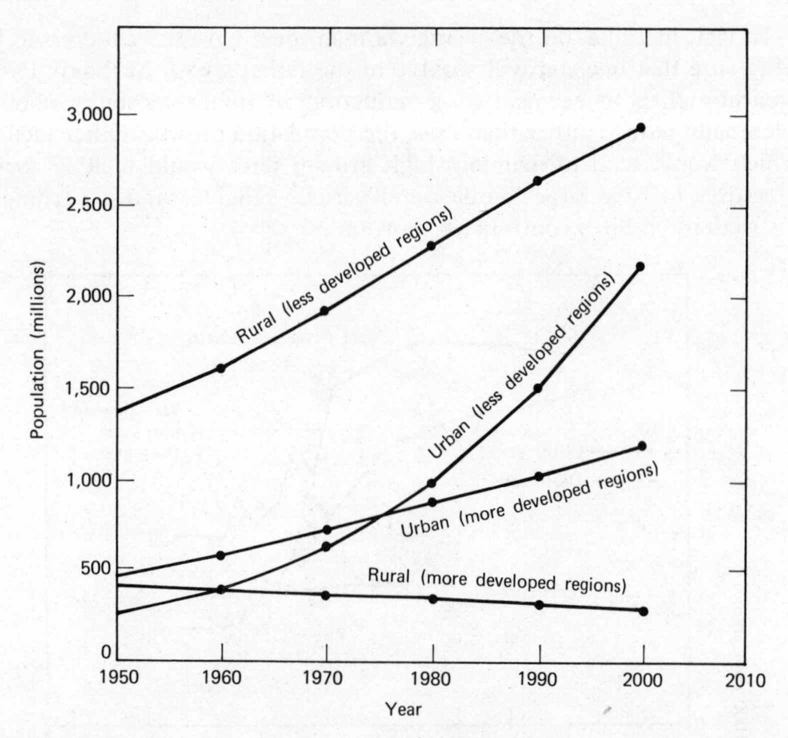

FIGURE 2.9 Urban and rural populations in the more developed and less developed regions of the world, 1950–2000 (United Nations 1971a).

Table 2.2 lists infant mortality rates for various DCs and UDCs. While this rate is relatively low in the United States compared to those of the UDCs, we rank eighteenth among the DCs which must be taken as an indictment of the substandard health care delivered to our poor.

What should be clear by now is that these tremendous growth rates cannot be sustained for long. Either the birth rates in the UDCs must drop in the very near future, completing the demographic transition, or death rates will rise again to close the gap.

2.4 *Age Structure*

If any kind of detailed population projections are to be attempted, they must be based on more than just past growth rates. It is necessary to take into account the distribution of ages in the population. Figure 2.10

TABLE 2.2 Annual Deaths of Infants under 1 Year of Age per 1000 Live Births for Selected Countries (Population Reference Bureau 1973)

Underdeveloped country	Mortality	Developed country	Mortality
Gabon	229	Sweden	11.1
Guinea	216	Finland	11.3
Niger	200	Netherlands	11.4
Angola	192	Norway	12.7
Tanzania	162	Japan	13.0
Pakistan	142	Iceland	13.2
India	139	France	13.3
South Africa	138	Luxembourg	13.6
Cambodia	127	Denmark	14.2
Indonesia	125	Switzerland	14.4
Turkey	119	New Zealand	16.5
Egypt	118	Australia	17.3
Zaire	115	Canada	17.6
Guatemala	88	Democratic Republic of Germany	17.7
Chile	88		
Colombia	76	Republic of China	18.0
Mexico	69	United Kingdom	18.0
Philippines	67	Taiwan	18.0
Argentina	58	U.S.A.	18.5

shows the age structure for a rapidly growing country, India, as compared to the nearly stable age structure for Sweden. Figure 2.11 shows the structure for the whole world. Notice these structures are drawn with male population sizes in each age group on the left and the corresponding female populations on the right. As this figure indicates, a rapidly growing population is characterized by an age structure with a very wide base which results when the number of children in each age category increases with decreasing age. From such an age structure it is easy to see that, even if each couple simply replaced itself with two children, the population would continue to grow as the larger and larger groups of already born children move into their reproductive years and then replace themselves.

Figure 2.12 shows the age structure for the United States in 1970. The figure shows a necking down in the age group around 30–45 which reflects the decrease in births in the depression and early war years. The postwar "baby boom" bulge is also clearly evident (ages 10–25 in the figure were born 1945–1960). Those babies are now moving into their reproductive years so that, even if the number of children per couple

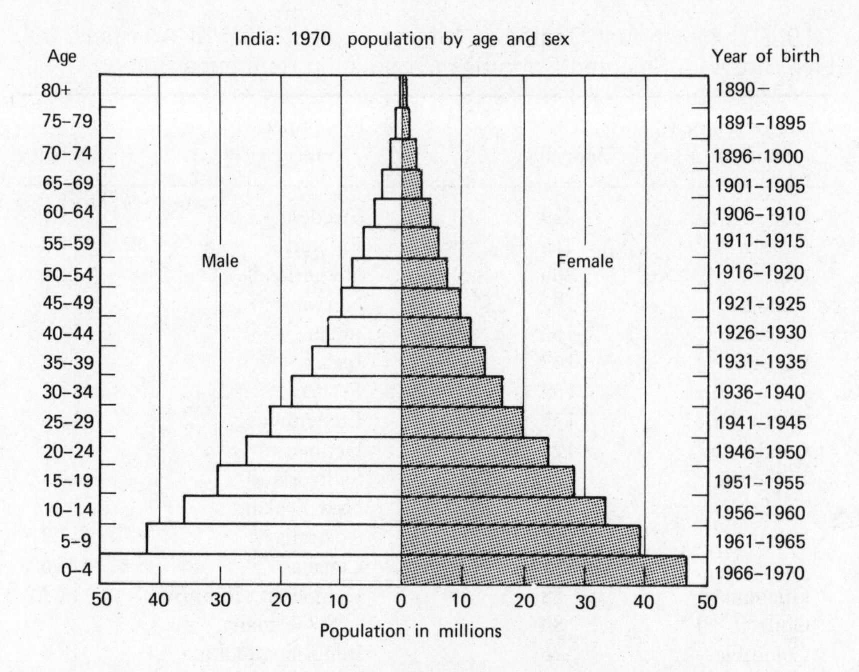

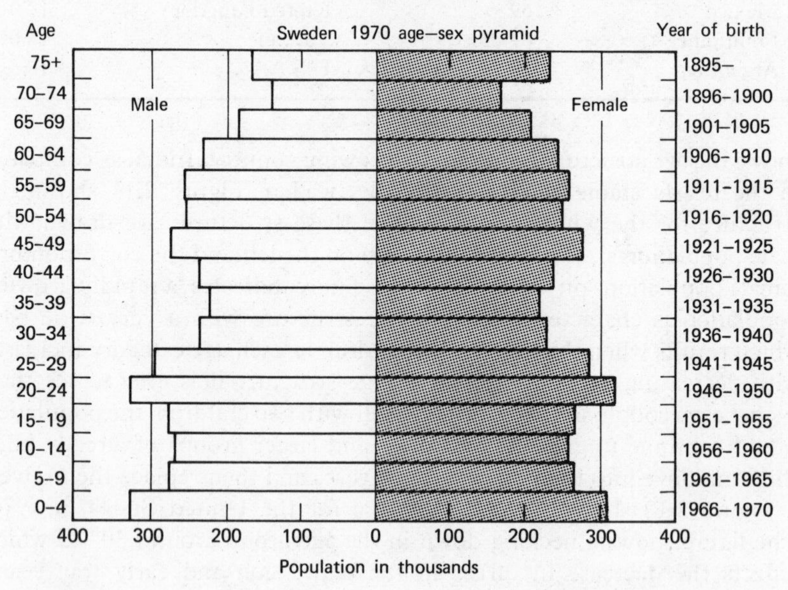

FIGURE 2.10 Age structures in 1970 for a rapidly growing population, India, and a nearly stable one, Sweden. (Population Reference Bureau 1970).

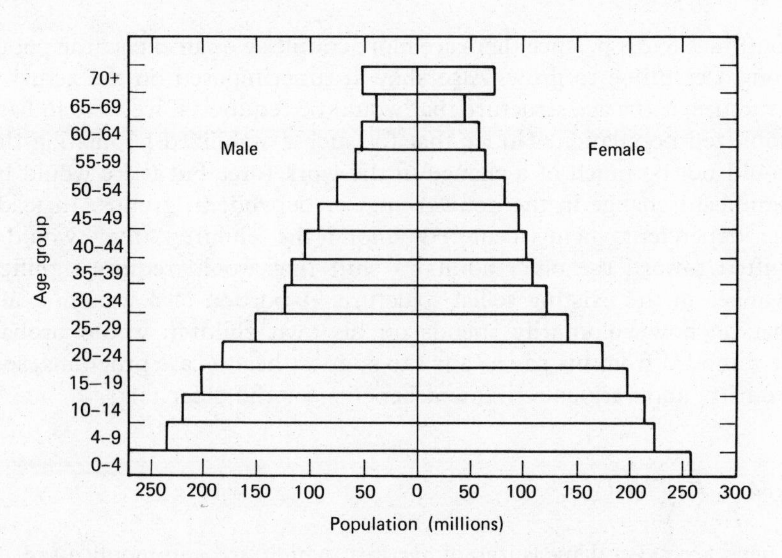

FIGURE 2.11 Age structure for the world, 1975 (U.S. Panel on World Food Supply 1967).

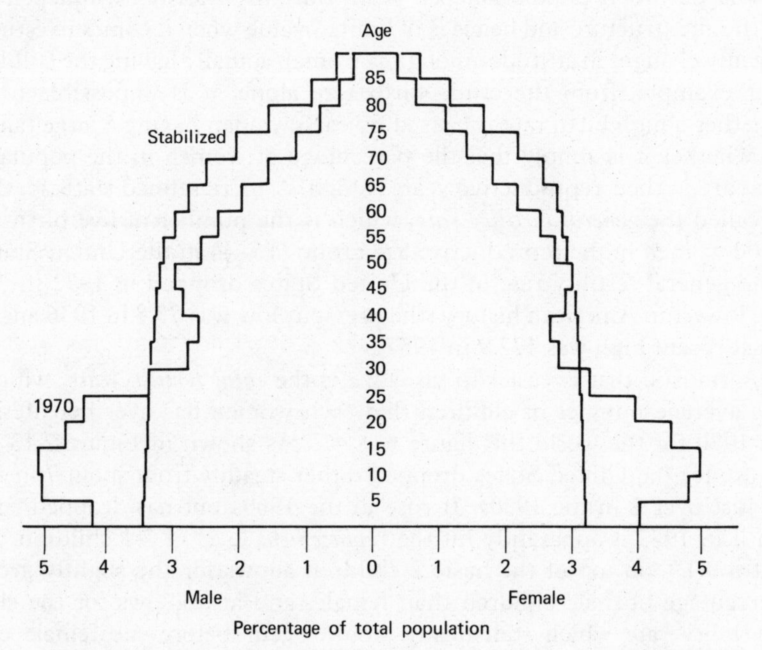

FIGURE 2.12 Age structure, 1970, for U.S. compared to that which would be required for a stabilized population. (Commission on Population Growth and the American Future 1972).

continues to drop, since there are more couples we can expect the population to continue to grow. Also shown, superimposed on the actual age structure, is the age structure that would be required if we were to have a stabilized population. Notice that for such a stabilized population there would not be much of a change in the work force but there would be a significant change in the nonworking, or dependent, groups. Instead of the dependents being centered among the children, they would be shifted toward the older adults, a shift that would require significant changes in the existing social structure. A portion of a worker's salary that he now voluntarily spends on his own children would probably be removed from his pay as a tax to support health care programs, social security, and various retirement benefits for the older folks.

2.5 *Fertility*

There are several measures of fertility which are commonly used. We have been using the *crude birth rate* so far, which is simply the number of births per 1000 population per year. But this statistic is independent of the age structure and hence is of limited value when it comes to estimating any changes in attitude among the women actually having the children. For example, from the crude birth rate alone, it is impossible to tell whether a high birth rate is caused by each woman having a large family, or whether it is simply that the percentage of women in the population that are in their reproductive years is high. A more refined statistic, then, is called the *general fertility rate*, which is the number of live births per 1000 women in the reproductive age group (15–44 in the United States). The general fertility rate in the United States dropped in 1972 to 73.1, the lowest in American history (the previous low was 75.8 in 1936 and the most recent high was 122.9 in 1957).

A statistic that is easier to visualize is the *total fertility rate,* which is the average number of children that each women has over her lifetime. In 1970 for the world this figure was 4.7. As shown in Figure 2.13 that figure for the United States dropped rather steadily from about 7 in 1800 to just over 2 in the 1930s. It rose in the 1960s but has dropped again, until in 1972 it apparently hit the *replacement* level of 2.1 children. The extra 0.1% on top of the basic 2 children allows for the slightly greater percentage of male children than female, and also allows for the slight mortality rate which statistically would occur before the female child reaches her reproductive years. In other words if each female had on the average 2.1 children, then, of these, exactly one female would survive to replace her mother and breed the next generation.

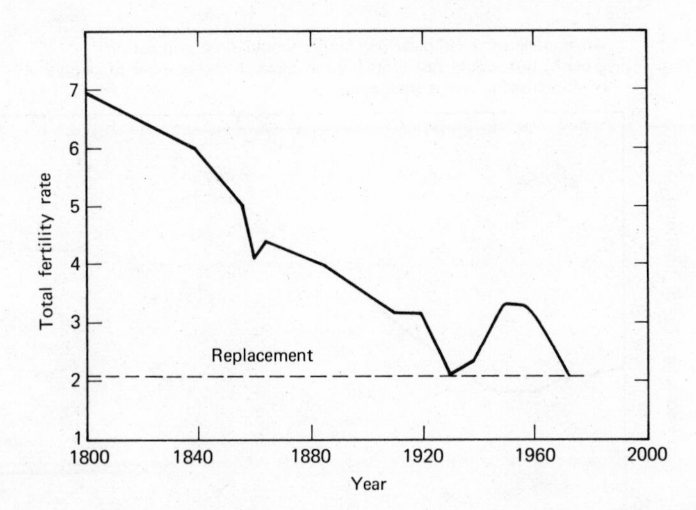

FIGURE 2.13 Total fertility in the U.S. dropped to replacement in 1972.

It is very important to realize that replacement does not mean zero growth in the population. Zero population growth only occurs (neglecting migration) when birth rate equals death rate. For most populations more people are entering their reproductive years than are leaving them, and replacement must be maintained for several generations before growth ceases.

2.6 *Momentum of Population Growth*

We have indicated that due to the age structure most populations would continue to grow for a number of years even if replacement is reached and held. Figure 2.14 demonstrates, for the United States, that due to the increasing numbers of females in the childbearing ages (females that have already been born, in fact), even if they each have only two children the population will continue to grow into the twenty-first century. Assuming continued immigration at the present rate of about 0.4 million each year, and continued total fertility at the replacement level, the population of the United States will have reached 300 million by 2015 and still be growing.

The analysis of this momentum is slightly complicated for individual countries because of migrations but it is straightforward when the world

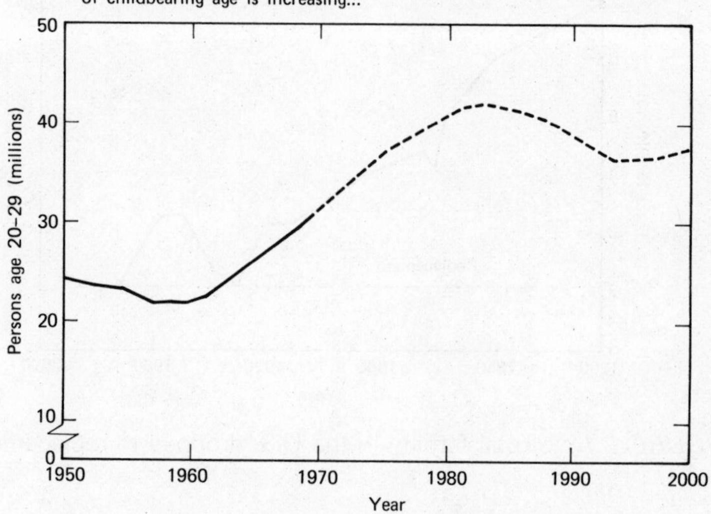

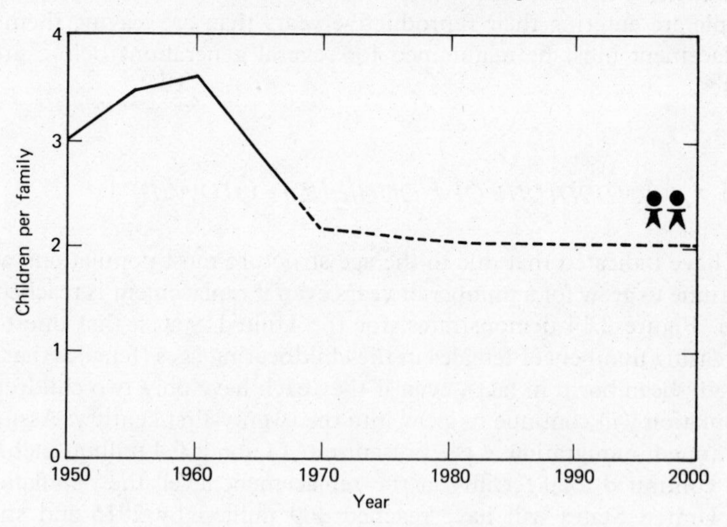

FIGURE 2.14 The continued growth in U.S. population even if replacement is maintained. Data assumes continued immigration at present levels (Commission on Population 1971).

... The resulting births will continue to exceed deaths for the rest of this century...

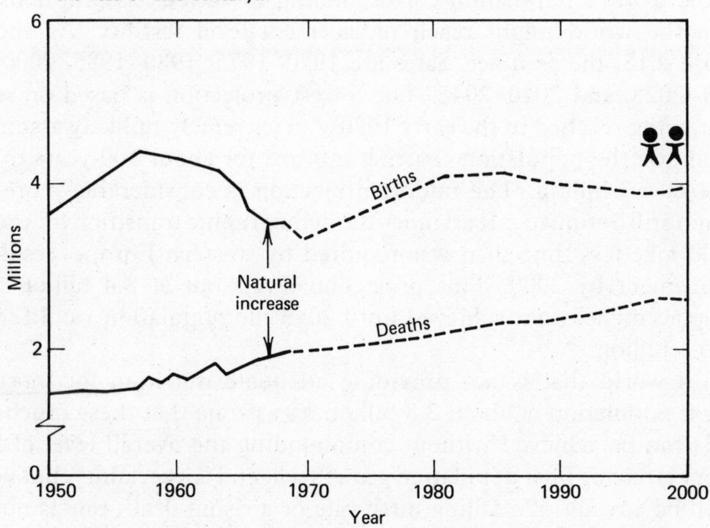

...so the population will still be growing in the year 2000, but at a decreasing rate.

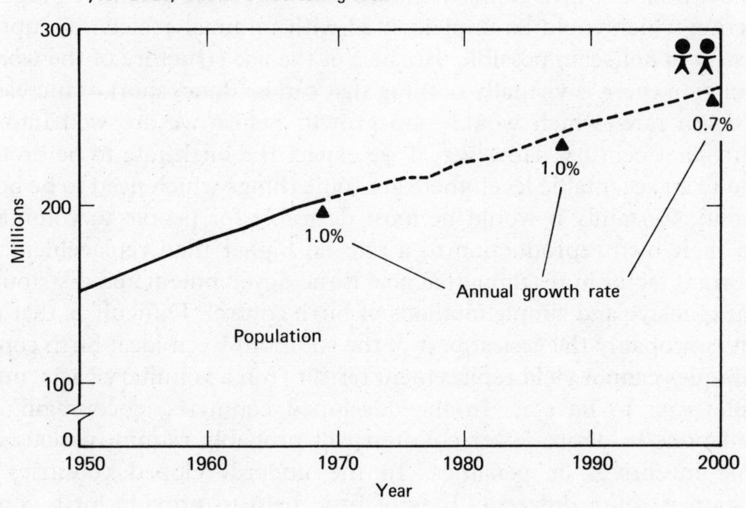

as a whole is considered. Frejka (1973) has generated a set of projections for the world's population corresponding to various assumptions about when the world might reach replacement-level fertility. As shown in Figure 2.15, the assumed dates are 1970–1975, 1980–1985, 2000–2005, 2020–2025, and 2040–2045. The lowest projection is based on replacement being reached in the early 1970s (an extremely unlikely assumption) and shows the population coasting upward for about 100 years to a final value of 5.7 billion. The middle projection is considerably more likely, though still optimistic. It assumes the demographic transition for the world would take less time than was required by western Europe, resulting in replacement by 2000. This projection levels out at 8.4 billion people. If replacement is not achieved until 2040 the population would increase to 15.1 billion.

In a world that is not providing adequate nutrition for most of its *present* population of about 3.8 billion, to assume that these much higher levels can be achieved without compounding the overall level of human misery is naive. That population growth will end is a certainty, but whether it will be a result of a falling birth rate or a rising death rate is not clear.

2.7 Conclusions

While it would be nice to finish this chapter with some definitive programs of action which could be implemented with minimal societal disruption, it just does not seem possible. Because of the age structure of the world's population there is virtually nothing that can be done, short of increasing the death rate, which would stop growth before we are well into the twenty-first century. However, if we expect the birth rate to be brought down to an acceptable level, there are some things which need to be borne in mind. Certainly it would be most desirable for people to voluntarily curb their own reproduction to a rate no higher than replacement. An important factor in reaching that goal is the development and distribution of inexpensive and simple methods of birth control. Difficult as that may be, it is probably the easiest part of the solution. Even ideal birth control techniques cannot yield replacement fertility (on a voluntary basis) unless people *want* to have it. In the developed countries, encouraging the populations to desire fewer children will probably require various economic incentives or penalties. In the underdeveloped countries the situation is quite different. It is of little help to provide birth control techniques when the people see their own survival linked to having a large family. For them, a significant amount of development will be required before the demographic transition can be expected to be com-

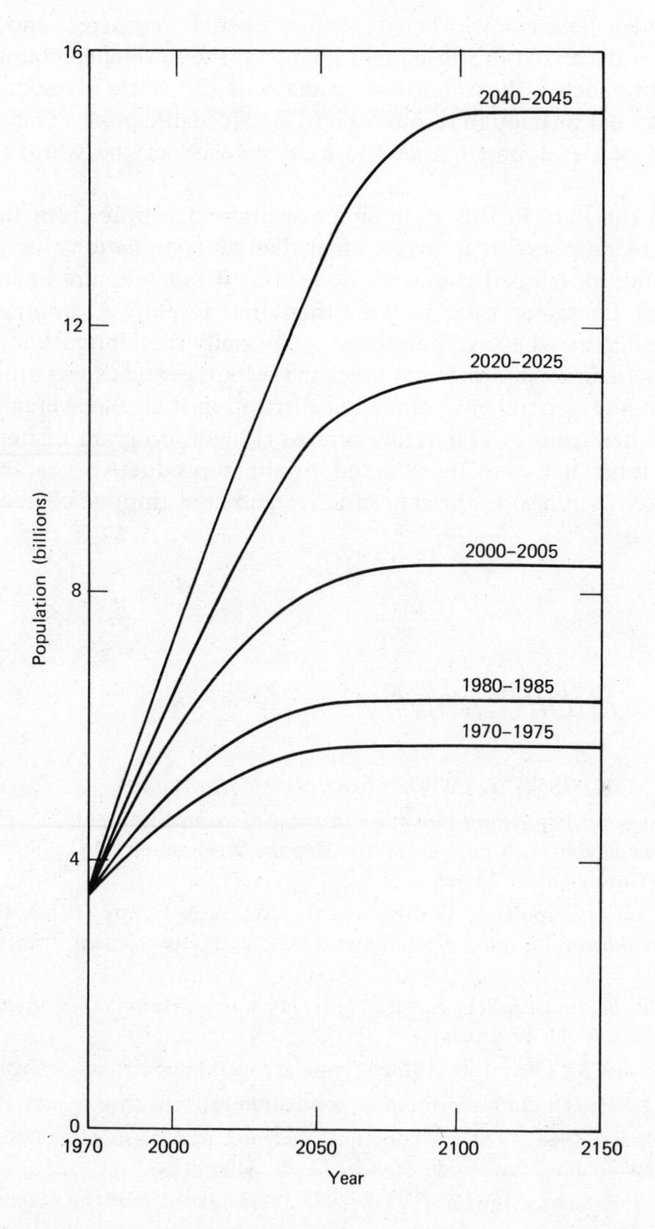

FIGURE 2.15 The momentum in world population growth under five assumptions of when replacement is reached. From Frejka, T. "The Prospects for a Stationary World Population."

pleted—development which will require energy, resources, and capital. As will be discussed in subsequent chapters, the developed countries are consuming such a disproportionate share of the world's resources that there may not be enough resources left, at affordable prices, to ever allow the kind of development that the vast majority of the world's people require.

When the data in this chapter are examined simply from the point of view of numbers, it is easy to feel that overpopulation is a problem in the underdeveloped countries alone. But if our measure includes the ecological effects of each person, then the developed countries must also be categorized as overpopulated—especially the United States, since each American consumes many times the resources and causes much more pollution and general environmental disruption than the average person of any other country. Therefore any worldwide program of population control must not only be directed to the reproductive characteristics of the poor, but also to the reproductive and consumptive characteristics of the rich.

Bibliography

Bennett, M. K. (1954). *The World's Food,* New York: Harper.

Commission on Population Growth and the American Future (1971). *Population Growth and America's Future.* Interim Report. Washington, D.C.: U.S. Government Printing Office. March.

Commission on Population Growth and the American Future (1972). *Population and the American Future.* Washington, D.C.: U.S. Government Printing Office. March.

Ehrlich, P. R., and Ehrlich, A. H. (1970). *Population, resources, environment.* San Francisco: W. H. Freeman.

Frejka, T. (1973). The prospects for a stationary world population. *Sci. Am.* March.

Keyfitz, N. (1971). On momentum of population growth. *Demography* Feb.

Klein, D. R. (1968). The introduction, increase, and crash of reindeer on St. Matthew Island. *Jour. Wildl. Manag.* 32 (2): 350–367.

Population Reference Bureau (1973). *1973 World Population Data Sheet,* Washington, D.C.

Population Reference Bureau (1970). *Population Bulletin,* Washington, D.C.

United Nations (1971a). *A Concise Summary of the World Population Situation in 1970.* Population Studies No. 48. New York.

United Nations (1971b). *The World Population Situation in 1970*. Population Studies No. 49. New York.

U.S. Panel on the World Food Supply (1967). *The World Food Problem*. Washington, D.C.: U.S. Government Printing Office. May.

Questions

1. Explain the differences between net reproductive rate, crude birth rate, general fertility rate, and total fertility. For the United States, what values of net reproductive rate and total fertility (TF) correspond to replacement? Would these values be different for different countries in the world? Explain.

   ans. $R_0 = 1$, $TF = 2.1$, only TF would change

2. Explain the difference between replacement and zero population growth.

3. Consider a population having 100 members with the following age structure: 50 members age 0–1, 30 members age 1–2, and 20 members age 2–3. In this population all reproduction occurs on a member's first birthday and everyone dies on their third birthday. If the population starts now with a net reproductive rate equal to 1 (replacement), draw the age structures corresponding to 1 year, and 2 years, from now. What will the total populations be then? What is the ultimate population?

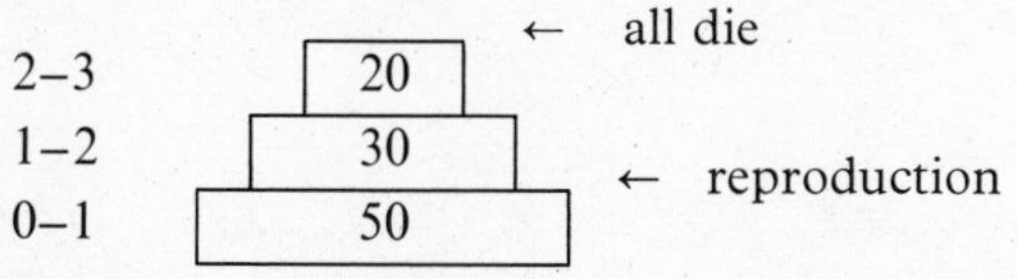

Age	Members	
2–3	20	← all die
1–2	30	← reproduction
0–1	50	

   ans. 130, 150, 150

4. The use rate of electricity in the United States is doubling every 10 years. What annual rate of growth, r, does this correspond to?

   ans. 7%

5. In 1972 the crude birth rate, crude death rate, and net immigration rate for Canada were 17.5, 7.3, and 6.8 per 1000. What is the annual growth rate r, and what is the corresponding doubling time?

   ans. $r = 1.7\%$
   $T_d = 41$ years

6. Which of the following circumstances would yield exponential growth?
   (a) Adding 100 books per year to a library.
   (b) Doubling the library's size every 10 years.
   (c) The accumulation of back issues of a library's periodicals.
   (d) Increasing the expenditures on new books by 10% each year.

   ans. b, d

7. Discuss the demographic transition and list some of the factors which you think caused it to take place in the developed countries. What changes would need to take place in the underdeveloped countries to help complete the transition?

8. The world's population in 1972 was 3.78 billion and growing at the rate of 2% per year. Assume that rate remains constant and calculate the world's population in the year 2000.
   (a) Using equation (2-3) directly.
   (b) Using the first three terms of the series approximation for the exponential (see Appendix).

   ans. 6.6 billion, 6.5 billion

Chapter 3

Food Production

The explosive growth in the world's population described in the last chapter is going to require an (at least) equally rapid increase in food production if even present, inadequate levels of nutrition are to be maintained. According to a 1963 estimate made by the Food and Agriculture Organization (FAO) of the United Nations (UNFAO 1963b), about 20% of the people in the developing areas of the world were *undernourished* (insufficient quantity of food, that is, calories) whereas about 60% were *malnourished* (inadequacy in the quality of the diet, mostly a lack of sufficient protein). As we shall see, conditions in these areas haven't improved much since then, which would imply that somewhere around half the world's population is presently suffering from hunger. This lack of proper quality and quantity of food is a principal contributor to the high morbidity and mortality rates in these areas which, as we have seen, is probably an important factor in their very high birth rates.

In this chapter we shall examine the world's food problem and try to evaluate the prospects for supply keeping up with the demand. We shall see that increased production must come mostly from higher yields on already cultivated land, rather than from an increase in land area. Increased yields are highly dependent upon a combination of proper irrigation, better seeds (the Green Revolution), increased use of fertilizers and pesticides, and higher consumption of energy (fossil fuels to run the tractors, harvesters, etc.). The ecological implications of our increasing utilization of this "package" of systems are very important. Towards the end of this chapter, the focus will be on the ecological dangers which result from our dependence on pesticides, while in Part II we shall study the water pollution problems that agriculture helps cause.

3.1 Human Food Requirements

Let us begin by considering overall food production statistics to see generally whether any progress is being made toward alleviating worldwide shortages. Figure 3.1 shows a history of food production since 1955, for the developing and developed countries, and for the world. As can be seen, the developing countries have increased their production by about 60% in that time period, but their population has correspondingly increased just over 50% so that the amount of food produced per person has gained only slightly. In the developed countries, where population is growing more slowly, food production per person has increased significantly. Some of this increase stays within the developed countries to satisfy increased per capita demands but some is exported to the less developed countries to help satisfy their needs. For example, in 1971, about 21% of U.S. crop acreage harvested was for exports. From the point of view of feeding the poor, it is unfortunate that a rather high percentage of these exports end up in other developed countries rather than in the more needy, less developed regions. For example, 25% of U.S. agricultural exports in 1971 were to the European community alone (U.S. Department of Agriculture, 1972). From another point of

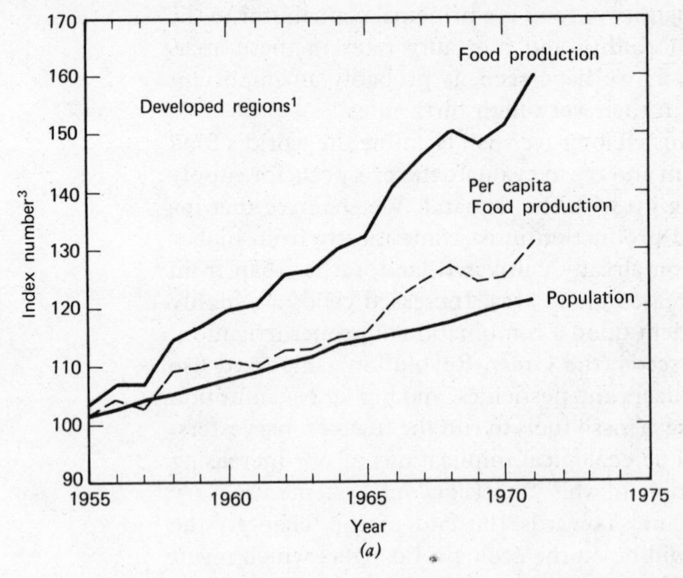

FIGURE 3.1 Food production, population, and per capita food production for developed and developing regions, and the world. Gains in production are keeping slightly ahead of population

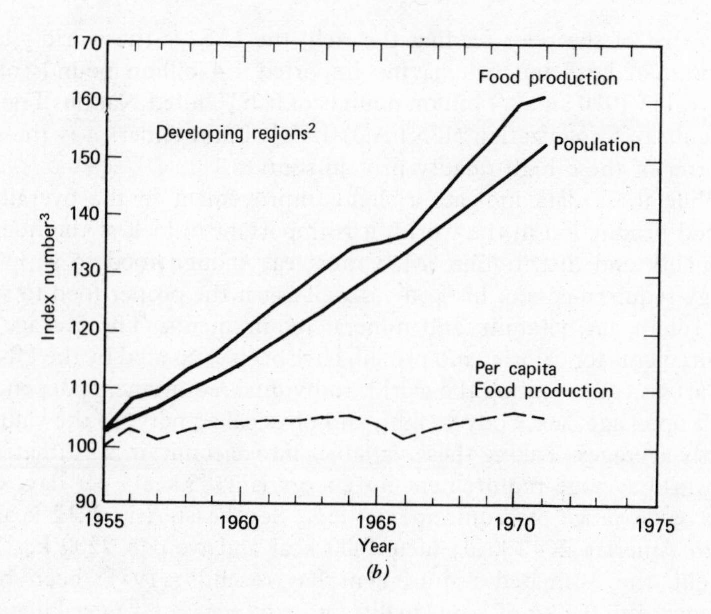

(b)

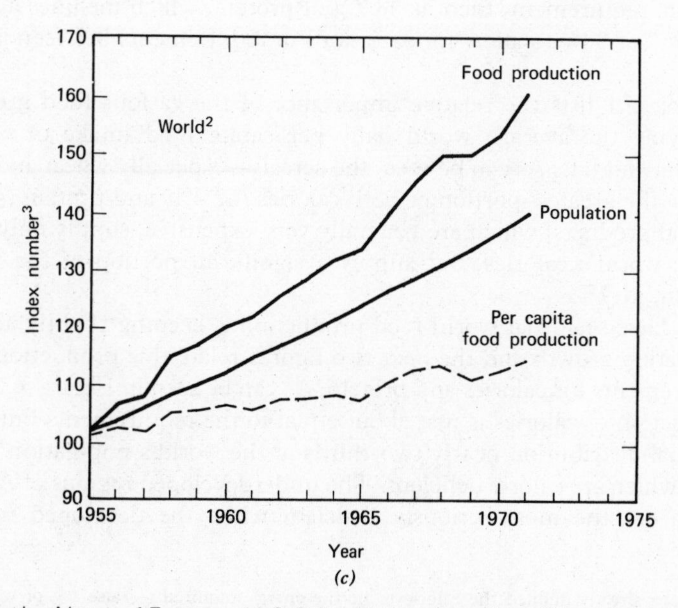

(c)

growth. *Notes:* [1]Europe, U.S.S.R., North America, Oceania, Israel, Japan, South Africa; [2]Excludes People's Republic of China; [3]1952–1956 = 100. (Data from UNFAO 1972).

view, that of the poor feeding the rich, the U.S. is the world's largest importer of beef and fish having imported 1.4 billion pounds of beef and veal in 1970 and 8.4 billion pounds of fish [United Nations Food and Agriculture Organization (UNFAO) 1971]. Latin America is the major exporter of these high-quality protein sources.

While these data indicate a slight improvement in the overall level of food production in the world, it is important to look at the questions of quality and distribution. Man must eat enough food to supply the energy requirements of his body as well as eat the proper food to supply his protein, fat, vitamin, and mineral requirements. The average daily requirements for calories and protein have been estimated by the UNFAO for various countries and the world. Individual requirements depend very much upon age, sex, body weight, and physical activity, so the values are strictly averages. Taking these variables into account, it is estimated that the world average requirement for energy is 2385 kcal* per day, with a fairly wide range of requirements (e.g., Southeast Asia 2212 kcal, and North America 2642 kcal; men 3000 kcal and women 2200 kcal). For protein, the estimated requirement has recently (1971) been revised downward to 0.57 g of high-quality protein (eggs, milk) per kilogram of body weight for men and 0.51 g per kilogram for women. The world average requirement, then, is 38.7 g of protein, which includes a "safety margin" of 30% to allow for variations in requirements between individuals.

Table 3.1 lists the relative importance of the various food groups in supplying the average world daily per capita food intake of calories, protein, and fats. As can be seen, the cereals—especially wheat and rice—supply the greatest portion of both calories (52.4%) and protein (47.4%). Animal products, which are generally very expensive, supply only 16.7% of the world's calories, but supply a significant portion of the world's protein, 31.7%.

We have seen that world food production is keeping slightly ahead of population growth and the next two figures relate this production to the requirements for calories and protein. As can be seen in Figure 3.2, world production of calories is just about equal to the requirements but due to unequal distribution nearly two-thirds of the world's population lives in areas which are calorie deficient. The underdeveloped regions of Asia and Africa are the most seriously deficient while the developed world is

*We have already defined the calorie to be the energy required to raise 1 g of water from 14.5 to 15.5°C. When referring to the energy content of food, the unit is one kilocalorie (kcal) which is usually called a Calorie with a capital C. (1 kcal = 1000 cal = 1 Cal.) To avoid confusion we will use the kilocalorie as the unit of food energy.

TABLE 3.1 Relative Importance of Various Food Groups in Average Daily per Capita Intake (UNFAO 1971)

	Calories		Proteins		Fats	
	Number	Percent	Grams	Percent	Grams	Percent
Cereals	1,245	52.4	31.1	47.4	5.1	9.3
Wheat	441	18.6	13.3	20.3	1.5	2.7
Rice	459	19.3	8.5	13.0	1.0	1.8
Maize	147	6.2	3.6	5.5	1.0	1.8
Millet and sorghum	119	5.0	3.5	5.3	1.2	2.2
Others	76	3.2	2.1	3.2	0.4	0.7
Roots and tubers	184	7.8	2.8	4.3	0.4	0.7
Sugar and sugar products	210	8.8	0.1	0.2	—	—
Pulses, nuts, and oilseeds	121	5.1	7.9	12.0	3.6	6.5
Vegetables	36	1.5	2.2	3.4	0.3	0.5
Fruits	47	2.0	0.6	0.9	0.3	0.5
Total, animal products	322	13.6	20.7	31.5	22.4	40.8
Meat	168	7.1	9.2	14.0	14.3	26.0
Eggs	18	0.8	1.4	2.1	1.3	2.4
Fish	19	0.8	3.0	4.6	0.6	1.1
Milk	117	4.9	7.1	10.8	6.2	11.3
Fats and oils	199	8.4	0.1	0.2	22.5	40.9
Vegetable oils	127	5.3	—	—	14.4	26.2
Animal fats	72	3.1	0.1	0.2	8.1	14.7
Total	2,374	100.0	65.6	100.0	55.0	100.0
Animal origin	396	16.7	20.8	31.7	30.5	55.5

supplied with more than is necessary. Figure 3.3 indicates that the world's supply of protein is considerably greater than minimum requirements would suggest is necessary. Interestingly, every area of the world has sufficient protein resources, but as the figure indicates the developed regions derive most of their protein from animal sources while the underdeveloped regions derive theirs from other sources, mostly cereals.

While it would appear from this figure that supplies of protein are quite adequate, the data are a bit misleading for several reasons. These are averages and do not show the grossly unequal distribution that exists in each area. Protein is expensive and there is considerable data to indicate that family income is directly related to protein consumption so that the story of malnutrition is largely a story of poverty. Further, if a person's calorie intake is not sufficient, then the deficit will be made up by burning proteins for energy, so both the calorie and protein inputs must be con-

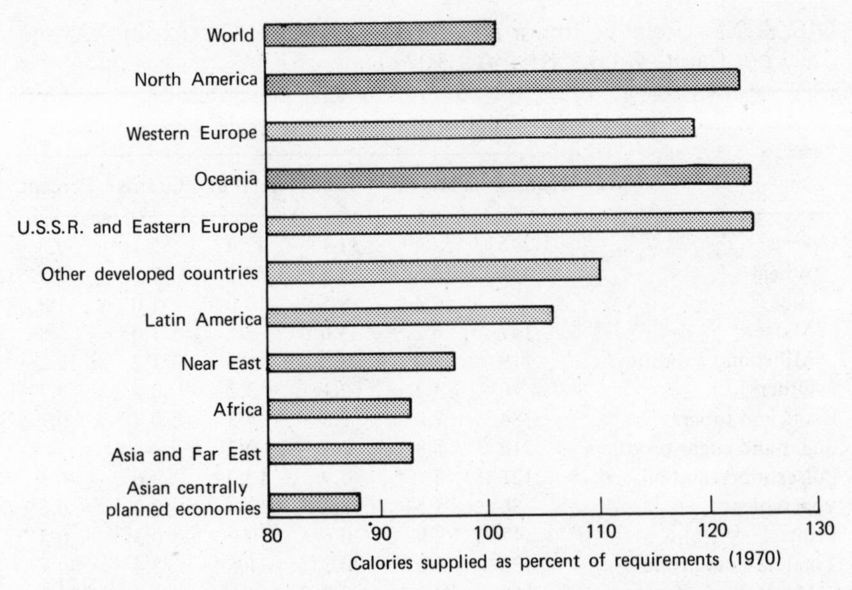

FIGURE 3.2 About two-thirds of the world's population lives in regions where calories supplied are less than requirements (UNFAO 1971).

sidered simultaneously. Since most people live in calorie deficient areas, a considerable portion of their protein intake is being wastefully used to meet their energy requirements. There is also a factor relating to the quality of the protein which we shall consider in the next section. That the world has a widespread protein malnutrition problem is clinically documented, so the figure must be viewed with caution.

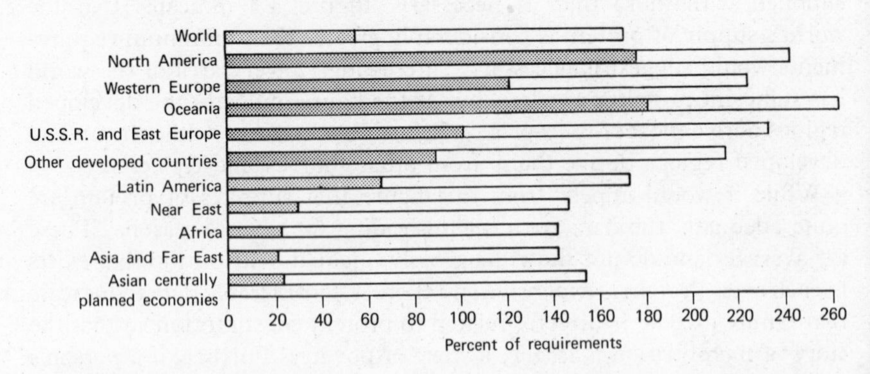

FIGURE 3.3 Protein supplied as a percent of requirements. Dark area is the protein supplied from animal sources. 1970 data (UNFAO 1971).

3.2 Protein

There are a number of diseases which are caused by malnutrition including anemia, goiter, beriberi, xerophthalmia, scurvy, and rickets, but it is the lack of protein which is the most widespread deficiency.

Proteins are the structural material of life; they are needed for growth, replacing tissues, and metabolism. Hormones, enzymes, and hemoglobin are all proteins. Protein malnutrition in infants and young children is often called *kwashiorkor*, which is an African term meaning the "displaced child" who is no longer breast fed and must fend for himself. Kwashiorkor has its own syndrome of characteristics including the distended stomach—a condition caused by fluids which accumulate in the interstitial spaces between the cells. But besides the disease itself, the weakened condition of the malnourished child causes the mortality rate from common infectious diseases to soar. In the developing countries the mortality rate in preschool children between 1 and 9 years of age is 10–30 times higher than that in the U.S., the large difference being attributed to malnutrition and complicating infections (President's Science Advisory Committee 1967, Vol. II, p. 13).

Proteins are essential not only to the growth of the body but also to that of the brain. During early childhood the brain is growing very rapidly and lack of protein will irreversibly impair the brain's development. Therefore, besides the stunted physical growth caused by malnutrition, large numbers of the world's children are being sentenced to irreparable mental deficiency.

About 20% of our body weight is protein, and while there are millions of different proteins they are all made up of different combinations of only about 20 amino acids. The amino acids are in turn made up primarily of atoms of carbon, hydrogen, oxygen, and *nitrogen* (hence the high protein content of nitrogen-fixing legumes). Nine of the amino acids are essential in our diets and the rest can be synthesized from these nine. When talking about the *quality* of a given protein, it is to the balance of these nine essential amino acids that we are referring. A high-quality protein, then, is one which has all of the essential amino acids in roughly the proportion required by man.

Proteins from animal sources (milk, eggs, and meat) are high quality; in fact, quite often the egg is used as the standard against which other protein sources are measured, since the egg comes very close to being an ideal protein source. Table 3.2 lists various sources of protein with their corresponding amino acid content measured relative to the egg. As noted, beef and fish are very good sources by themselves, whereas the cereals, for example, are generally deficient in at least one of the essential amino acids. However, by eating a combination of foods it is possible to come up

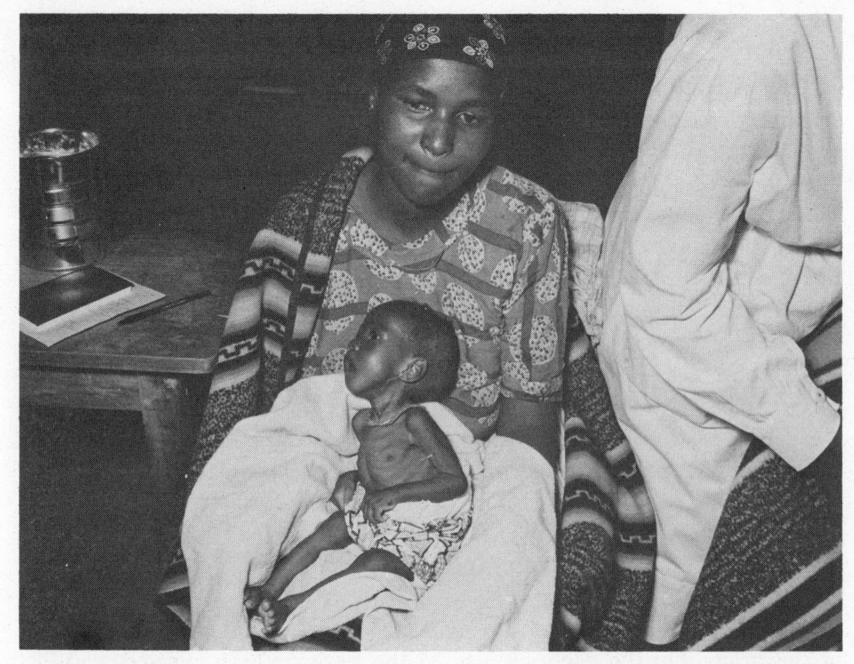

Malnutrition is extremely common among children in the developing nations of the world. (Courtesy UNFAO.)

with a balanced protein diet—for example, the lysine deficiency of whole rice or whole wheat could be made up by consuming dried roast beans or soybeans, both of which are high in lysine. In fact, one of the most promising ways of meeting the protein needs of cereal-consuming populations is to fortify the grains with the lacking amino acids, such as with lysine and tryptophan for corn, before distributing them to the population. Not only is it an economical, relatively simple solution, but it wouldn't require a shift in the eating habits of the consumers.

3.3 *Food Chain Losses*

In Chapter 1, the flow of energy through the biosphere was traced from the initial capture of solar energy on through the various trophic levels of the food chain. It was indicated there that a reasonable rule-of-thumb is that there is only about a 10% efficiency of energy transfer from one trophic level to the next. Thus when man consumes meat rather than plants we would expect this additional trophic level to result in about a 90% loss in available calories. In areas where the food supply is minimal, then, it is necessary for people to consume the crops directly rather than

TABLE 3.2 Percentage of Ideal Concentration of Essential Amino Acids Observed in Typical Proteins (Using Egg as 100%) (President's Science Advisory Committee 1967)

Foodstuffs	Histidine	Threonine	Valine	Leucine	Isoleucine	Lysine	Methionine	Phenyl-alanine	Tryptophan
Beef	157	90	73	87	84	141	84	70	92
Fish muscle	124	96	86	106	105	148	100	79	109
Soybean meal, low fat	138	80	76	89	97	111	53	95	127
Whole rice	81	78	88	91	84	52	106	89	118
Whole wheat	100	67	62	78	64	44	78	91	109
Cottonseed meal	128	61	69	67	64	57	53	107	118
Whole corn	119	76	76	167	103	38	97	89	55
Peanut flour	100	57	66	79	66	57	25	88	72
Dried roast beans	104	79	78	78	89	106	62	89	73
Sesame meal	106	81	67	70	63	38	53	78	93

allow the luxury of the additional trophic level required when animals are consumed. This is apparent in Figure 3.3 where the less developed regions of the world are shown to be consuming quite small amounts of animal protein.

In the U.S., where animal products are significant sources of calories and protein, it takes roughly 10,000 kcal of food production to supply the 3100 kcal that are typically consumed per day. As Table 3.3 indicates, 6300 kcal goes to feed for the animals, from which an American derives about 900 kcal of energy. Handling and waste account for another 1500 kcal.

Not only is there a loss of calories when man acts as a carnivore, but it is also inefficient in terms of protein. Figure 3.4 indicates that 1 acre of land could supply about 6 men with their required protein if soybeans are raised but only about 0.22 men could be supported if the land is used to raise beef which are subsequently utilized by the men. Soybeans could be a much more important source of protein in the future, if they would be used directly as food for humans instead of being used as feed for animals. Soybeans, peas, and beans are seen to be extremely efficient producers of protein, while among animals poultry is more efficient than hogs which are in turn more efficient than beef.

It is important to point out that while these data indicate that animals are inefficient sources of food, this is only important when the animals are consuming food that could be consumed directly by man, or when they are grazing on land that could otherwise be used to grow crops. Quite often neither of these is the case. Animals consume many foods which could not be utilized directly by man such as forages, wastes, by-products, and even urea; and often they graze on lands that are not suitable for anything else. In such cases animals are extremely important because they

TABLE 3.3 In the U.S. about 10,000 kcal are required to supply the 3,100 kcal that are consumed per person per day. Most of the loss is for feed for animals (Cook 1971)

Caloric distribution	Total, kcal	To man, kcal
Handling and waste	1,500	0
Plant material in diet	2,200	2,200
Food for animals	6,300	900
Total	10,000	3,100

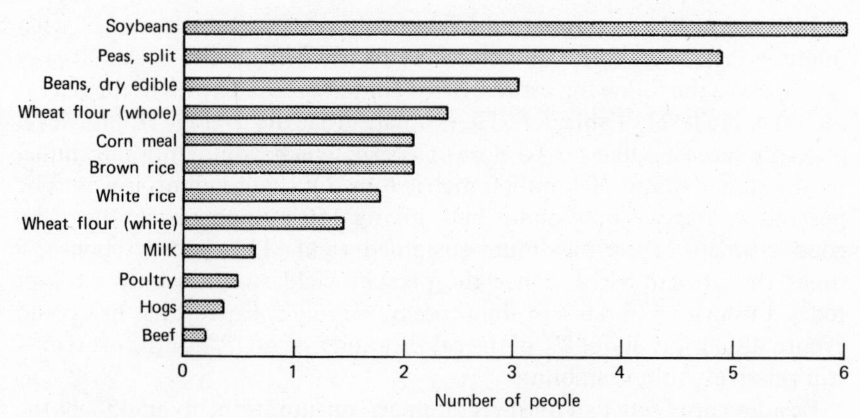

FIGURE 3.4 Number of people whose protein requirements could be met by one acre of selected food products (President's Science Advisory Committee 1967)

supply man with very high quality foodstuffs without competing with him. For example, in India cattle are considered sacred and so are not slaughtered for food. While on the surface this appears irrational in a country where the food problem is so severe, it is in reality rather sensible. The cattle obtain a good portion of their food by scavenging from roadsides, ditch banks, or crop residue wastes and their milk provides high-quality protein. Besides their food value, cow dung is rural India's main cooking fuel and cow manure is an important source of fertilizer. Bullocks also provide most of the draft power for agriculture.

3.4 *Food from the Sea*

So far the discussion has centered around the food resources of the land, so let us briefly shift our attention to the sea since it is often erroneously thought to be potentially a source of vast amounts of food. Referring back to Table 3.1 we see that, as of 1965, fish supplied only about 0.8% of the world's calories and about 4.6% of the protein so that, for the world as a whole, it was a relatively minor food source. However, for some countries, such as Japan, it is a very important source of protein. World consumption of fish is projected to increase from the 1970 value of 66.5 million metric* tons to about 83.3 million metric tons in 1980. This corresponds to an annual growth rate of about 2.3% per year, a value slightly greater than the world population growth rate (UNFAO 1971).

*A metric ton is 1000 kg, or 1.1 tons.

It is of interest to compare these annual yields with estimates of what might be the maximum annual sustained yield from the ocean. Ricker (1969) gives the following estimates for annual productivities in the sea for each trophic level (Table 3.4). He estimates that the average trophic level of man's harvest is likely to be close to level 4, which would imply an annual production of about 300 million metric tons. Of that amount, it would be possible to harvest only about half, giving 150 million metric tons as a good estimate of the maximum sustained yield. This is only about 2.5 times the present yield. Since the present yield supplies only 0.8% of today's calories and 4.6% of the protein, we might expect that fish could eventually supply about 2% of the calories and maybe 12% of the protein—still relatively minor amounts.

Besides supplying fish for direct human consumption, about 35% of the total catch is used to produce fish meal which is used in high-protein feeds for pigs and poultry. About 5.5 tons of protein in the feed are required to produce 1 ton of protein in poultry and about 8 tons are required for 1 ton of pork (President's Science Advisory Committee, Vol. II p. 338). So we can see that this is an inefficient way to use these valuable proteins.

What about harvesting plants from the sea? This appears to be unattractive because much of the plankton is of such low density that more energy would be required to gather it up than would be gained by the harvest (as Ricker states, if it were all concentrated in the top 10 m of water, it would have a density of about 1/2 g/m^3). It appears to be better to let the fish gather it up for us and then harvest the fish. Plants such as kelp are being commercially harvested but since they are restricted to shallow water (in order to be attached to the bottom and still reach the surface for photosynthesis), the total potential harvest is not great.

The difficulty with the sea as a major food source is that, even though it covers about three-fourths of the earth's surface, the majority of the

TABLE 3.4 Estimated Annual Productivities at Different Trophic Levels (Ricker 1969)

Trophic level	Annual production, millions of metric tons
1	130,000
2	13,000
3	2,000
4	300
5	45

Bags of fish meal ready for export from Peru to North America and Western Europe. This high-quality protein will be used as the main component of animal feeding stuffs. (Courtesy, UNFAO).

productivity is concentrated in certain very localized areas. The open sea, which comprises about 90% of the oceans, is sometimes referred to as a biological desert, the productivity is so low. It is in the coastal areas and certain regions where there are natural upwelling currents that nutrients are available in sufficient quantities to support a large biomass of producers (phytoplankton) which in turn support the higher trophic levels that we are interested in exploiting. Unfortunately it is precisely here in these coastal areas where most of man's toxic pollutants end up as streams and rivers carry these wastes to the sea. So there is the growing danger that pollution, in such forms as pesticides or heavy metals, may seriously damage the coastal ecosystems which are absolutely essential to life in the sea. We will have more to say about this in later chapters.

3.5 *The Land*

Since it appears that the sea has rather limited potential to become a major food source, it is going to be necessary to increase food production on the land in order to keep up with the increasing demand. There are two ways that this can be done—increase the acreage being cultivated or increase the yield per acre. Of course, both approaches will be necessary so let us briefly consider the possibilities.

The President's Science Advisory Committee (1967) has estimated the potentially arable land in the world to be about 7.88 billion acres. This includes all the land then under cultivation, about 3.43 billion acres, plus any other land which could potentially grow crops. To bring these not-yet cultivated lands into production would probably require such measures as irrigation, drainage, stone removal, clearing of trees, new roads, and a certain amount of resettlement—all of which would require capital expenditures. Estimates by the Committee of the costs per acre to bring new land under cultivation, range from $32 to $973, so it is a rather expensive proposition. As can be seen in Figure 3.5, most of the remaining arable land which is not yet under cultivation is in Africa and South America. Asia and Europe have very little land left while North America has the potential to nearly double its cultivated land. Since the majority of the world's population growth is going to take place in Asia, where there is very little unused land remaining, either the yields per acre there must increase to keep up, or a much better world trade and food distribution system must be established to move the food from these other continents into Asian mouths. Both will be required, but since it will not

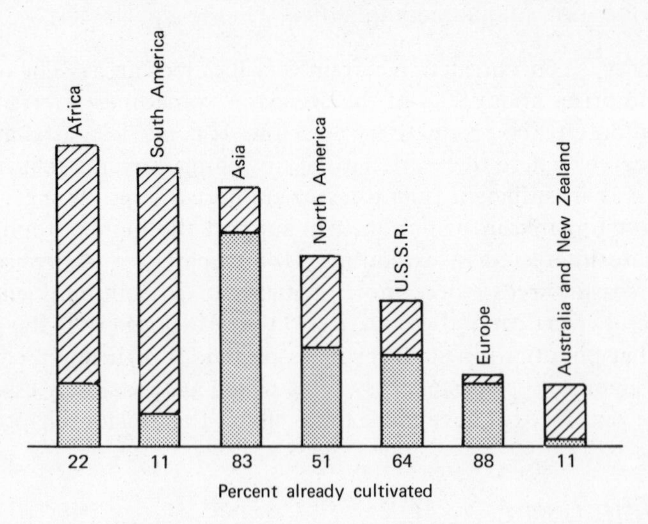

FIGURE 3.5 Potentially arable land (total bar) compared to land already under cultivation (shaded). Most of the remaining arable land is in Africa and South America (President's Science Advisory Committee 1967).

be possible for new lands to keep up with population growth anyway, all efforts need to be devoted to increasing yields per acre. However, as we shall see, there are many severe ecological problems associated with increasing yields.

It is useful to understand a few concepts relating to soils when considering the possibility of increasing yields or bringing new land under cultivation. We have already pointed out that plants get their energy from the sun and their carbon (for production of new organic material) from the carbon dioxide in the air. The nutrients that are needed are absorbed from the soil through the roots. The soil itself must be porous, not only so that water can drain to the root system, but also so that oxygen from the air can be available for respiration by the roots. The oxygen supply can easily be cut off if the ground becomes water logged from excessive irrigation, rendering the land unusable.

For a soil to be fertile it must have a rich content of organic material called *humus*, consisting of dead and decaying plant and animal material, which is being decomposed by bacteria and fungi. This decomposition recycles the nutrients and makes them available for reuse by the plants. The humus gets mixed with the mineral components of the soil by the burrowing activities of countless soil organisms, which helps to distribute these essential nutrients throughout the soil. As living plants absorb the nutrients, the soil's fertility is maintained by the leaf fall and general accumulation of organic litter, or in some cases, by annual flooding with nutrient-rich silt. In a temperate forest the annual leaf fall results in a buildup of slowly rotting organic material covering the rich topsoil. After thousands of years a relatively thin layer of rich soil builds up, held in place by the complex root systems of the covering plants. When such an area is cleared and planted, not only are the nutrients removed from the soil when the crops are harvested, but the protective cover is removed subjecting the land to one of the world's most pressing problems, that of soil erosion. Each year millions of acres of cropland must be abandoned when improper care of the soil leads to erosion. The conversion of North Africa from the fertile granary for the Roman empire into a desert, and the more recent "dust bowl" experience in the U.S. are two important examples. Not only are invaluable soil resources lost during erosion, but the resulting sedimentation in waterways can lead to such harmful effects as the silting in of dams, the smothering of coral reefs, and excess fertilization of the water leading to algal blooms and eutrophication.

Much of what has been discussed refers to the buildup of topsoil in ecosystems which exist in the temperate zones of the world. The situation is quite different in the tropics which, significantly, contain most of the arable land that is not yet being cultivated. What is the problem in the tropics? It would appear from the lush vegetation that the land would be eminently suitable for intense cultivation, but just the opposite happens to be true. Tropical soils are usually extremely poor because the heavy rainfall leaches away nutrients from the soil before they have a chance to accumulate. Most of the nutrients are therefore locked up in the vegetation. When plants die and decay their nutrients must be quickly returned to the living vegetation or else they will be lost.

When a tropical rain forest is cleared for any length of time (or defoliated as in Viet Nam) the recycling of nutrients is halted and what small supply of nutrients that exist in the soil are quickly leached away by the heavy rainfall. Sometimes what remains is a soil that is high in iron and aluminum oxides which, when baked in the sun, produce a rocklike substance called *laterite* (from Latin *later* for brick). In fact, the ruins at Angkor Wat in Cambodia were built of lateritic bricks and have lasted some 800–1000 years so far. McNeil (1964) describes the problems of laterization including an example from Brazil in which an agricultural colony was attempted in the Amazon basin. The forests were cleared and crops were planted but in less than 5 years the soil irreversibly hardened into virtual pavements of rock.

3.6 *Increasing the Yield*

Increasing the yields on already cultivated lands is going to be a key factor in meeting the increasing demand for food. Figure 3.6 shows, as an example, the progress which has been made on farms in the U.S. where in the past 20 years yields per acre have increased by about 60%. Much

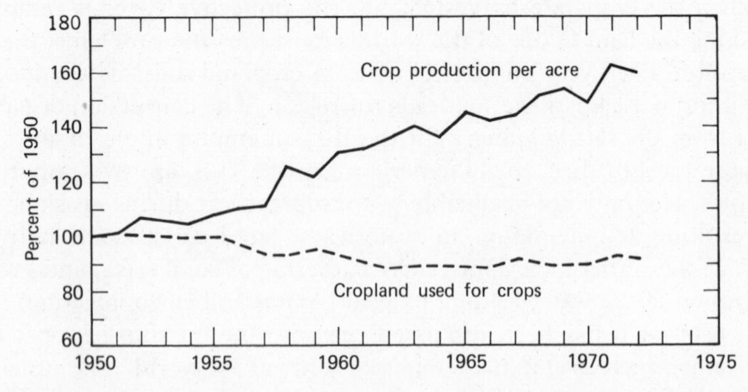

FIGURE 3.6 Crop production per acre in the U.S. has increased about 60% in a 20 year period. (USDA 1972.)

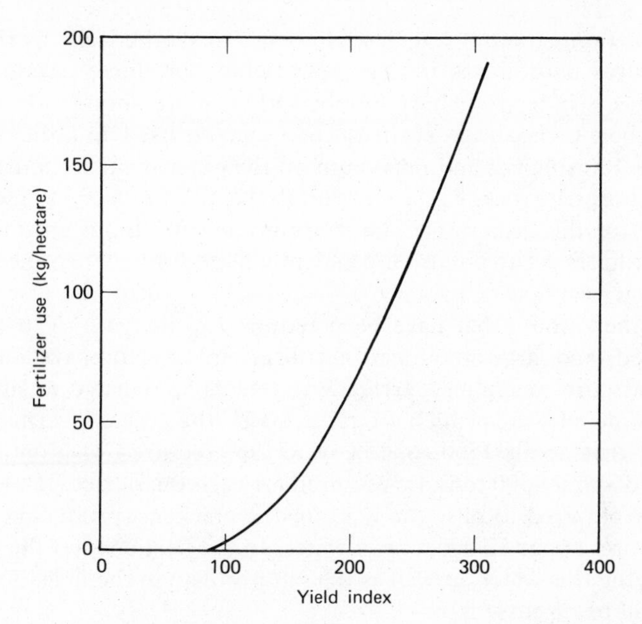

FIGURE 3.7 Doubling the food yield per unit area requires nearly a ten-fold increase in fertilizer as well as better seed and increased pesticide use. (President's Science Advisory Committee 1967, Vol. II, p. 381).

of the progress made by modern agriculture can be attributed to advances in plant genetics which have made possible the development of cereals and other plant species that according to Brown (1970) are "more tolerant to cold, more resistant to drought, less susceptible to disease, more responsive to fertilizer, higher in yield and richer in protein." In turn, these advances are dependent on mechanization, irrigation, fertilization and pesticides. While each of these inputs has helped bring about increased yields, they have also all perturbed the natural cycles of the biosphere.

It has been mentioned that as crops are harvested, the nutrients which were incorporated into the plants from the soil are removed. In order to maintain the soil's fertility (or indeed to build it up in the first place) it is necessary to replace those nutrients either by returning the waste products after consumption, by encouraging nitrogen fixation by planting legumes, or by manufacturing and utilizing artificial fertilizers. It is the latter that concerns us here. As Figure 3.7 indicates, doubling crop yields per acre will require about a ten-fold increase in fertilizer use.

The three nutrients that are most needed are nitrogen, phosphorus, and potassium, in that order. Nitrogen is obtained from the air so there is no danger of running out of reserves. However, to synthetically fix the nitrogen requires natural gas in a reaction which produces ammonia, and natural gas is indeed in short supply. Other, more abundantly available hydrocarbon compounds such as coal, can be used but they are less desirable. Phosphorus and potassium on the other hand are mined in the form of phosphate rock (P_2O_5) and potash (K_2O), the reserves of which are adequate for the near future. Of more immediate importance for all of these fertilizers is the pollution problem which results when they end up in our waterways (see Chapter 6).

The other inputs that have been required to increase yields, besides better seeds and large amounts of fertilizer, are irrigation systems, pesticides, and farm machinery. Irrigation systems are needed to supply the vast amounts of water which are required by the crops. Overman (1969) indicates that about 1500 tons of water are required to grow 1 ton of wheat and about 4500 tons are required for each ton of rice. If we include the water required to raise the food for the chickens, producing 1 ton of eggs requires about 12,000 tons of water. In the next chapter the problem of supplying this water, as well as the salt problem in the irrigation runoff water, will be discussed.

Farmers in Honduras display results of a fertilizer experiment. Corn grown without fertilizer is on the right, while other bags of corn were grown with varying fertilizer formulas. (Courtesy UNFAO).

The importance of machinery is generally thought of in terms of its labor-saving function, but besides that, by speeding up such processes as seedbed preparation and harvesting, and by doing a better job in the placement of seed and fertilizer, machinery contributes to increasing yields per acre. The manufacture and use of this machinery depends upon a resource base of metals and fuels—a resource base that is fast diminishing as will be discussed in Chapter 15.

The last of these important inputs to increasing yields is pesticides. Our growing dependence on pesticides has such serious ecological implications that we will treat this problem now in some detail.

3.7 *Pesticides*

Pesticides are used for both the protection of agricultural crops and for control of such dreaded insect-borne diseases as typhus, malaria, encephalitis, and bubonic plague. The introduction of DDT during World War II was credited with stemming a potentially serious outbreak of typhus in Europe by controlling body lice. By 1953, estimates indicate that perhaps 5 million lives had been saved from malaria by DDT control of mosquitoes. The incidence of malaria in India alone dropped spectacularly from 75 million cases in 1952 to about 100,000 in 1964 (President's Science Advisory Committee 1967, Vol. II, p. 135). Encephalitis, another mosquito-borne disease, has no cure and can only be controlled by eliminating the mosquitoes. The remarkable achievements of DDT earned for its discoverer, Paul Mueller, a Nobel Prize.

In the area of crop protection, the requirement for pesticides can be thought of as a direct consequence of the reduction in species diversity that characterizes agricultural monocultures. Conditions are created in agriculture that favor the crop pest, which upsets the balance between it and its natural enemies so that some kind of external control is necessary Chemical controls have typically been relied upon and there has been a good correlation between the amount of pesticide used and crop yields (again, realize that the pesticides are just one component of a "package" of techniques for increasing yields). The data in Table 3.5 indicate that a doubling of yields has been accompanied by roughly a ten-fold increase in pesticide use.

The word *pesticides* is a general term that refers not only to insecticides but also herbicides, fungicides, fumigants, and rodenticides. The use of herbicides to control weeds which compete with the crops often helps increase yields substantially. In carefully controlled farmer demonstrations in the Philippines, West Pakistan, and Brazil, rice yield increases averaging 46% were noted. And it has been estimated that maize yields are reduced

TABLE 3.5 Pesticide Use and Yields of Major Crops[a]

Area or nation	Pesticide use, grams per hectare	Yields, kilograms per hectare
Japan	10,790	5,480
Europe	1,870	3,430
United States	1,490	2,600
Latin America	220	1,970
Oceania	198	1,570
India	149	820
Africa	127	1,210

[a]Source: UNFAO 1963a.

an average of 30% by weeds in subtropic and tropical climates. All told, insects, weeds, rodents, birds, and other organisms contribute substantially to crop losses which, for example, in the U.S. are estimated at somewhere around 30%, in spite of pesticides (President's Science Advisory Committee 1967).

In view of the need to control disease vectors and limit crop losses, and considering the outstanding successes which pesticides have achieved, why then is there a controversy over their use? Since most of the controversy centers around insecticides, DDT in particular, we will concentrate on them.

3.8 *Insecticides*

There are three main groups of synthetic organic insecticides: chlorinated hydrocarbons, organophosphates, and carbamates. All three groups kill by disrupting the transmission of nerve impulses. For organophosphates and carbamates the mechanism is better understood being one which causes the repetitive firing of nerves which leads to convulsions and death. The insecticides in these three groups also share the common property of being highly soluble in fatty tissues. This property makes it possible to kill on contact by allowing the insecticide to penetrate the thin layer of hard fatty material that covers the body of insects. Table 3.6 lists some of the more important insecticides in each of the three groups (for the exact nomenclature and structural diagrams for each of these, see for example, American Chemical Society 1969).

One of the most important properties initially designed into insecticides was nondegradability. It was thought to be highly desirable for an insecticide to be persistent so that repeat applications need not be made very frequently. Another property that is desirable from the point of view of both the pesticide manufacturer and the casual user, is that the insecticide should be effective against a broad spectrum of organisms so that a single product can be prescribed for many pests. As we shall see these two properties are responsible for much of the ecosystem disruption caused by pesticides.

In terms of ecosystem disruption the chlorinated hydrocarbons are the most damaging. The key distinction is that they are highly persistent, while organophosphates and carbamates are broken down rather quickly in the environment. Since the chlorinated hydrocarbons do not degrade very quickly, their effects will be felt far into the future. Figure 3.8, for example, shows computer predictions of the length of time that DDT would remain in the environment if all usage were suddenly stopped, and also if the usage rate were to be gradually decreased to zero. As can be seen, even if all applications of DDT were to be stopped immediately (as happened in the U.S. in 1973), the concentration in fish would still be sizable well into the next century. It is also significant to note that DDT breaks down into two other biologically active compounds, DDD and DDE, which are themselves long-lived and biologically disruptive.

Combining the long-lived property with the fact that chlorinated hydrocarbons are highly mobile (DDT, in fact, codistills with water) results in their spread throughout every ecosystem in the world. For example, DDT residues have been found in Alaskan Eskimos as well as in seals and penguins in Antarctica. Figure 3.9 shows DDT moving from the fields into the air and water and finally ending up in the oceans where it is picked up in the food web. As it passes through the food chain the con-

TABLE 3.6 Common Insecticides

Chlorinated hydrocarbons	Organophosphates	Carbamates
DDT, DDD, DDE	Methyl parathion	Carbaryl
Aldrin	Parathion	Zectran
Dieldrin	Malathion	
Endrin	Diazinon	
Heptachlor	Fenthion	
Chlordane	TEPP	
Lindane	Azinphosmethyl	

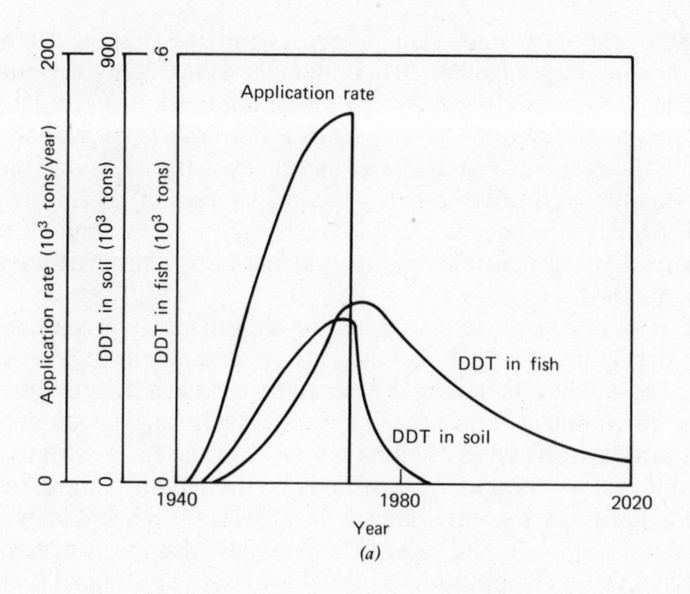

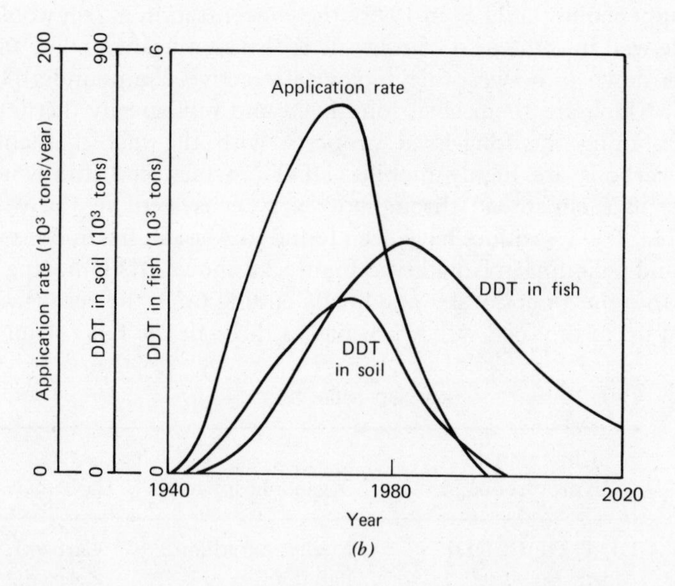

FIGURE 3.8 Computer predictions for the length of time DDT would remain in the environment if all applications were to cease: (*a*) immediately; (*b*) gradually. From Meadows, D. L., and Randers, J., "Adding the time dimension to environmental policy," in D. A. Kay and E. B. Skolnikoff eds., *World Ecocrisis*, (© 1972 by the Regents of the University of Wisconsin).

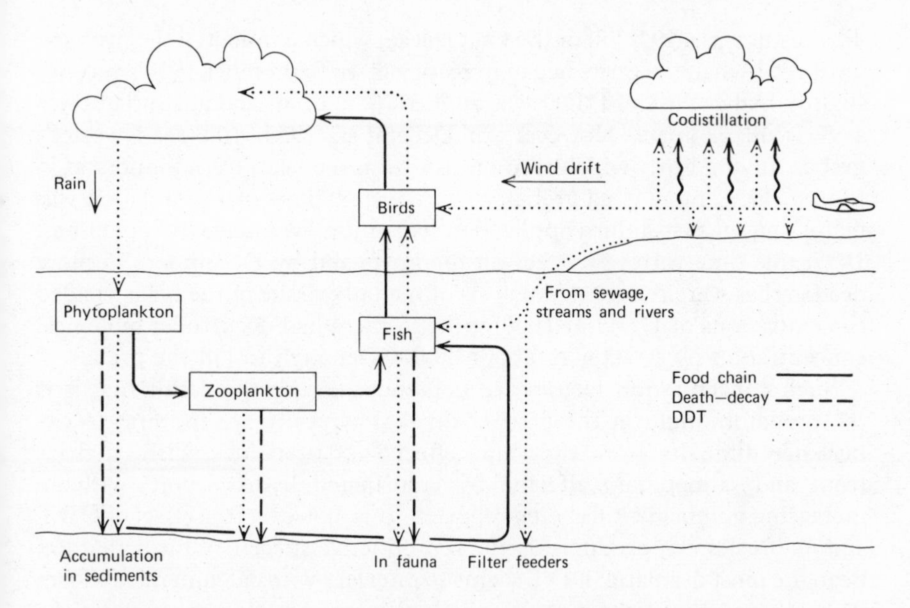

FIGURE 3.9 DDT codistills with water and can drift far from its point of application, concentrating as it moves through the food chain.

centration of DDT in each organism generally increases with each trophic level (see Figure 1.4). This phenomenon, called *biological concentration,* results because less DDT is excreted by an organism than is consumed, the difference being stored in the fat. Organisms at successively higher trophic levels are consuming food that has successively higher concentrations of DDT, which increases the DDT burden at higher trophic levels. The consequences of this biological concentration are far-reaching. While the DDT concentration in water may be quite low, it can be multiplied by a factor of thousands high up the food chain.

Effects on Nontarget Organisms

"It is impossible to do just one thing" is a common ecological expression. Pesticides not only affect the particular pest that they are used on, but they always have unwanted effects on the rest of the ecosystem. As a particular example, consider the case of Clear Lake, California, where DDD was used to try to control gnats (Hunt and Bischoff 1960). In 1949, Clear Lake was treated with an amount of DDD which would result in a concentration of 0.014 parts per million (ppm) of DDD in the water.

The result was a 99% kill of the gnat larvae, which eliminated the problem until 1951 when the gnats began to reappear. In September 1954 a second, slightly higher dose of DDD was applied (0.02 ppm) and again there was a 99% kill of gnats. However, in December, 1954, 100 dead western grebes (diving birds which feed on fish from the lake) were found and in March, 1955, more dead grebes were reported. The gnat population was increasing again so a third application (0.02 ppm) was made in September, 1957, this time with less effect on the gnats and by December, 75 more dead grebes were reported. Analysis of the fatty tissue of the birds yielded concentrations of DDD of 1600 ppm; so there had occurred a biological concentration by a factor of about 80,000—enough to kill the grebes.

Such concentration factors are not unusual with the result that it is the predators high on the food chain that typically are the first to experience difficulty from the "side effects" of pesticides. Fish, such as trout and salmon, are affected by chlorinated hydrocarbons with an increasing number of kills being reported. It is, however, the effect of DDT and its breakdown products on the reproductive success of birds that has been the most dramatic. DDT seems to interfere with calcium metabolism in birds, resulting in eggs with shells that are too thin to support the weight of the parent. The result is an inability to reproduce, which is affecting such species as the peregrine falcon, the bald eagle, ospreys,

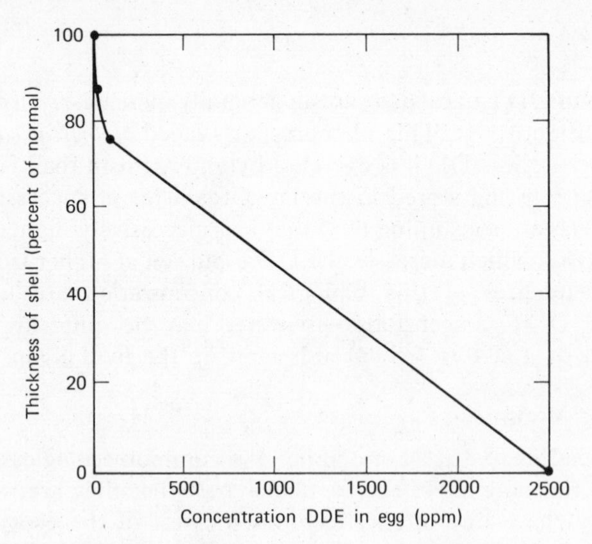

FIGURE 3.10 Severe effect of the concentration of relatively small amounts of DDE: 25 ppm in the egg reduces shell thickness by 15%; after 20% shells are usually broken. From Peakall, D. B., "Pesticides and the Reproduction of Birds." Copyright © 1970 by Scientific American, Inc. All rights reserved.

and brown pelicans. Figure 3.10 shows the relationship between eggshell thickness and concentration of DDE in the egg indicating that even very small doses can cause enough thinning to cause eggshell breakage.

It should be pointed out that another pollutant, polychlorinated biphenyls (PCB), also cause thinning of eggshells, perhaps even more effectively than DDT. PCBs are used commercially in plasticizers, dielectrics, and lubricants, are similar to DDT and can cause similar physiological reactions. They can be released into the environment when certain plastics weather or are burned.

Very disturbing is the discovery that DDT reduces photosynthesis by marine phytoplankton (microscopic floating plants), as shown in Figure 3.11. The concentrations at which significant decreases in photosynthesis occur, are higher than usual background levels of DDT (roughly 1 part per billion), but are comparable to the levels near the site of DDT application (recall the concentrations applied to Clear Lake were about 20 ppb). Besides the fact that marine phytoplankton, through photosynthesis, supply about 70% of the world's oxygen production, any decrease in productivity at the base of the food chain could have drastic effects on the entire marine system. Further, recall that most of the photosynthesis is occurring in the shallow coastal areas, where nutrients and sunlight are abundant, and that is where the DDT is being washed into the sea.

When it comes to assessing the damage which pesticides may be causing to man directly, the picture is not yet very clear. The low-level concentrations to which man is generally exposed have not been shown conclusively to cause any adverse effects. The current average daily dose of DDT in the U.S. is about 0.0004 mg/kg/day (milligrams per kilogram of body weight per day). It is ten times higher in India, and some workers in DDT factories have been receiving about 0.25 mg/kg/day for 19 years with apparently no ill-effects (WHO 1972). Obtaining meaningful data on low-level effects is extremely difficult since, for example, if a pesticide is a carcinogen (cancer-causing), it may take anywhere from 5 to 30 years to induce the cancer. Also, experimentation is complicated by the fact that there are no control groups to make comparisons with since essentially everyone has already been exposed.

Peakall (1970), however, reports that the National Cancer Institute has determined that 46 mg of DDT per kilogram of body weight causes a four-fold increase in tumors in liver, lungs, and lymphoid organs of animals. Human cancer victims have been found to have 2.5 times more DDT in their fat than occurs in normal people.

High-level applications of pesticides to humans as can occur during an accident in handling or perhaps from improper usage, can result in sickness or death. Agricultural workers are particularly vulnerable. The agricultural industry has the highest general occupational disease rate of

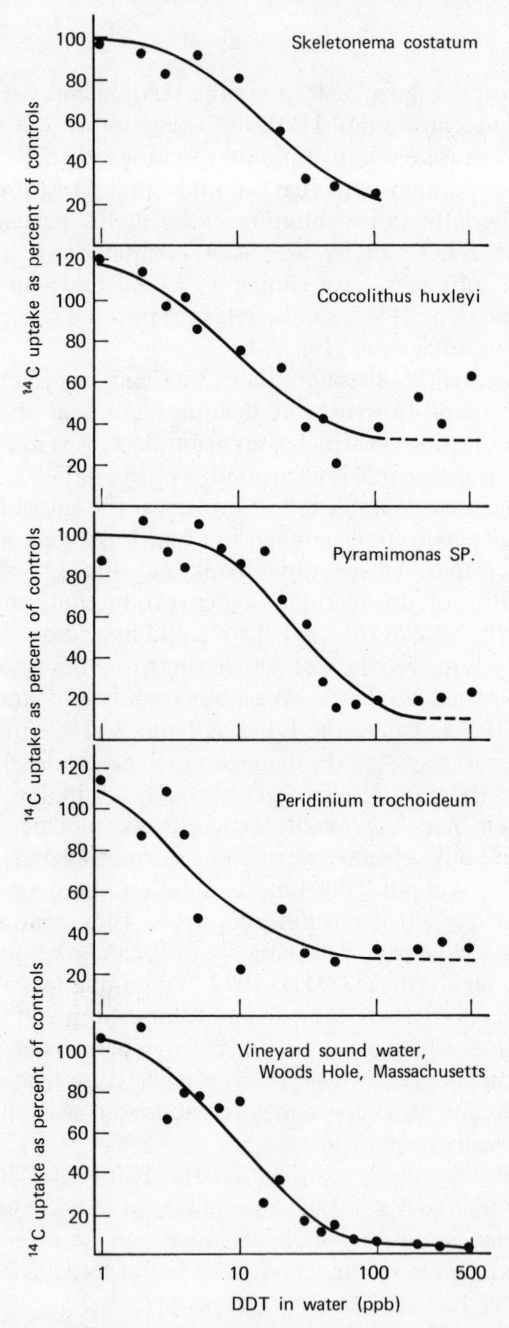

FIGURE 3.11 Reduction in photosynthesis as measured by uptake of ^{14}C relative to controls, for various species of marine phytoplankton. (Wurster 1968). Copyright 1968 by the American Association for the Advancement of Science.

all California industries—three times the average. From 1951 to 1967 there were 151 accidental deaths attributed to agricultural chemicals in California and in 1967 there were 692 reported cases of disease among California farm laborers attributable to agricultural chemicals (Gibson and Watson 1970). This danger is increasing since the less toxic chlorinated hydrocarbons are being replaced by the extremely toxic organophosphates.

Biological Resistance

While evidence of the destructive effects of pesticides mounts, it has long been apparent that insects very quickly evolve resistance to pesticides making the pesticides less and less useful—or on the other hand, necessitating higher and higher doses to accomplish the same task. The buildup of resistance is a classic example of Darwinian selection. Of those insects that survive the first spraying, some do so because they are genetically different in such a way as to make them less susceptible to the effects of the pesticide. This genetic resistance is then passed on to their offspring, so the next generation of insects has a higher percentage of resistant individuals.

The usual spraying program accelerates this effect. When it is seen that the initial dosage does not seem to be working anymore, the usual thing to do is to increase the dosage. This backfires because it ensures that the only bugs that will survive and continue to reproduce are those that are really resistant. Hence you are selectively breeding "super bugs." Also, by spraying as broad an area as possible, the farmer ensures that no non-resistant insects will migrate into the area, which allows the resistant insects to firmly establish themselves.

These effects were noticed as early as 1946 when some populations of houseflies were reported to be resistant to DDT. By 1969 there were 224 species of insects that had developed resistance to one or more groups of insecticides, including 127 agricultural pests and 97 pests of medical and veterinary importance [U.S. Department of Health, Education, and Welfare (HEW) 1969]. Among malaria-carrying mosquitoes, 38 species have developed resistance to various insecticides including dieldrin, DDT, malathion, and propoxur. The widespread use of insecticides for agriculture increases the danger that their effectiveness as a control for insect-borne diseases such as malaria, encephalitis, yellow fever, and bubonic plague, will be diminished.

The development of resistance can never be completely avoided as long as pesticides are used, however it is possible to reduce the problem by using the pesticides wisely and sparingly. Not only would this cut down on the increasing resistance being built-up, but the polluting effects of pesticides on the rest of the environment could be minimized.

Effects on Predator-Prey Relationships

Consider the simple food chain of an agricultural crop being attacked by a pest which is in turn being preyed upon by a predator:

$$\text{crop} \rightarrow \text{pest} \rightarrow \text{predator}$$

Pesticides not only kill off the pests, but they also kill the pest's natural enemies, oftentimes doing a better job on the latter. This is especially true when broad-spectrum poisons are used. Then, with no natural enemies around, the pests which have survived can flourish.

In Kormondy (1969) the statement is made that "any factor which is moderately destructive to both predator and prey will increase the average *prey* population and decrease that of the predator." There are several reasons for this. One is simply that there are many more pests than predators and the probability of some pests surviving is therefore higher. Also a few pests can easily be hidden by the environment making it difficult for the predators to find them.

Another reason is that biological concentration causes species higher up the food chain to be subject to higher doses of poison. And finally pests are more used to being poisoned because plants, for defense, have been evolving their own poisons (e.g. nicotine, poison oak, quinine, and pepper) and in response the pests have evolved defense mechanisms which the predators have not.

There is another aspect of broad-spectrum pesticides which often causes unanticipated results—that is the elevation of a hitherto harmless species to pest status. Consider a food web having two pests being held under control by one predator. If a pesticide is used to eliminate one of the pests, the predator may inadvertently also be eliminated, in which case the second pest species suddenly increases in number. An example of this occurred at Lake Tahoe where spraying for mosquitoes resulted in a large increase in a scale insect on pine trees. The natural control of the scale insect was the wasp, which inadvertently was being affected by the pesticide being used for mosquitoes.

The overall effect of the use of pesticides is thus, most often, an upsetting of the natural balances between pests and their natural enemies which, in some cases, results in higher pest damage than would have occurred had the pesticide not been used at all. Figure 3.12 shows such an example which took place in California, wherein a citrus pest, the cottony cushion scale *Icerya purchasi*, was accidentally introduced in 1868 from Australia and soon grew to such numbers that it threatened disaster

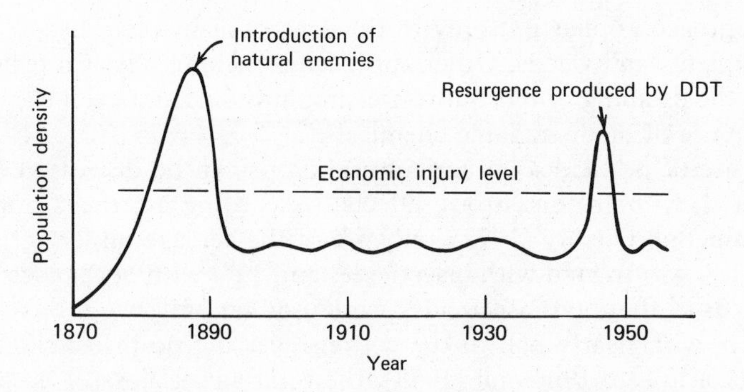

FIGURE 3.12 A scale insect pest was kept under control by its natural enemies until DDT upset the balance.

to the entire California citrus industry. Two of the pest's natural enemies were then introduced and quickly control over the scale was established and held for some 60 years until 1946 when DDT began to be used in the area. The DDT killed off the natural enemies so the scale once again became a major pest. The withdrawal of DDT then gradually allowed the natural enemies to regain control (Doutt 1958).

Integrated Control

Against this background of initial spectacular success followed by progressively worsening results, the question is what approach can be used? Clearly, if pesticides will no longer protect us against the spread of disease and if we have become dependent upon them to protect the food supply for the mushrooming population, then what can be done?

There are a wide variety of approaches which can be taken depending upon the particular pest that is involved. The term "integrated control" refers to a combination of chemical, biological, and physical techniques for controlling pests without the use of long-lived pesticides. It requires a much more detailed knowledge of the life cycle of the pests so as to know the most effective way to control them.

An example of the approach is the program which effectively removed the screwworm fly as a pest in the Southeast United States. Male screwworm flies were sterilized by irradiation and then released into the general population. Since the females mate only once, if she happens to mate with one of the large number of sterile males, there will be no offspring. The reproduction of the screwworm was effectively stopped with this technique without adverse side effects to other species.

As was suggested by Figure 3.12, one of the important control techniques is to introduce and encourage the natural enemies of the pest.

This approach applies not only to the control of insects but also to unwanted plants and weeds. Other approaches include physical techniques such as the draining of mosquito-breeding grounds and even the careful, selective use of biodegradable chemicals.

The use of pesticides on agricultural crops can be decreased or even eliminated without disastrous effects. According to the Council on Environmental Quality (1972), only 5% of the acreage under agriculture in the U.S. was treated with insecticides and 12% with herbicides. About two-thirds of the total insecticides were used on only two crops—cotton and corn, with nearly half of the cotton receiving no insecticides at all. And, according to Pimental (1973), the total annual loss of crops from pest damage in the U.S. would rise only about 7% if pesticides were eliminated altogether.

3.9 *Conclusions*

"Most children who are born in our era will face malnutrition in their formative years, and an unknown minority of human life suffered and will suffer severe malnutrition. In some areas half will die. Diseases which a healthy child survives can easily prove fatal for the undernourished. There is a great deal of hard evidence from rat experiments and much (perforce) circumstantial evidence from investigations on humans that those who survive severe malnutrition in early infancy have permanently, deleteriously affected, physical and mental capacities and are much more vulnerable to diseases" (Palmer 1972, p. 83).

The world is presently producing just enough food to feed everyone adequately, but due to the inequities of distribution and the invariable wastage, probably half the world's population is suffering from hunger. With the population expected to double in 35 years the prospects for widespread famine by the turn of the century are extremely high.

Theoretically, it will be possible to meet the increasing demand for food by a combination of approaches including cultivating the remaining arable land, continuing to improve yields, improving the amino acid balance in grains, utilizing foods such as soybeans directly for human consumption rather than as animal feed, and very importantly, to provide for better food distribution. In addition, there are several more exotic schemes for meeting the needs for high-quality protein which were not discussed since it appears they will not be significant in the near future. Included in this category are protein additives made from fish called Fish Protein Concentrate (FPC) and single cell proteins which are produced by the culture of yeasts or bacteria on hydrocarbon substrates such as coal, oil, and natural gas.

The most important method being used to keep the food supply up to demand is to increase the yield per acre. This requires a package of technological techniques which includes increased consumption of fertilizers and pesticides, improved irrigation systems, and more horsepower. While each component of the package increases the ecological impact of agriculture, the emphasis in this chapter has been on the use of pesticides.

Pesticides have been credited with saving millions of lives through the control of insect-borne diseases, and they have been helpful in the improvement of agricultural yields, but there are strong arguments for severely curtailing their use. The long-lived, broad-spectrum pesticides, especially DDT, have many deleterious effects on nontarget organisms which range from a decrease in phytoplankton photosynthesis to reproductive failures among predatory birds. Especially alarming is the increase in biological resistance to pesticides which insects are acquiring. Much more work needs to be done in the area of integrated control to help decrease the growing dependence on pesticides.

Bibliography

American Chemical Society (1969). *Cleaning our environment: The chemical basis for action.* Washington, D.C.

Brown, L. R. (1970). Human food production as a process in the biosphere. *Sci. Am.* Sept.

Cook, E. (1971). The flow of energy in an industrial society. *Sci. Am.* Sept.

Council on Environmental Quality (1972). *Environmental Quality—1972.* Washington, D.C.: U.S. Government Printing Office.

Doutt, R. L. (1958). Vice, virtue, and the Vedalia. *Entomol. Soc. Am. Bull.* 4 (4): 119–123.

Ehrlich, P. R., and Ehrlich, A. H. (1970). *Population, resources, environment.* San Francisco: W. H. Freeman.

Gibson, M., and Watson, A. (1970). *Pesticide exposure and protection of California farm workers.* Stanford Workshop on Political and Social Issues, Stanford, Ca. Nov.

Hardin, C. M. ed. (1969). *Overcoming world hunger.* Englewood Cliffs, N.J.: Prentice-Hall.

Hunt, E. G., and Bischoff, A. I. (1960). Inimical effects on wildlife of periodic DDD applications to Clear Lake. *Calif. Fish and Game* 46: 91–106.

Kormondy, E. J. (1969). *Concepts of ecology.* Englewood Cliffs, N.J.: Prentice-Hall.

Meadows, D. L., and Randers, J. (1972). Adding the time dimension to environmental policy. In D. A. Kay and E. B. Skolnikoff eds. *World eco-crisis.*

McNeil, M. (1964). Lateritic soils. *Sci. Am.* Nov.

Overman, M. (1969). *Water: Solutions to a problem of supply and demand.* New York: Doubleday.

Palmer, I. (1972). *Food and the new agricultural technology.* Geneva: United Nations Research Institute for Social Development.

Peakall, D. B. (1970). Pesticides and the reproduction of birds. *Sci. Am.* Apr.

Pimental, D. (1973). Realities of a pesticide ban. *Environment* Mar. 19–31.

President's Science Advisory Committee (1967). *The World Food Problem.* 3 vols., Washington, D.C.: U.S. Government Printing Office. May.

Ricker, W. E. (1969). Food from the sea. In National Academy of Sciences. *Resources and man.* San Francisco: W. H. Freeman.

United Nations Food and Agriculture Organization (UNFAO) (1963a). *Production Yearbook 1963*, Vol. 17, Rome, Italy.

UNFAO (1963b). *Third World Food Survey.* Freedom from Hunger Campaign, Basic Study No. 11, pp. 49–50.

UNFAO (1971). *Agricultural Commodity Projections, 1970–1980.* CCP 71/20. Rome.

UNFAO (1972). *Monthly Bulletin of Agricultural Economics and Statistics.* Vol. 21, Jan.

U.S. Department of Agriculture (USDA) (1972). *1972 Handbook of Agricultural Charts.* No. 439. Washington, D.C. Oct.

U.S. Department of Health, Education, and Welfare (HEW) (1969). *Report of the Secretary's Commission on Pesticides and Their Relationship to Environmental Health.* Washington, D.C. Dec.

Woodwell, G. M. (1967). Toxic substances and ecological cycles. *Sci. Am.* Mar.

Wurster, C. F. Jr. (1968). DDT reduces photosynthesis by marine phytoplankton. *Science.* 158: 1474–1475.

World Health Organization (WHO) (1972). *Health Hazards of the Human Environment.* Geneva.

Questions

1. Why does calorie deficiency aggravate protein malnutrition?
2. Explain how malnourishment contributes to the high population growth rates typical of the UDC's.
3. Why are legumes important to soil fertility?
4. Discuss the relationship between species diversity, stability, and agricultural systems.

5. Why does the addition of lysine to wheat improve the wheat's nutritional characteristics?
6. Discuss how many of the improvements in food production can aggravate the food distribution problem.
7. What is biological concentration? Biological resistance?
8. Why is biodegradability of a pesticide sometimes considered an asset and sometimes a liability?
9. How is it possible for crop damage to actually increase when pesticides are used?

Part II

Water Pollution

Chapter 4

Water Resources

As the human population grows, our interaction with the water resources on which we are completely dependent becomes more and more critical. Even though water is the most abundant substance on the surface of the earth, its uneven distribution leads to disastrous flooding in some areas and acute shortages in others. In our attempts to redistribute water to meet the growing demands of agriculture, industry, and municipalities, it is becoming increasingly important to understand the environmental impact of what seem to be necessary water projects.

In this chapter the natural distribution and movement of water through the biosphere will be discussed, followed by a description of some of the environmental effects that are being experienced as a result of our attempts to accelerate the development of our water resources.

4.1 *The Hydrologic Cycle*

In an earlier section we traced the flow of the sun's energy through the atmosphere to the earth's surface and saw that a very small but extremely important percentage is utilized by plants for photosynthesis. A much larger percentage, roughly one-fourth of the sun's energy that reaches the surface of the earth, is used to evaporate water and hence propel what is known as the *hydrologic cycle.* Water is not only evaporated from any wet surface, but plants release considerable amounts of water to the atmosphere through their leaves during transpiration. The combination of evaporation and transpiration is referred to as *evapotranspiration.* The water which passes into the atmosphere is eventually returned as precipitation, but not necessarily to the same region. In fact, on a worldwide basis, more water is evaporated from the oceans than is returned to them by precipitation; and to maintain a balance, on the land there is more precipitation than evaporation. The excess water that the land gains from precipitation, some 10×10^{15} gal/yr, is returned to the oceans as *runoff*, both from streams and from the slow movement of underground water. This movement of water is summarized in Figure 4.1.

As the water evaporates, most of the impurities are left behind (a notable exception being DDT), making the evaporative process one of nature's great water purifiers. The pure water which subsequently falls back onto the land is destined to either be returned to the atmosphere or become part of the runoff. As the data in Figure 4.1 indicate, about 66% is evaporated. The rest can be considered to be the potential water supply and is a renewable resource—that is, it represents the amount of water which can be used year after year without causing any depletion of the resource.

Most of the world's water, some 94.2%, is contained in the oceans. Most of the accessible remainder, about 1.65%, is locked up in icecaps and glaciers. While it is only a small percent of the total, it has been estimated that if all of the ice melted, the earth's oceans would rise about 200–250 feet, enough to inundate the entire atlantic and southern seaboard of the United States (USDI). Table 4.1 shows the distribution of the earth's water, including estimates of the amount which exists in the ground.

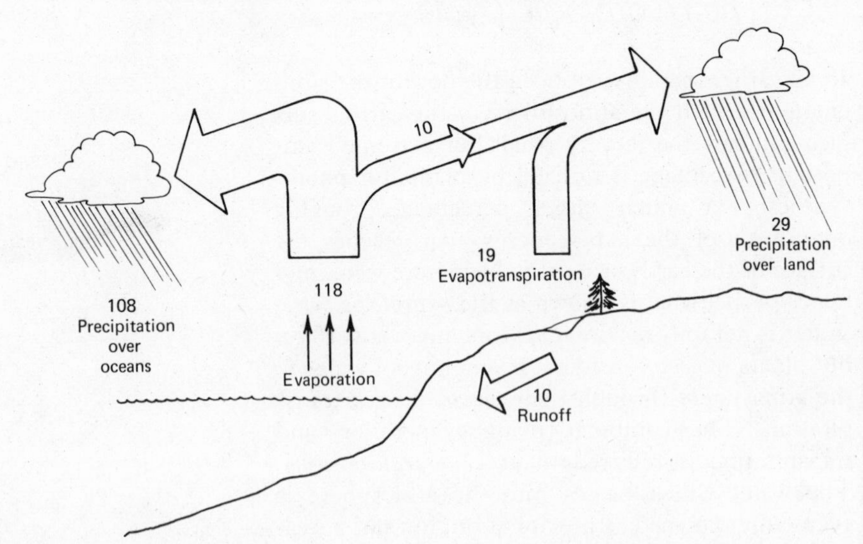

FIGURE 4.1 World hydrologic cycle, units are 10^{15} gal/yr. (Lvovitch 1972).

TABLE 4.1 Distribution of the Earth's Water Resources (Lvovitch 1972)

Location	Water volume, 10^{15} gal	Percent of total
World ocean	362,000	94.2
Glaciers	6,350	1.65
Lakes	60.5	0.016
Soil moisture	21.6	0.006
Atmospheric vapor	3.7	0.001
River (channel) waters	0.32	0.0001
Total groundwater	15,800	4.13
Groundwater less than .5 mile down	1,160	0.28

4.2 *U.S. Water Resources and Uses*

The average annual rainfall on the conterminous United States is about 30 inches, which is equivalent to a daily precipitation of 4200 billion gallons of water. About 70% of this, or 3000 bgd (billion gallons per day), is lost to evapotranspiration, leaving 1200 bgd as the daily runoff. Of this amount, Kruse (1969) indicates that at most we may be able to develop about 315 bgd as a sustained, dependable, water supply for irrigation, industry, cooling water, and domestic requirements. The rest is either required for channel uses such as navigation, hydroelectric power, fish and wildlife, or could not be economically developed into a dependable supply.

Let us compare present and projected requirements for water to the estimated ultimate supply of 315 bgd. First, realize that most of the water which man uses is returned to the overall water supply and, if it is not too polluted, may be used over and over again. So it is necessary to be careful about the terms which are used to describe water usage. *Withdrawal* refers to the water which is taken up in some process, such as water which is used to cool atomic power plants. Some of the water withdrawn is *consumed*, which means either it becomes part of some larger entity or it is returned to the atmosphere making it unavailable for reuse until it rains again. Agriculture consumes a large amount of water through evapotranspiration. Water which is withdrawn but not consumed is *returned* to the water base and is potentially available for reuse.

Figure 4.2 shows water usage in the U.S. in 1970, indicating that the principal user of fresh water, both in terms of withdrawal (41%) and consumption (84%), was agriculture. Municipal and industrial water consumption is far lower. Notice that lots of water, both fresh and saline,

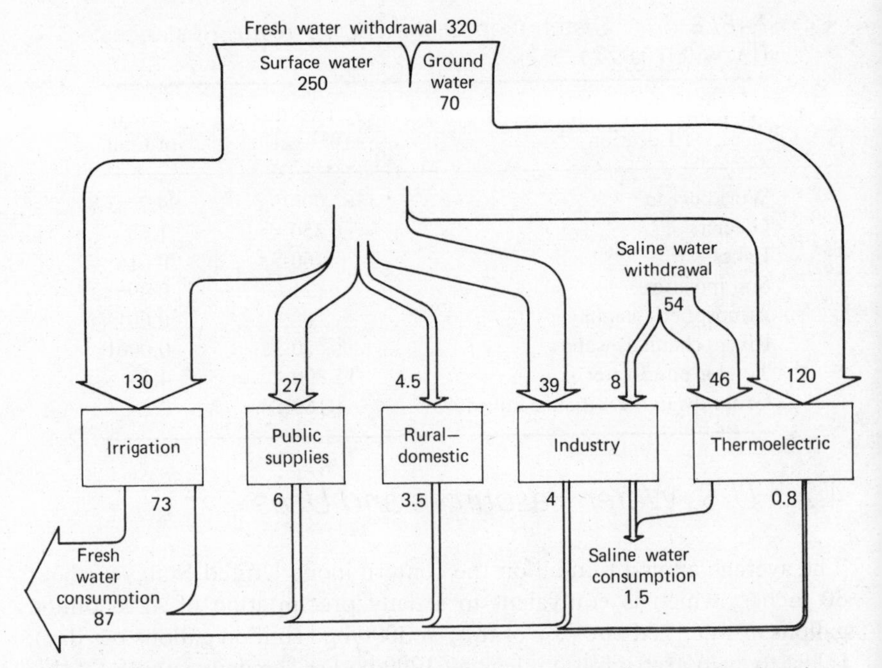

FIGURE 4.2 Estimated water withdrawals and consumption in U.S. in 1970; units are billions of gallons per day. (Murray and Reeves data, 1972).

was withdrawn for cooling steam-electric power plants but that almost all of it was returned (albeit somewhat heated; this thermal pollution will be discussed in a later chapter). In the future, the amount of water required for cooling electric power plants is going to increase dramatically, and as shown in Figure 4.3, will push the total amount of water withdrawn above the daily runoff before the year 2020. As the figure indicates, about half of that cooling water will be saline.

Even though withdrawal already exceeds the estimated dependable supply of 315 bgd, and by 2020 will exceed the daily runoff, it is the *consumption* of water which is of most importance. As shown in Figure 4.4, consumption is estimated to remain well below the estimate of what is available, far into the future. Thus, without yet considering the uneven distribution in the country, overall water resources are more than adequate. However, these data also indicate that if the returned water is of such poor quality that it cannot be reused, then we are already at a critical usage rate. Thus the emphasis needs to be put on water quality and water distribution rather than on quantity.

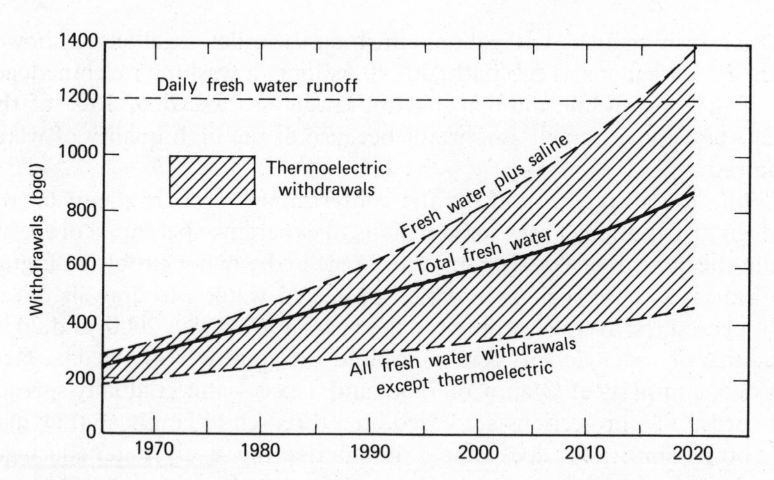

FIGURE 4.3 Estimated future U.S. water withdrawals. Cross hatched area indicated withdrawals for cooling thermoelectric power plants.

If we divide the total withdrawals of 320 bgd of fresh water, by the 1970 population, we obtain a figure of about 1600 gallons a day per capita usage of water in the U.S. Of that, municipal usage is about 130 gallons per person per day. Typical home activities consume a large percentage

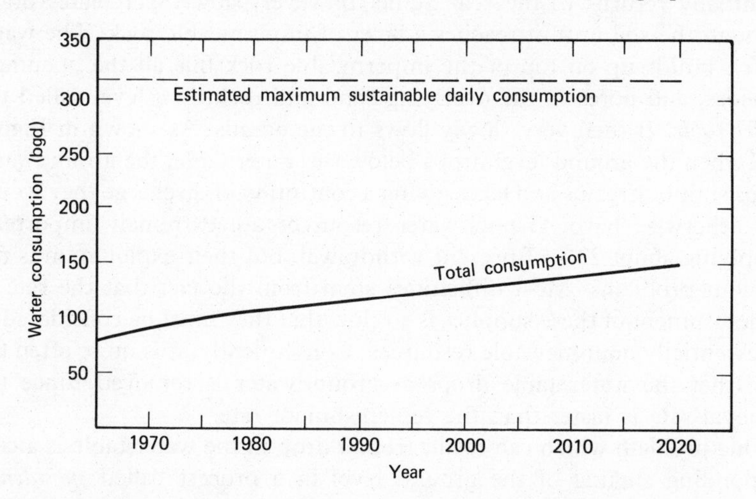

FIGURE 4.4 Estimated total freshwater consumption for U.S. is considerably lower than maximum sustainable level (U.S. Water Resources Council 1968; estimated maximum sustainable consumption, Kruse 1969).

of this: washing dishes, 10 gallons; flushing the toilet, 4 gallons; a shower bath, 20–30 gallons; a tub bath, 30–40 gallons; a washing machine load, 20–30 gallons. While municipal withdrawals are less than 10% of the total, they are extremely important because of the high quality of water required.

While the average runoff for the conterminous U.S. is adequate, the uneven distribution causes some regions to experience perennial droughts while the control of flooding in other areas is the major problem. Figure 4.5 indicates those regions in which projected water withdrawals do, or will, exceed the average supply for the years 1960, 1980, 2000, and 2020. The area of water deficits centers around the southwest—Arizona, New Mexico, and parts of Utah, Colorado, and Texas—and gradually spreads outwards. The projections are based on a three-child family so they may be a bit pessimistic. It does appear though that the severe water shortages already being experienced in some areas are going to soon spread to other regions.

4.3 *Groundwater*

We have seen that about 70% of the rainfall in the U.S. is lost to evaporation. Most of the rest runs off the surface into streams and lakes, and eventually returns to the sea. Some, however, slowly percolates down through the soil until it reaches a layer of impermeable rock. The water which builds up on top of the impermeable rock fills all the openings, crevices, and pores in the overlying sand and rock, to a level called the *water table*. It then very slowly flows to the oceans. As shown in Figure 4.6, when the ground level drops below the water table, the groundwater helps supply streams and lakes giving a continuity of discharge they would not otherwise have. Groundwater resources are extremely important, supplying about 22% of present withdrawal, but their exploitation is not without problems. Most difficulties stem from the fact that the rate of replenishment of these supplies is so slow that they must be considered to be essentially nonrenewable resources. Consequently, it is quite often the case that the water table drops as groundwater is removed, since the removal rate is faster than the replenishment rate.

One problem which can result from a drop in the water table is a corresponding sinking of the ground level in a process called *subsidence*. For example, the U.S. Geological Survey estimates that the city of San Jose sank more than 5 feet during the 14-year period from 1948 to 1962 with a total drop of about 10 feet from 1912 to 1962. The water table was dropping precipitously due to the demands put on it by the (at that time)

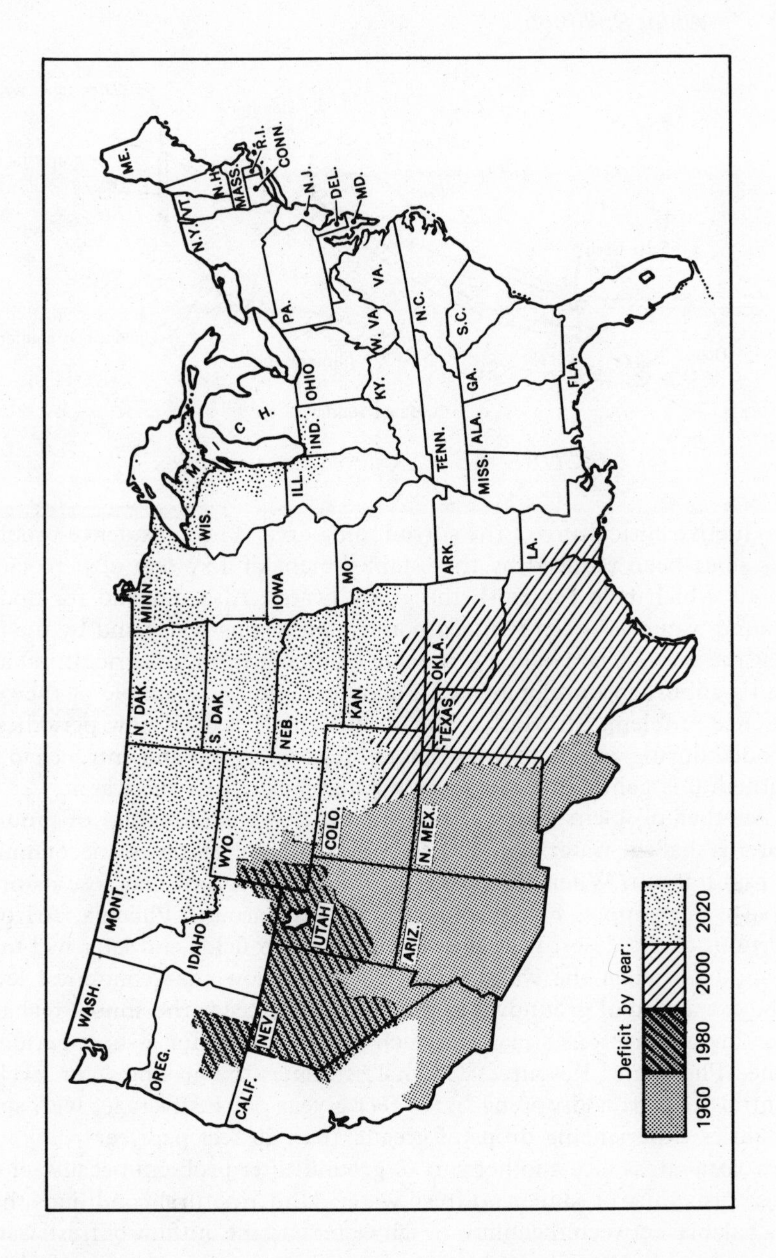

FIGURE 4.5 Water deficit regions. Projections based on three-child families, rapid economic growth, maximum development of water storage facilities, and tertiary treatment (Commission on Population Growth and the American Future 1972).

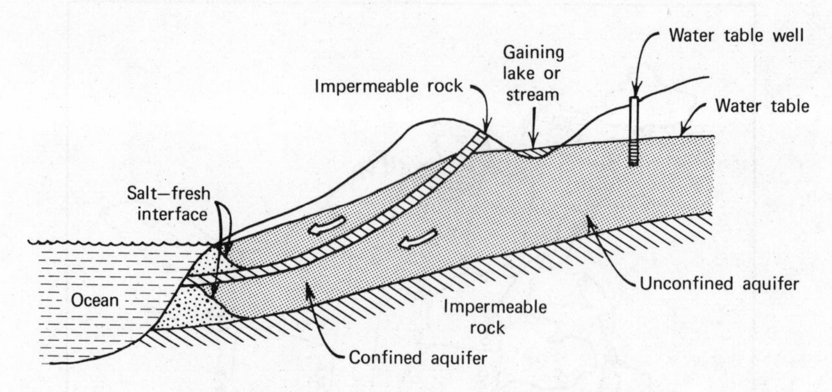

FIGURE 4.6 Groundwater resources.

productive agriculture of the surrounding area. The subsidence problem has since been reduced by the establishment of a system of percolation ponds which help replenish the groundwater. Also demand for underground water has been reduced both by importing water and by the fact that the Santa Clara valley is quickly changing from an agricultural area into a suburban development. In another dramatic example of the subsidence problem, St. Mark's Square in Venice, Italy is now periodically flooded during very high tides. Shops are sandbagged and entrance to the Cathedral is gained by a temporary gangplank across the water.

Another problem which can result from overexploitation of groundwater is that the water table can drop so low that it becomes uneconomical to pump it out. When that happens, the area involved may lose its principal water supply. For example, in the area around Phoenix, Arizona, portions of the desert were converted to cotton fields and later had to be abandoned when the water table dropped below the economical level. The extraction of groundwater there greatly exceeds the annual recharge creating a dangerous situation which can exist only for a short period of time. The Water Resources Council estimates the groundwater level in central Arizona is dropping by 10 feet a year on the average, with some localities experiencing drops of greater than 20 feet per year.

Coastal areas face another sort of groundwater problem because of the close proximity of saline and fresh water. Under natural conditions there is a balance between the inflow of salt water and the outflow of fresh water, which establishes a boundary between the two regimes. However, as fresh water is pumped out of the area, the water table drops and the boundary moves inland. As suggested in Figure 4.7, a coastal freshwater well may have to be abandoned, after a period of time, because of this saltwater intrusion. The urbanization and industrialization of coastal areas has resulted not only in a greater usage of groundwater but the

St. Mark's square, Venice. The city is sinking approximately ¼ inch per year and periods of high-water flooding are occurring ever more frequently.

accompanying improvements in drainage and sewer systems in these areas has decreased the groundwater recharge rate. Water which would normally collect in puddles and pools and then seep into the soil is now efficiently collected and released directly into the sea. The result is that

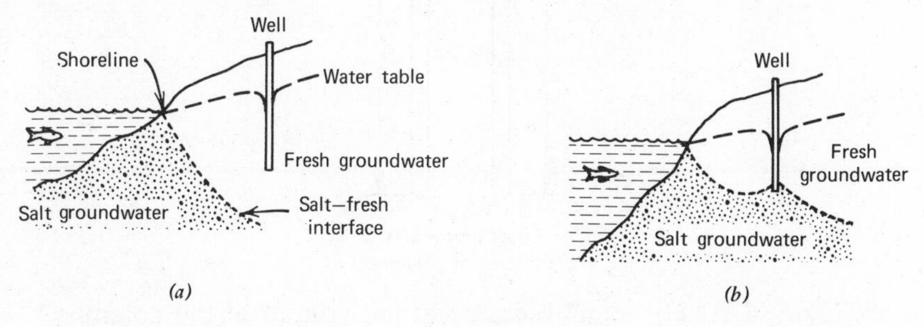

FIGURE 4.7 Saltwater intrusion into a fresh water well due to overpumping. (a) before, (b) after.

every state on the Atlantic coast, and California in the west, has problems with seawater intrusion. So far, the only way to alleviate these problems, other than to simply refrain from exploiting groundwater sources, is to inject fresh water or treated waste water back into the ground to recharge the aquifer.

There is a simple model which helps explain the movement of the boundary inland as the water table drops. The model is based on the fact that the density of seawater is about 2.5% greater than that of fresh water. This fact implies that a column of fresh water must be 2.5% taller than a column of seawater for the pressures to be the same at the bottoms of each column (e.g. as shown in Figure 4.8 a 41 foot column of fresh water produces the same pressure as a 40 foot column of seawater).

At the interface between the seawater and the fresh water, the hydrostatic pressure is the same on either side which means the freshwater side must be under a column of water 2.5% higher than the saltwater side. The extra 2.5% of height on the freshwater side comes from the rise above sea level of the water table as shown in Figure 4.9. As this figure indicates, the interface will exist 40 feet below sealevel for every foot above sea level that the water table rises. What is interesting about this conclusion is that it suggests for every foot the water table drops because of overpumping, the interface will rise by 40 feet. And the higher the interface rises, the more likely the wells will pump saline water as was shown in Figure 4.7.

There are other groundwater problems relating to contamination from wastes which percolate down through the soil to the water table. As will be discussed in Chapter 7, something like 50 million people in the U.S. rely on septic tank systems, such as is shown in Figure 4.10, for their

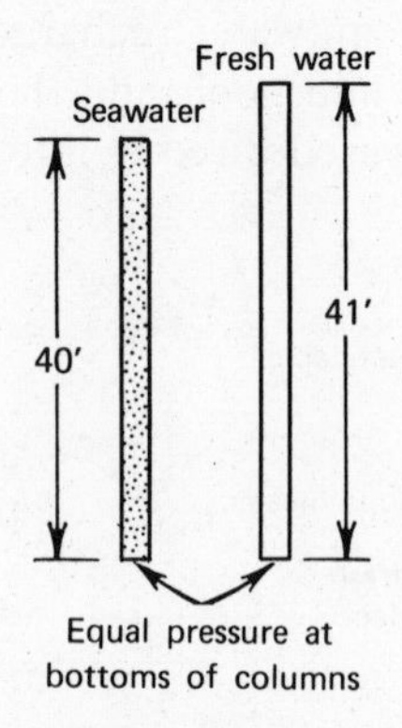

FIGURE 4.8 For equal pressure at the bottom of the columns, the freshwater column must be 2.5% higher than the seawater column.

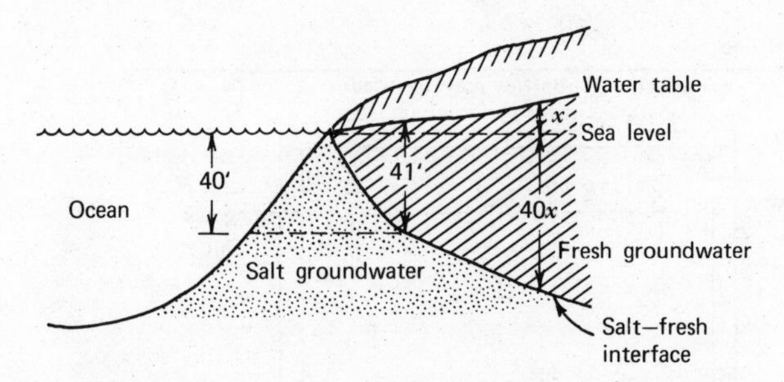

FIGURE 4.9 A simple model predicts that the seawater–fresh-water interface will exist 40 feet below sea level for every foot that the water table exists above sea level.

waste disposal. Some decontamination of the wastes occurs in the soil as the effluent filters down but sometimes it is not sufficient and the groundwater can become polluted. Hepatitis has been known to be spread this way. Refuse dumps and landfills also can cause contamination but the major contributor is agriculture. As was discussed in the last chapter, modern agricultural practices include heavy use of fertilizers and pesticides as well as the concentration of livestock in feedlots. The tremendous quantities of nutrients, poisons, and organic wastes that result not only cause pollution of surface waters but groundwater as well.

A particular problem associated with the use of nitrogen fertilizers is an increase in the nitrate (NO_3) content of groundwater supplies. Nitrates themselves are not especially dangerous, but they can be converted by certain bacteria into nitrites (NO_2) which are highly toxic. These bacteria occur in the digestive tract of babies and when combined with high levels of nitrate in the baby's water can result in a disease called methemoglobinemia. The disease reduces the blood's ability to carry oxygen and can cause suffocation and death.

The U.S. Public Health Service drinking water standards recommend a maximum of 45 mg/l of nitrates. Samples taken around Delano, California measure 70+ and in Illinois, 30% of all wells less than 25 feet deep supply water which exceeds the recommended limit [U.S. Environmental Protection Agency (EPA) 1972]. In such areas it has been recommended that babies be given only pure bottled water.

4.4 *Dams and Reservoirs*

So far in this chapter the main concern has been with the quantities and distribution of water resources. We shift our attention now to some water resource management techniques which are designed to equalize

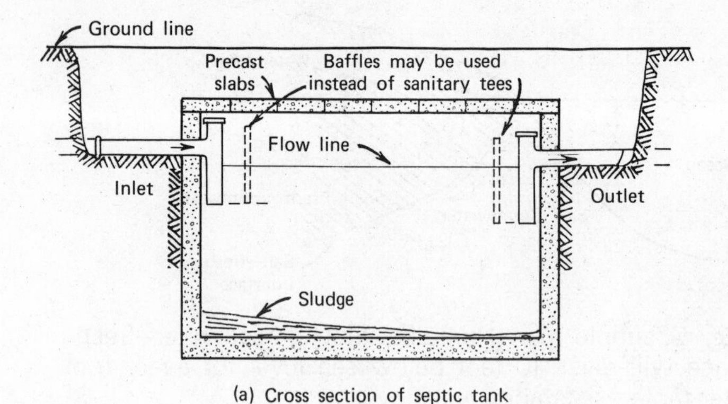

(a) Cross section of septic tank

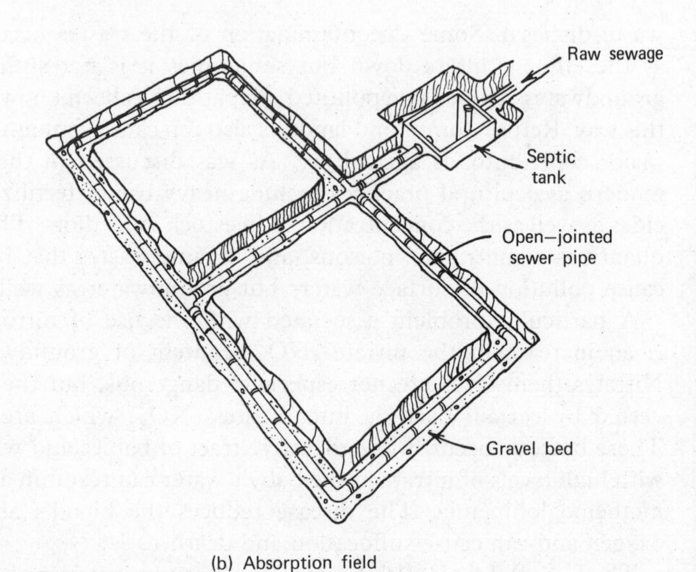

(b) Absorption field

FIGURE 4.10 Septic tank disposal systems can contaminate the groundwater.

distribution and smooth the irregular flow rates of nature. The emphasis will be on the environmental effects of such projects to once again make the point that it is impossible to do "just one thing."

The principal component of any multipurpose water resource development project is the dam. By capturing high winter flows, the danger of flooding is reduced and a reliable source of water can be provided throughout the year to supply the agricultural and domestic needs of the area. Moreover, by releasing the water through hydroelectric turbines, virtually pollution-free electrical power can be generated. According to the U.S. Water Resources Council there were in 1963 nearly 1400 major reservoirs (over 1.6 billion gallons each) in the U.S. with a combined capacity of 117 trillion gallons of water. To get some idea of how large a number that is, it is equal to about 100 days worth of the total U.S. runoff.

Irrigation, municipal water supply, flood control, hydroelectric power—these are the most obvious benefits of reservoir projects. Other benefits are not quite so clear-cut. For example, the increased recreational potential of the reservoir must be weighed against the loss of a free-flowing river, the destruction of what may have been a beautiful canyon, and the complete change of character of the area. In the process of providing a flood-free valley, many acres behind the dam are permanently flooded. It has been estimated, for example, that should the huge Pa Mong Dam be built, as part of a proposed Mekong River Development program in Southeast Asia, that upwards of 500,000 people may have to be displaced by the impounded water.

It should also be brought out, that while dams permanently destroy a natural area, their benefits are not permanent. The silt which flows with a river settles when it reaches the still water behind the dam, reducing the storage capacity. Sterling (1972) reports that the storage capacity of Lake Austin in Texas was reduced by 95% by silting within 13 years, while Pakistan's $600 million Mangla Dam is expected to silt up completely in 50 years. Not only does the silt fill in the reservoir, but the flood plain below the dam suffers from a lack of silt, with several consequences.

The Aswan High Dam in Egypt, completed in 1971, provides a good example of the kinds of unwanted "side effects" which may accompany large water management projects. The dam has produced 7 billion kilowatt-hours of electricity and brought 900,000 acres of land under cultivation, but it has ruined the eastern Mediterranean sardine industry, has led to a drastic increase in the incidence of a severely debilitating disease called schistosomiasis, and is capturing about 110 million tons of silt annually (Turner 1971).

The loss of silt below the dam has upset the balance which previously existed between the erosion caused by the river and the deposition of silt along the river banks. Erosion now prevails: bridges are being undermined, the Mediterranean shoreline is receding, threatening coastal cities such as Alexandria; land barriers are being broken down allowing the seawater to flood into freshwater lakes.

Further, the lack of nutrient-rich silt entering the eastern Mediterranean has decimated the bottom of the food chain resulting in a 95% loss in the area's sardine catch. Also, the annual flooding, which is now controlled, used to provide a natural, free, form of fertilization. Typically, silt carries many nutrients and the periodic flooding which used to occur spread those nutrients over the land. Artificial fertilizer production and distribution is now necessary.

Most serious is the rapid rise in a painful, energy-draining disease called bilharzia or *schistosomiasis*. The disease is spread by blood flukes which, during one stage of their life cycle, require snails as an intermediate host. The snails thrive in the still waters of the reservoir or in the many new irrigation canals. A person can become infected with the disease by merely standing in the water, no cut is necessary. The infection rate in areas with new canals has gone from 0 to 80 %, and on the order of half of Egypt's population has the disease, which essentially has no cure.

Getting a little closer to home, we might consider the problems caused by the roughly 150 dams on the Columbia River Basin in the Pacific Northwest. "Roll on Columbia" is no more, as it is virtually possible to water ski from tide water to the Canadian border. There are only about 50 miles of free-flowing Columbia left. The salmon runs have been severely disrupted, the Grand Coulee Dam itself caused about 70% of the nesting and nursery areas for spring runs of chinook salmon on the Columbia to disappear. Another serious problem is nitrogen supersaturation which can kill fish by giving them the bends. When water falls over a spillway excess atmospheric nitrogen may be driven into the water. If the water quietly flows into another reservoir the supersaturation may remain long enough to kill the fish.

4.5 *Irrigation and the Buildup of Salts*

In this brief discussion of water resource management, it is of interest to consider the salt problem which accompanies irrigation projects and which is often compounded by other massive water-diversion projects. Water naturally accumulates a large variety of dissolved matter as it passes through soils and rocks on its way to the sea, typically including such ions as calcium, magnesium, sodium, sulfate, chloride, and bicarbonate. These plus minor amounts of other dissolved constituents are commonly referred to as salinity and are generally measured in terms of parts per million (ppm) of dissolved solids in the water.

To see how the concentration of salts in a river is increased by agricultural practices, consider Figure 4.11. A portion of the upstream water which is low in salts is diverted for use in irrigation. As the water is spread over the soil a high percentage of it is lost to evaporation and transpiration—processes which leave the salts behind. The concentration of salts in the remaining water is therefore increased. These salts must

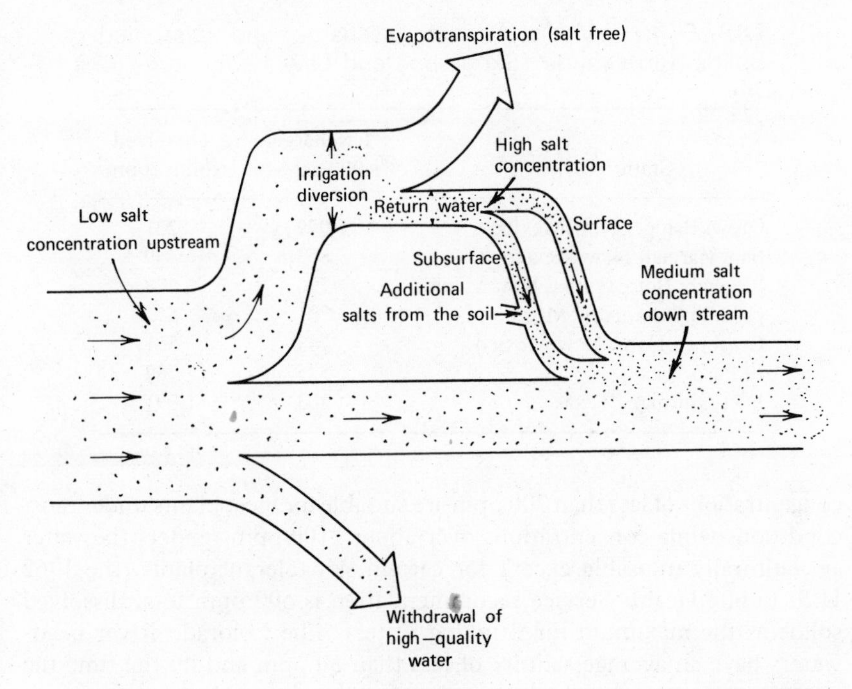

FIGURE 4.11 The salt concentration of a river increases due to evapotranspiration of pure water and withdrawal of high-quality water.

not be allowed to build up in the soil or they will damage the crops, so a certain portion of the irrigation water (as much as one-third) must be used simply to wash away salts. Some of this irrigation return water, which is very high in salts, drains off of the surface back into the river and some percolates through the soil, picking up even more salt, before returning to the river. As this process is repeated over and over again down the length of the river, the salt concentration gets progressively higher.

Also shown in Figure 4.11 is the withdrawal of high-quality water which may not be returned at all. With less water remaining in the river for dilution, the irrigation return flow problem is increased. The Colorado River in particular is subject to large diversions. Besides the water which is taken by California, the authorized Central Arizona Project will divert large quantities of water to the Salt River Valley and in addition, the huge power plant complex at the Four-Corners region will divert even more for cooling water, with no return flow.

Table 4.2 shows the decreased flows and increased salt concentrations along the Rio Grande as it travels from New Mexico to Texas. Salt

TABLE 4.2 Mean Annual Discharge and Dissolved Solids, Rio Grande (Skogerboe and Law 1971)

Station	Discharge, 1,000 acre-ft	Dissolved solids, ppm
Otowi Bridge, New Mexico	1,079	221
San Marcial, New Mexico	853	449
Elephant Butte Outlet, New Mexico	790	478
Caballo Dam, New Mexico	781	515
Leasburg Dam, New Mexico	743	551
El Paso, Texas	525	787
Fort Quitman, Texas	203	1,691

concentrations of less than 700 ppm are suitable for most plants under most conditions while concentrations over about 2100 ppm renders the water agriculturally unusable except for certain salt tolerant plants (the 1962 U.S. Public Health Service recommendation is 500 ppm total dissolved solids as the maximum for drinking water). The Colorado River headwaters have an average salinity of less than 50 ppm and by the time the water reaches the Imperial Dam just north of the Mexican border it ranges in salinity from about 750 to 1060 ppm (EPA 1972). Water thus delivered to Mexico from the U.S. is so salty that it has ruined some of that country's richest cotton-growing land, constituting probably the biggest problem between the two countries.

4.6 *Grandiose Water Projects*

While reservoirs make it possible to efficiently utilize the water resources of a given area, if the local annual demand exceeds the annual supply, then it is necessary to import water from some outside area. Systems of canals and reservoirs to accomplish this task have been common for a long time but what is unusual is the growing scale of these projects. As the size of the project increases, the more difficult it is to accurately predict the hidden consequences, and some proposals presently being considered are truly colossal.

The most amazing plan being proposed calls for the creation of a huge system of canals and reservoirs stretching from the Yukon River in Alaska all the way down to the interior of Mexico. Called the North American Water and Power Alliance (Nawapa), the project would cost more than

$100 billion and would supply at least 120 million acre-feet of water per year (about 110 bgd). Many dams would be constructed, the largest being over 1700 feet high—twice as high as any dam ever built; 15 reservoirs would be larger than Lake Mead, the largest man-made lake in North America. Part of the water would be transported by a canal 630 feet wide and 35 feet deep. Everything about the project would be spectacular, including, one might confidently predict, the ecological consequences.

Compared to Nawapa, the California Water Plan is trivial, and yet it is one of the most complex and ambitious water development projects ever attempted by man. By studying the California project, the initial stages of which are nearing completion now, we may gain some insight into some of the ecological consequences of such large-scale water transfers. As in all water transport projects the basic philosophy is the same, namely to move water from an area of "surplus," in this case northern California, to an area of "deficiency," southern California. Ecologically speaking, there is no such thing as "surplus" since everything exists as a balance. When water is removed there will be changes in the affected ecosystems and what really is at question is the severity of those changes.

In the case of the California Water Plan, the most seriously endangered ecosystem is the Sacramento Delta–San Francisco Bay estuary in northern California. Fresh water which is deposited in the Delta from the Sacramento and San Joaquin Rivers normally flows out through the Bay into the Pacific Ocean, but as the water project progresses, more and more of this fresh water will be diverted to southern California. As shown in Figure 4.12, present flows out of the Delta are already significantly reduced, and projected decreases in the year 2020 are drastic. To anticipate the possible effects we should first consider the importance of an estuary.

An estuary is a semienclosed coastal body of water with a free connection to the open sea. It is a region characterized by the mixing action from saltwater tidal movements and freshwater drainage. Their extreme importance as the nursery for many species of fish is demonstrated by the fact that 85% of the total U.S. fisheries catch in 1962 was of estuary-dependent species (MacIntyre and Holmes 1971). Estuaries are especially vulnerable to destruction by man's activities. They are the sink that catches the runoff of our polluted streams, and their "useless" appearance makes them tempting to land developers. Already San Francisco Bay has lost about 40% of its surface area due to filling.

As the flow of fresh water into the Delta-Bay estuary decreases, not only will the boundary between fresh and salt water move inland, but the flushing action in San Francisco Bay will be decreased. The decreased input of fresh water, especially in the southern part of the bay, will mean pollutants will remain in the bay for a longer period of time, so their

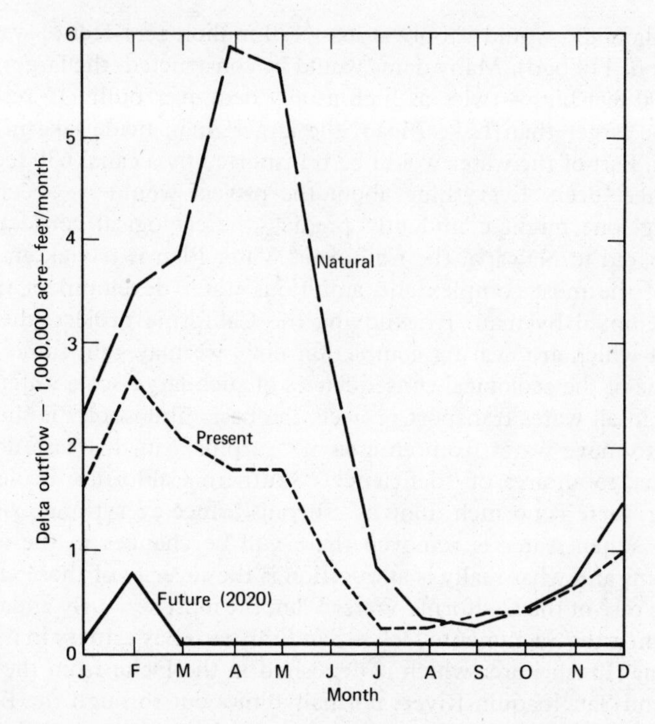

FIGURE 4.12 The median outflow of fresh water from the Sacramento Delta into San Francisco Bay will be severely decreased because of the California Water Plan. (USDI 1967).

concentrations will increase. Moreover, recall that evaporative processes always increase the salinity of the remaining water. With less freshwater input to compensate for the evaporation, the bay will become more saline which may upset the biota.

These are only a few of the problems which may result from the diversion of "surplus" water. The actual consequences cannot be predicted with any accuracy since not enough is known about the Bay-Delta system. There does seem to be cause for concern though, lest south San Francisco Bay turn into another Lake Erie.

There are other massive water projects being proposed, for example, the Texas Water Plan which calls for the diversion of part of the Mississippi River into West Texas and New Mexico (Price 1971), but the new environmental awareness, typified by the new Federal laws requiring a thorough environmental impact analysis before a project can begin, are decreasing the likelihood of such projects ever being built.

4.7 *Conclusions*

This chapter has pointed out the necessity to treat water-supply problems in terms of distribution and quality. The total quantity is adequate, providing the water is of sufficient quality after being used to enable it to be reused. There are major problems of distribution, especially in the southwestern U.S. where present use patterns, which depend on exploitation of groundwater, cannot continue for long. The dropping water table is going to require major importations of water, strict water conservation measures, and, very likely, increased construction and use of desalination facilities. Moreover, the overexploitation of the major rivers in the area has led to deteriorating relations with Mexico, as they are put on the receiving end of progressively poorer quality water.

One of the important themes of the chapter is that growth is not without its price. As population grows, the need for increased agricultural production results in increasing demands placed on the environment. The difficulties associated with pesticides were discussed in the last chapter and the problems of water supply were discussed in this one. The decrease in water quality caused by the use of fertilizers will be brought out in Chapter 6.

It is not only population growth but also the growth in our overall material well being which contributes to the demand for water. The complexity of the interrelationships which are affected by this growth makes it imperative to carefully analyze the possible environmental effects of proposed engineering solutions, before project approval is given. As the size of these projects increases and the awareness of our lack of detailed knowledge of ecosystem behavior becomes more evident, it is apparent that in the future as much effort should be devoted to decreasing our demands as is presently given to the technological aspects of increasing supplies.

Bibliography

Ballentine, R. K., Reznek, S. R., and Hall, C. W. (1972). Subsurface pollution problems in the United States. U. S. Environmental Protection Agency, Report TS-00-72-02, Washington, D.C. May.

Berghinz, C. (1971). Venice is sinking into the sea. *Civil Eng.–ASCE* 41: (3).

Commission on Population Growth and the American Future (1972). *Population and the American Future*. Washington, D.C.: U.S. Government Printing Office.

Kruse, Cornelius W. (1969). Our nation's water: Its pollution control and management. In J. N. Pitts and R. L. Metcalf eds., *Advances in environmental sciences,* Vol. 1. New York: Wiley-Interscience.

Linsley, R. K., and Franzini, J. B. (1972). *Water-resources engineering.* San Francisco: McGraw-Hill.

Lvovitch, M. I. (1972). World water balance (general report). In *World Water Balance, Proceedings of the Reading Symposium.* Vol. 2. UNESCO, Geneva.

MacIntyre, F., and Holmes, R. W. (1971). Ocean pollution. In W. W. Murdoch ed., *Environment resources, pollution and society.* Stamford, Conn.: Sinauer.

McKee, J. E., and Wolf, H. W. (1963). *Water quality criteria.* 2d ed. California State Water Quality Control Board, Pub. 3-A.

Murray, C. R., and Reeves, E. B. (1972). Estimated Use of Water in the United States in 1970, U.S. Geological Survey Circular 676.

Overman, M. (1969). *Water: Solutions to a problem of supply and demand.* Garden City, N.Y.: Doubleday.

Poland, J. F., and Green, J. H. (1962). Subsidence in the Santa Clara Valley, California, A Progress Report, U.S. Geological Survey Water-Supply Paper 1619-C.

Price, B. (1971). Possible diversion of Mississippi river water to Texas and New Mexico. *Water Resources Bull.* 7 (4): 676–683.

Seckler, D. (1971). *California water, a study in resource management.* Berkeley: Univ. of Calif. Press.

Sewell, W. R. D. (1967). Pipedream or practical possibility. (Nawapa.) *Bull. Atom. Sci.* Sept.

Skogerboe, G. V., and Law, J. P., Jr. (1971). Research needs for irrigation return flow quality control. U.S. Environmental Protection Agency, Project No. 13030, Nov.

Sterling, C. (1972). Superdams, the perils of progress. *Atlantic* June: 35–41.

Turner, D. J. (1971). Dams and ecology, can they be made compatible? *Civil Eng.–ASCE,* Sept.: 76–80.

U.S. Department of the Interior (USDI). *River of Life.* Conservation Yearbook Series, vol. 6.

USDI (1967). *San Joaquin Master Drain, Effects on Water Quality of San Francisco Bay and Delta.* Federal Water Pollution Control Administration, San Francisco.

U.S. Environmental Protection Agency (EPA) (1972). *Reconvened Seventh Session of the Conference in the Matter of Pollution in the Interstate Waters of the Colorado River and its Tributaries,* Denver, Colo. April 26, 27.

U.S. Water Resources Council (1968). *The Nation's Water Resources.* Washington, D.C.: U.S. Government Printing Office.

Questions

1. What does the term "nonrenewable resource" mean? List some resources which can be considered renewable and some which are nonrenewable.

2. If a faucet drips at the rate of 1 drop per second and it takes 25,000 drops to equal 1 gallon, (a) How much water is lost per day? If water costs 30 cents per 1000 gallons, how long would it take to equal the cost of fixing the leak: (b) If you do it yourself for 2 cents? (c) If a plumber does it for $10?

   ans. (a) 3.4 gal/day; (b) 20 days; (c) 10,000 days

3. Explain the statement "it is impossible to do just one thing." Include examples relating to the production of food.

4. Monitor your daily activities to try to estimate your own personal water consumption. Is it anywhere near the daily residential average of 55 gallons?

5. Suppose the water table near the seacoast drops by .5 foot, how much would you expect the seawater–freshwater interface to rise?

   ans. 20 feet

6. Explain how it would be possible to withdraw more water per year than the annual runoff.

7. The storage capacity of Lake Mead behind Hoover Dam is about 10,000 billion gallons. (a) If it is used for municipal purposes only, how many people could it supply for 1 year? (b) If it is used for all purposes how many could it supply? Make any simplying assumptions that seem necessary.

   ans. (a) 210 million; (b) 17 million

8. Suppose water having a salt concentration of 700 ppm is used as irrigation water. If two-thirds of the water is lost to evapotranspiration, what would be the salt concentration in the return water? Assume no extra salts are picked up in the soil.

   ans. 2100 ppm

Chapter 5

Water Pollutants

Much of the water which is withdrawn for use by man has already been used in one application or another, and some of it in fact will be reused again before it finally returns to the sea. With each use of the water, various forms of pollution contribute to a degradation of its quality. Sometimes the degradation is only temporary—natural self-purification being sufficient to eventually restore the quality—but oftentimes either the pollutant is one which does not degrade naturally or the sheer volume of pollution is sufficient to overload the self-purification mechanisms, in which cases the water quality is more permanently degraded.

In this chapter various forms of water pollution will be examined with an emphasis on biodegradable organic wastes, thermal pollution, and toxic heavy metals. Other pollutants will be discussed in Chapters 6 and 7.

5.1 *Types of Pollutants*

There are several ways to categorize water pollutants, the following being one which is used by the Environmental Protection Agency.

1. *Oxygen-demanding wastes* are biodegradable organic compounds contained in domestic sewage or certain industrial effluents. When these compounds are decomposed by bacteria, oxygen is removed from the water. If the oxygen level drops low enough, the fish will die.
2. *Disease-causing agents* are various pathogenic microorganisms which usually enter the water with human sewage. Contact with these microbes can be made by drinking the water or through various water-contact activities.

3. *Synthetic organic compounds* include detergents and other household aids, pesticides, and various synthetic industrial chemicals. Many of these compounds are toxic to aquatic life and may be harmful to humans.
4. *Plant nutrients* such as the nitrogen and phosphorus which drain from fertilized lands, as well as the effluent from most sewage treatment plants, stimulate the growth of algae and water weeds.
5. *Inorganic chemicals and mineral substances* includes the acids which form when water drains from abandoned mines, as well as the heavy metals such as mercury and cadmium.
6. *Sediments* are particles of soils, sands, and minerals washed from the land. They can smother bottom life such as shellfish and coral, as well as fill in reservoirs and harbors. Improper soil management leading to erosion is a major contributor.
7. *Radioactive substances* can enter the water from the mining and processing of radioactive ores, from various nuclear power operations, from medical facilities, and from nuclear weapons testing. Radioactivity typically concentrates in the food chain in much the same way as was shown for DDT.
8. *Thermal discharges* from steam-electric power plants raise the temperature of the receiving water by as much as 20°F, resulting in various changes in the local ecosystem.

Pollution of our waterways is often caused by a combination of the above eight sources which can severely compound the problem.

5.2 *Aquatic Ecosystems*

Before going on to a discussion of the effects of these various pollutants, let us examine the normal operation of an aquatic ecosystem as shown in Figure 5.1. The producers in the figure are considered to be of two types: large plants that may be rooted to the bottom, and the free-floating minute plants, usually algae, called *phytoplankton*. Phytoplankton are not usually visible, though they give the water a greenish color. Along with the attached algae, phytoplankton are the main producers in the system, storing energy and liberating oxygen during photosynthesis.

The primary consumers are indicated to be benthos, or bottom forms, and zooplankton. Zooplankton are minute animal plankton with little or no swimming ability. Other consumers in the system are fish, insects, frogs, man, etc., and a category called detritivores which live off the organic debris.

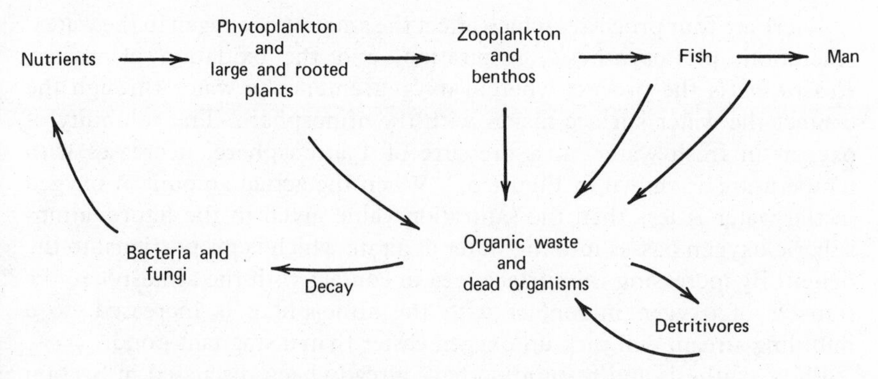

FIGURE 5.1 An aquatic ecosystem.

The decay organisms, bacteria and fungi, recycle the nutrients back into the ecosystem, thus closing the circle. This is a balanced ecosystem, capable of keeping the water clean and healthy automatically, with no help from man, provided it is not *overloaded* with nutrients or wastes. Unfortunately, as man concentrates his activities in cities, and as his technology becomes more sophisticated, quite often these aquatic ecosystems are forced out of balance and no longer function properly.

5.3 *Dissolved Oxygen*

The Environmental Protection Agency reports that 73.7 million fish were killed by water pollution in the U.S. in 1971; 81% more than in any previous year on record (beginning 1960). The highest single source of kills was sewerage systems (21.3 million fish) with the majority of these reports indicating low dissolved oxygen in the water as the immediate cause of death.

The amount of dissolved oxygen (DO) in water is an important parameter of water quality. Fish, for example, require certain minimum amounts of DO depending upon their species, stage of development, level of activity, and the water temperature. For example, for a well-rounded, warm-water fish population, McKee and Wolf (1963) recommend that the DO remain above 5 ppm* for at least 16 hours of the day and during the other 8 hours it should not drop below 3 ppm. In general the more desirable species such as trout require more oxygen than the coarser species, such as carp.

*Again, ppm means "parts per million" by weight. This is approximately equal to milligrams of oxygen per liter of water, mg/l.

There are four processes which affect the amount of oxygen in the water: reaeration, photosynthesis, respiration, and the oxidation of wastes. *Reaeration* is the process wherein oxygen enters the water through the contact the water surface makes with the atmosphere. The solubility of oxygen in fresh water, at a pressure of 1 atmosphere, decreases with temperature as shown in Figure 5.2. When the actual amount of oxygen in the water is less than the saturation value given in the figure, atmospheric oxygen passes into the water at a rate which is proportional to the deficit. By increasing the surface area in contact with the atmosphere the transfer of oxygen in contact with the atmosphere is increased, so a bubbling stream will pick up oxygen easier than a stagnant pond.

Photosynthesis and respiration have already been discussed in Section 1.5. During photosynthesis, which of course occurs only during the daylight hours, oxygen is liberated, thereby increasing the DO level in the water. Respiration, however, is a process which continuously removes oxygen from the water. Combining the three effects of photosynthesis, respiration, and reaeration produces a diurnal variation of DO as shown in Figure 5.3. In this figure, photosynthesis is assumed to occur from 6 A.M. to 6 P.M., bringing the DO level to slightly above saturation in the

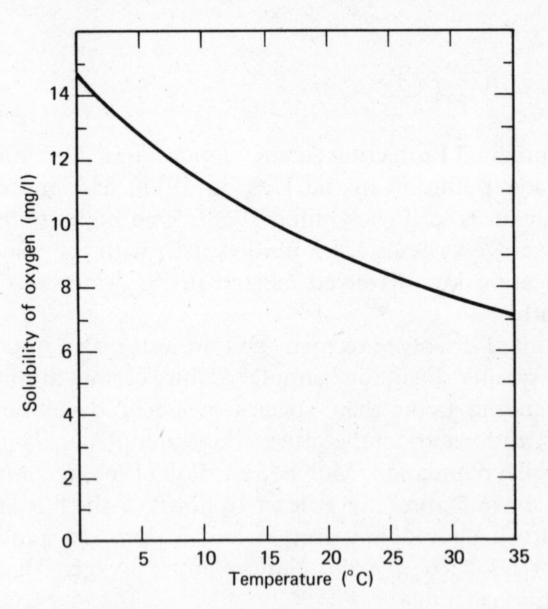

FIGURE 5.2 Solubility of oxygen in fresh water exposed to water saturated air at 1 atmosphere pressure (American Public Health Association 1971).

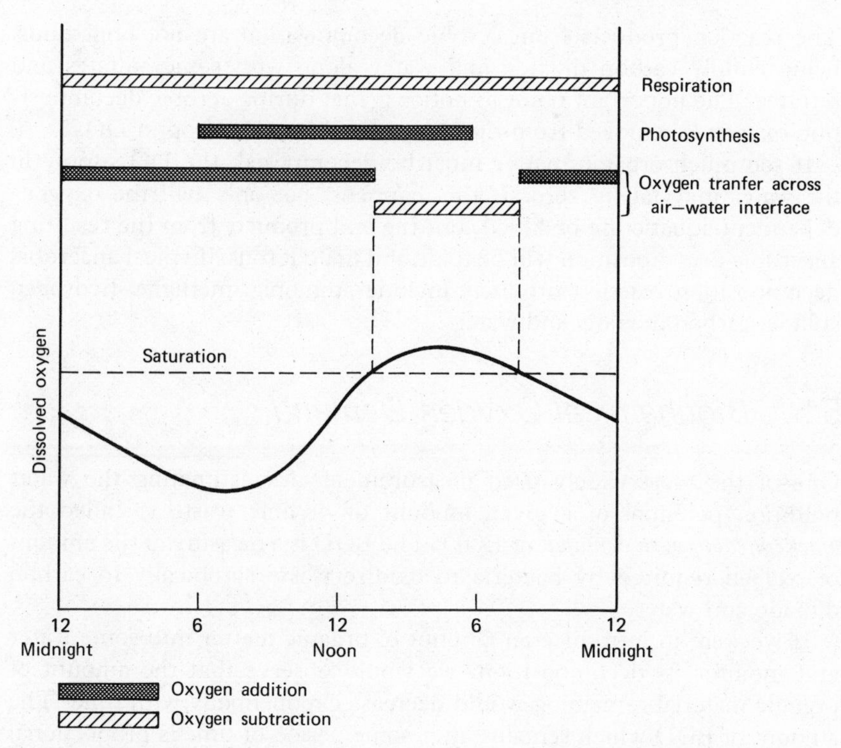

FIGURE 5.3 Diurnal variation of DO assuming photosynthesis from 6 a.m. to 6 p.m.

afternoon. While the water is supersaturated, oxygen diffuses out of the water instead of into it.

5.4 *Decomposition*

The fourth process which affects the amount of oxygen in the water is the oxidation of wastes. Microorganisms, especially bacteria, use organic wastes as food and in the process break down the complex organics into simple organic and inorganic materials. This decomposition may occur in the presence of oxygen, in which case it is called *aerobic* decomposition, or it may occur in the absence of oxygen in which case it is called *anaerobic* decomposition.

The general form of the equation for aerobic decomposition of organic material is

$$\text{organic matter} + \text{bacteria} + O_2 \rightarrow CO_2 + H_2O + \text{new bacterial cells}$$

The reaction products from aerobic decomposition are not obnoxious, being simply carbon dioxide and water along with some sulfates and nitrates. The important point to notice is that during aerobic decomposition oxygen is removed from the water, resulting in a drop in DO.

If too much organic matter must be decomposed, the DO supply in the water may fall to zero. If this happens, not only will the oxygen-dependent aquatic life be killed, but the end products from the resulting anaerobic decomposition will be toxic and malodorous. Typical anaerobic decomposition reaction products include ammonia, methane, hydrogen sulfide, carbon dioxide, and water.

5.5 *Biochemical Oxygen Demand*

One of the most widely used measurements for estimating the water pollution potential of a given amount of organic waste is called the *biochemical oxygen demand*, or BOD. The BOD is a measure of the amount of oxygen required by bacteria to oxidize waste aerobically to carbon dioxide and water.

If we were to introduce an amount of organic matter into some water and monitor its decomposition, we would observe that the amount of organic material remaining would decrease exponentially with time. The amount of BOD which remains after some period of time is proportional to the remaining organic material, which results in a curve as shown in Figure 5.4. The initial value of BOD will of course be the total oxygen requirement to oxidize the organic material. This quantity is called the *ultimate BOD*, or BOD_L. An equation describing this curve is as follows:

$$BOD_{remaining} = BOD_L\, e^{-Kt} \qquad (5\text{-}1)$$

where K is a reaction-rate constant (which increases with temperature) and t is time.

Usually when curves of BOD versus time are drawn, they are actually made for BOD *utilized* and not BOD remaining. BOD utilized will be proportional to the amount of organic matter which has been oxidized and hence will have the shape shown in Figure 5.5. This distinction between BOD utilized and BOD remaining can cause confusion, so be aware of the difference.

The equation for BOD utilized is

$$BOD_{utilized} = BOD_L\,(1 - e^{-Kt}) \qquad (5\text{-}2)$$

A standard way to measure BOD is to determine the amount of oxygen required by the bacteria during the first 5 days of decomposition (at

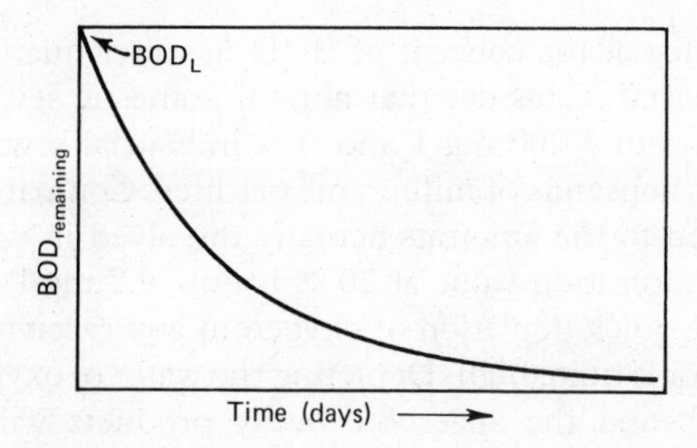

FIGURE 5.4 BOD remaining as a function of time with initial value BOD$_L$.

20°C). This is the 5-day BOD called BOD$_5$ which is shown in Figure 5.5. As the following example points out, the measured value of the 5-day BOD may be considerably lower than the ultimate BOD.

EXAMPLE 5.1 The 5-day BOD for some waste has been found to be 200 mg/l. With $K = 0.22$/day, find the ultimate BOD.

Solution: From 5.2 with $t = 5$:

$$\text{BOD}_\text{L} = \frac{\text{BOD}_5}{1 - e^{-5K}}$$

$$= \frac{200}{1 - e^{-5 \times 0.22}} = \frac{200}{1 - 0.33}$$

$$= 300 \text{ mg/l}$$

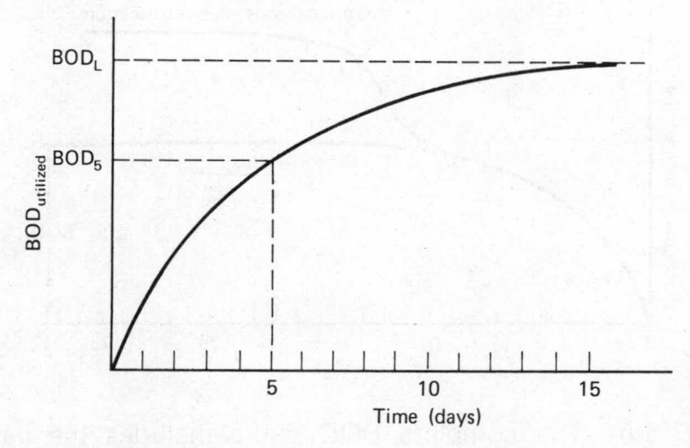

FIGURE 5.5 This is the usual curve for BOD, and is actually BOD utilized. Also shown is the 5 day BOD.

The importance of the concept of BOD in water quality management must be emphasized. Consider that normal domestic sewage may have a 5-day BOD of around 200 mg/1 and that industrial sewage may have a BOD of several thousands of milligrams per liter. Compare these requirements for oxygen to the amounts actually dissolved in water (e.g., from Figure 5.2, the saturation value at 20°C is only 9.2 mg/l). It is thus easy to anticipate the quick depletion of oxygen in any receiving water unless the dilution factor is quite high. Depleting the water of oxygen, remember, will kill the fish and the anaerobic decay products will be extremely objectionable.

It should be mentioned that there is a secondary effect which causes the demand for oxygen to suddenly increase after about 8–10 days. The organic nitrogen in the wastes is converted to ammonia during decomposition, and the subsequent oxidation of the ammonia to nitrite (NO_2) and then nitrate (NO_3) requires oxygen. This process of *nitrification* was shown as part of the nitrogen cycle in Figure 1.7.

$$\underset{\text{ammonia}}{NH_3} \xrightarrow[\text{nitrite bacteria}]{O_2} \underset{\text{nitrite}}{NO_2} \xrightarrow[\text{nitrate bacteria}]{O_2} \underset{\text{nitrate}}{NO_3}$$

The total oxygen demand curve showing both the carbonaceous demand and the nitrification requirement is shown in Figure 5.6.

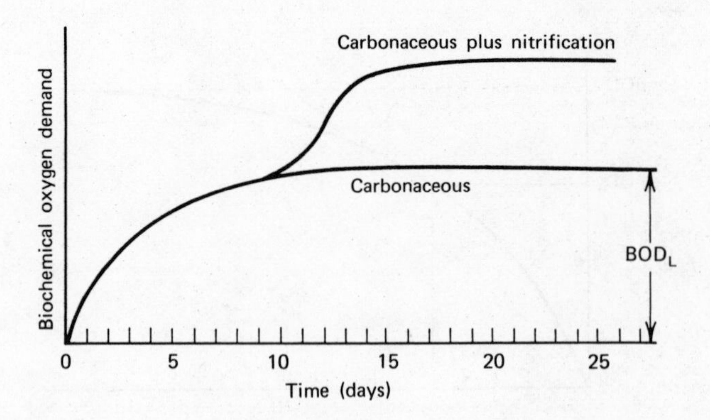

FIGURE 5.6 The complete BOD curve includes the carbonaceous and nitrification demands (Sawyer and McCarty 1967).

5.6 *The Oxygen Sag Curve*

We have seen that when wastes are discharged into a body of water, the amount of dissolved oxygen will decrease due to oxidation by bacteria. Opposing this drop in dissolved oxygen is reaeration which replaces oxygen through the surface, at a rate which is proportional to the depletion of oxygen below the saturation value. The simultaneous action of deoxygenation and reaeration produces what is called the *oxygen sag curve,* as shown in Figure 5.7. The DO curve initially drops as the wastes deplete the oxygen faster that it can be replaced. At the point where the DO is a minimum, the rate of reaeration becomes equal to the rate of oxygen utilization. Beyond that point the rate of reaeration exceeds the rate of utilization and the DO level eventually returns to normal. This sequence is referred to as the natural self-purification ability of water.

The horizontal axis of the oxygen sag curve may be either time or distance. If for example, a certain amount of waste is released all at one time into some impounded water, the DO level will be a function of time. If, however, there is a continuous discharge of wastes into a stream, then the oxygen sag curve will be a function of the distance downstream from the point of discharge.

As we move downstream from the point of discharge, and the dissolved oxygen begins to drop, there will be a corresponding change in the biota. Normal game fish will be replaced by fishes tolerant of the turbid, low DO waters. At the point downstream where the DO reaches its lowest value, conditions will be at their worst. If the stream goes anaerobic, there will be no fish and the only organisms present will be those able to obtain their oxygen from the surface, or those which are tolerant of low oxygen conditions.

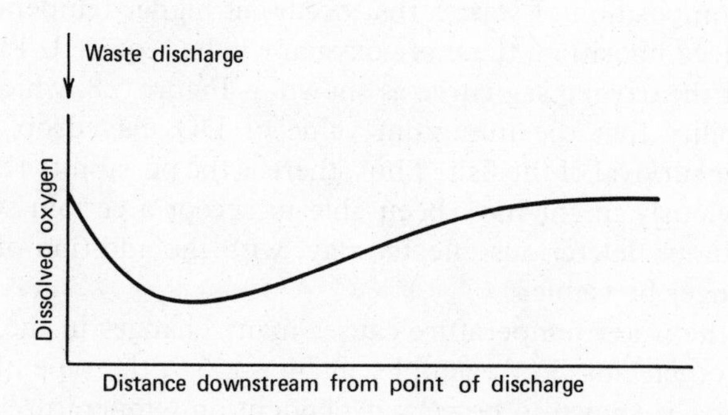

FIGURE 5.7 The oxygen sag curve.

5.7 *Thermal Pollution*

In the last chapter we saw that more water is withdrawn for cooling steam-electric power plants than for any other purpose. A single large nuclear power plant, for example, requires about 1500 cubic feet of cooling water per second—equivalent to a stream of water 10 feet in diameter flowing at the rate of 15 miles per hour. This water is extracted from a stream, lake, or ocean, run through the power plant cooling system where its temperature is raised about 20°F, and returned to its source. In Chapter 12, the thermodynamic details of this heat transfer will be explored but for now we are interested in the effects that this thermal discharge may have on the receiving body of water.

There are some circumstances when warmed water might be considered desirable. Within certain limits, thermal additions will promote fish growth and some suggestions have been made concerning the possibility of utilizing the heat in warm-water fish farms. Often times fishing is improved in the thermal plume from a power plant, but there is always the danger that the fish, which have become acclimated to the warmer temperature, will be killed when the temperature drops back to normal during periodic shutdowns of the plant.

Many of the deleterious effects to aquatic life associated with thermal pollution stem from the increased rate of metabolism that occurs as temperature increases. Generally, the metabolic rate doubles for every 10°C (18°F) rise in temperature. This causes an increased demand for oxygen by the organism. At the same time, the dissolved oxygen in the water decreases with increasing temperature, as was shown in Figure 5.2. Thus, as the organism's demands for oxygen are increasing the amount available is decreasing.

A second factor that decreases the dissolved oxygen is the increased rate of decomposition of wastes that occurs at higher temperatures. The faster the decomposition, the more oxygen will be required. This causes a lowering of the oxygen sag curve as shown in Figure 5.8, which increases the probability that the minimum value of DO may drop below that required for survival of the fish. Thus, there is the possibility that a stream which previously might have been able to accept a certain sewage load (BOD) without deleterious effects, may, with the addition of a thermal load, no longer be viable.

Raising the water temperature causes many changes in the local community of organisms. For example, in Figure 5.9, the type of algae that predominates is shown to be very dependent on temperature. The blue-green algae, characteristic of higher temperatures, are in many respects the least desirable food organisms for aquatic life. They are the algae

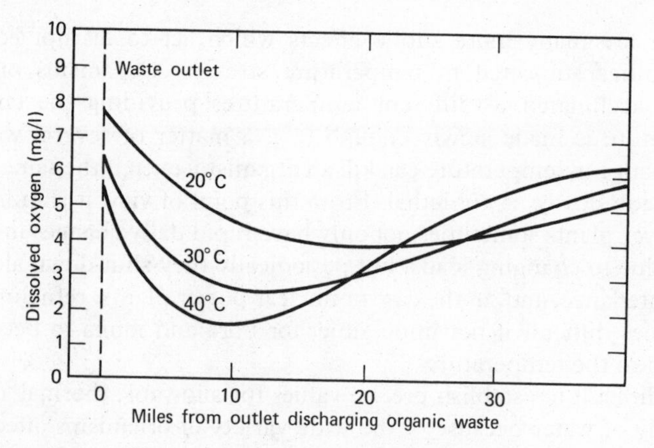

FIGURE 5.8 Relation between temperature and oxygen profile, showing the minimum point is lower at higher temperatures (USDI 1968).

most responsible for taste and odor problems in water and they are even toxic to some organisms. Ecosystems, recall, are sensitive to changes in the food chain, so the shift represented here, at the bottom of the chain, can be very important.

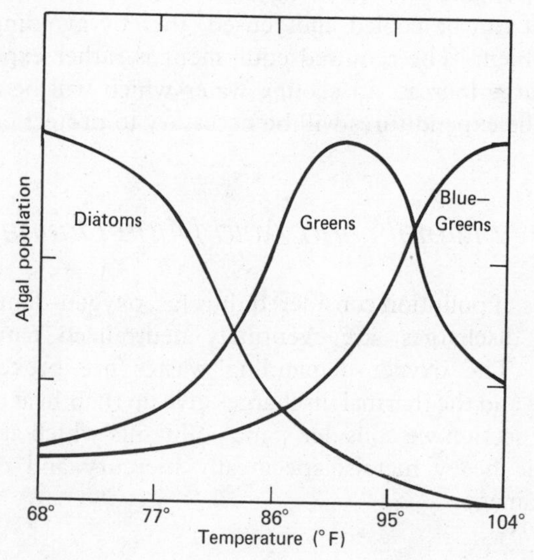

FIGURE 5.9 Effect of temperature on types of phytoplankton. Higher temperatures yield the less desirable blue-greens. (Cairns 1956).

There are many more subtle effects which act to disrupt ecological communities subjected to temperature stresses. Organisms often can become acclimated to different temperatures providing the change in temperature is made slowly enough (e.g. a matter of several weeks). A rapid change in temperature can kill an organism, even if the same change, when made slowly, is not lethal. From this point of view it is unfortunate that power plants sometimes not only have rapid daily changes in thermal output due to changing loads, but periodically they shut down altogether for maintenance, and, in the case of nuclear power plants, refueling. Thus it becomes difficult if not impossible, for flora and fauna to become acclimated to the temperature.

It is difficult to establish precise values for allowable thermal additions to a body of water because of the wide variety of organisms affected and because each organism can tolerate different amounts of temperature change during different stages in their development. The Environmental Protection Agency recommends a maximum water temperature of 90°F, with a maximum permissible rise above the naturally existing temperature of 5°F in streams and 3°F in lakes. It recommends that trout and salmon waters not be warmed at all. For estuarine and marine waters it recommends their temperatures not be raised by more than 4°F from September through May, not by more than 1.5°F from June through August.

In Chapter 12 we shall consider ways in which the heated discharge from a power plant can be cooled and reused, thereby avoiding the thermal pollution problem. The required equipment is rather expensive, but in view of the large increase in cooling water which will be needed in the near future, the expenditures will be necessary to protect our waterways.

5.8 *Heavy Metals and Acid Mine Drainage*

The two types of pollution considered thus far, oxygen-demanding wastes and thermal discharges, are eventually neutralized naturally by the environment. The oxygen-demanding wastes are broken down into simpler forms and the thermal discharges give up their heat to the environment. In this section we consider some pollutants which are not so easily degraded—the heavy metals, specifically mercury and cadmium, and acid mine drainage.

Mercury

In 1953 in Minamata, Japan, the first major incident of poisoning by mercury began to unfold. By 1960, 116 people had been irreversibly poisoned, that is, suffered such extensive brain cell damage that they will remain mentally defective for the rest of their lives. There were 43 deaths. Later, in 1965, in Niigata, Japan, 26 were irreversibly poisoned, including 5 fatalities. In both incidents the poisonings were traced to effluents from local plastics (vinyl chloride) manufacturing plants. The mercury, which is biologically concentrated as it passes along the food chain, was contaminating the fish which were being consumed by the local residents, leading to the poisonings.

Mercury enters the biosphere as a waste product from a number of industrial activities. A major source is the production of chlorine where mercury is used as an electrode in the electrolysis of brine. Other sources include "long-life" alkaline batteries, the effluent from the manufacture of paper, and as already mentioned, the plastics industry. Besides these unintentional releases of mercury, it is used extensively in agriculture as a fungicide to prevent seeds from rotting and to protect various plants, fruits and vegetables. Its use as a pesticide has been severely curtailed in Sweden (but not in the U.S.) after the discovery that certain bird populations there were suffereing severe population declines attributed to mercury poisoning from the methylmercury fungicide.

But most serious of all the mercury poisoning incidents is one which occurred in 1972 in Iraq. Mercury-treated seed arrived too late for planting and so was consumed directly by large numbers of people, resulting in an estimated 500 deaths and 7000 injuries [Council on Environmental Quality (CEQ) 1972]. In North America the concern has risen dramatically since about 1970, when elevated levels of mercury were found in swordfish and tuna. The concentration of mercury in some fish is considered by the U.S. Food and Drug Administration to be too high for consumption.

It is important to note that the toxicity of mercury is dependent on its form. Elemental liquid mercury itself is not toxic, but as a vapor it is known to produce detrimental effects in the central nervous system. However, it is the organic form of mercury, most especially methylmercury, which is the most injurious and poses the greatest environmental threat. Since the mercury in most industrial effluents is in an inorganic form, there was not much concern in the past over the release of these wastes. However, it has been discovered that microorganisms, especially bacteria in detritus and sediments, can convert inorganic mercury into the highly toxic form, methylmercury. Thus the inorganic mercury wastes which previously were considered harmless are potentially very dangerous, so even if industry should begin to do a much better job of managing its wastes, the large quantity of inorganic mercury already in the environment will be a significant threat far into the future.

Mercury contamination at the Calero Reservoir, San Jose, California.

Cadmium

As with mercury, the first indications that another heavy metal, cadmium, was becoming a serious environmental pollutant arose in Japan. In Toyama prefecture, on the west coast of Japan, incidences of an extremely painful cadmium-poisoning disease called "itai-itai" (literally ouch-ouch) have been reported ever since 1946. The disease most seriously affects the bones, gradually causing them to disintegrate. As the disease progresses "not only the bones of the extremities but also the ribs and other bones are susceptible to multiple fractures after very slight trauma such as coughing" (Friberg 1971). It has been estimated that nearly 100 deaths attributable to itai-itai had occurred by the end of 1965.

The source of cadmium in Toyama prefecture was drainage from a lead-zinc-cadmium mine. The drainage contaminated a river which supplied irrigation water for the local rice paddies. Cadmium ore is always found in intimate association with zinc.

In the U.S. the major end uses of cadmium are: electroplating, 50%; plastics, 26%; pigments, 14%. The electrical industry also uses cadmium for electrical contacts, alloys, and nickel-cadmium batteries. The Council on Environmental Quality (1972) reports that the daily average intake in the U.S. is between 0.02 ppm and 0.1 ppm, and there is some evidence that a reduction in lifespan may occur with continuous exposure at 0.1 ppm.

Acid Mine Drainage

One of the most significant causes of the degradation of water quality in and around the major mining areas of the United States, such as in Appalachia and the Ohio Basin States, is mine drainage. Besides the problem of the heavy metals which find their way into the streams, and which are generally toxic to fish and wildlife, there is the special problem of acid drainage from coal mines. Sulfuric acid is formed when water and air react with the sulfur-bearing minerals which are commonly associated with coal. This acid then enters streams either from surface runoff from the exposed tailings, or from groundwater percolation through the mines or refuse piles. The resulting acidic condition kills fish and most plant life and can make the water virtually useless. Around 75% of the coal mine drainage problem occurs in Appalachia, resulting in some 10,500 miles of streams with substandard water quality (USDI 1970).

Approaches at control have ranged from the sealing of abandoned underground mines to the reclamation of the land disturbed by surface mining. Sometimes the approach taken is to divert the drainage so that it can be treated, perhaps with lime to neutralize the acidity. Unfortunately, there is very little to be gained by the mining company from such control measures; so old mines are usually simply abandoned. For example, of the 3.2 million acres of land in the U.S. that had been disturbed by surface mining operations prior to 1965, approximately 2 million acres are either unreclaimed or only partially reclaimed.

5.9 *Conclusions*

The water-pollution examples of this chapter illustrate several important points. The biodegradable organic wastes are seen to be relatively harmless to an aquatic ecosystem providing the wastes are not so concentrated as to significantly decrease the dissolved oxygen level. Man, however, can easily overload these systems as he concentrates his numbers and activities, with the result that his wastes may be decomposed anaerobically. Anaerobic decay not only releases obnoxious reaction products, but it completely changes the character of the aquatic ecosystem. To avoid these problems, wastes may be treated before discharge to remove their oxygen demand, but, as we shall see in succeeding chapters, this is sometimes only a partial solution.

The problem of thermal discharges is similar in several respects to that of oxygen-demanding wastes in that both affect the oxygen relationships in the water, and both are relatively harmless in small doses. But more importantly, they both are degradable so the effects of a given discharge will diminish with time.

The nondegradable pollutants such as the heavy metals, some pesticides, and long-lived radioisotopes are quite different in that their effects diminish only very slowly with time, if at all. Thus there is the danger that their concentrations will keep increasing as man continues to pollute the environment. As the DDT and mercury examples have illustrated, we run the special risk with nondegradable wastes that we may discover their detrimental environmental effects after it is too late to do anything about them. Once these pollutants are discharged into the environment they are generally dispersed beyond recovery, so it is vital for us to make every effort possible to minimize their release in the first place.

Bibliography

American Public Health Association (1971). *Standard Methods for the Examination of Water and Waste Water*. 13th ed. Washington, D.C.

Cairns, J., Jr. (1956). Effects of increased temperature on aquatic organisms. *Industrial Wastes* 1(4): 150–153

Clark, J. (1969). Thermal pollution and aquatic life. *Sci. Am.* Mar.

Council on Environmental Quality (CEQ) (1972). *Environmental Quality*. Washington, D.C.: U.S. Government Printing Office.

Eckenfelder, W. W. (1970). *Water quality engineering for practicing engineers*. New York: Barnes and Noble.

Eliassen, R. (1952). Stream pollution. *Sci. Am.* 186: (3).

Friberg, L. et al. (1971). *Cadmium in the Environment*. Cleveland, Ohio: CRC Press.

Goldwater, L. J. (1971). Mercury in the environment. *Sci. Am.* May: 15–21.

Irukayama, K. (1967). The pollution of Minamata Bay and Minamata disease. *Advances in Water Pollution Research*, vol. 3. Water Pollution Control Federation, Washington, D.C.

Krauer, G. A., and Martin, J. H. (1972). Mercury in a marine pelagic food chain. *Limnol. Oceanog.* 17(6): 868–876.

Krenkel, R. A., and Parker, F. L. eds. (1969). *Biological aspects of thermal pollution*. Proc. Nat. Symp. Therm. Pollu. Nashville: Vanderbilt University Press.

McKee, J. E., and Wolf, H. W. (1963). *Water quality criteria.* The Resource Agency of California, State Water Quality Control Board, Pub. No. 3-A.

Mitchell, R. ed. (1972). *Water pollution microbiology.* New York: Wiley-Interscience.

Sawyer, C. N., and McCarty, P. L. (1967). *Chemistry for sanitary engineers.* 2d ed. New York: McGraw-Hill.

U.S. Department of the Interior (USDI) (1966). Federal Water Pollution Control Administration. *Handbook of pollution control costs in mine drainage management.* Washington, D.C. Dec.

USDI (1968). Federal Water Pollution Control Administration. *Industrial Waste Guide on Thermal Pollution.* Corvalis, Ore. Sept.

USDI (1970). Federal Water Quality Administration. *Clean Water for the 1970's.* Washington, D.C. June.

U.S. Environmental Protection Agency (EPA) (1971a). Office of Water Programs. *Temperature.* Washington, D.C. March.

U.S. EPA (1971b). *Toxic Substances.* Washington, D.C. April.

U.S. EPA (1972). *Fish Kills Caused by Pollution in 1971.* Washington, D.C.

Wood, J. M. (1971). Environmental pollution by mercury. In J. N. Pitts and R. L. Metcalf eds., *Advances in Environmental Science and Technology*, Vol. 2. New York: Wiley-Interscience.

Questions

1. Explain the influence that each of the following factors has on dissolved oxygen: photosynthesis, respiration, reaeration, decomposition, temperature.

2. For the wastes in Example 7.1, what would be the remaining BOD after 5 days?
   ans. 100 mg/l

3. Sketch the BOD-utilized curve for the sewage in Example 7.1. Suppose a temperature rise causes the reaction rate K to increase to 0.25/day. Sketch the new BOD curve. What is the new value of BOD-utilized after 5 days?
   ans. 214 mg/l

4. Show the diurnal variation of an oxygen sag curve by sketching curves corresponding to the afternoon and late night.

5. Sketch an oxygen sag curve for a stream having two sources of sewage pollution. Assume the downstream source is located slightly beyond the point at which the sag curve from the first source is at its minimum.

6. Suppose the stream described in Figure 5.8 has a flow rate of 1000 cubic feet per second and an average cross section of 2000 square feet. Calculate the time required for the wastes to reach the minimum point in the 40°C oxygen sag curve.

   ans. 1.2 days

Chapter 6

Eutrophication

All lakes are subject to an aging process wherein the gradual accumulation of silt and organic matter causes a transformation from lake into marsh and then marsh into field. A young lake which, for example, might have been formed during a glacial retreat, is characterized by a low nutrient content and low plant productivity. Such lakes, which are called *oligotrophic* ("few foods"), gradually acquire nutrients from the streams in their drainage basins, thereby enabling an increasing growth of aquatic organisms. As the resulting organic debris gradually builds up in the sediments, the lake begins to get shallower and warmer; more plants take root in the bottom; the aquatic life changes; and the lake slowly changes into a marsh.

The term *eutrophication* refers to this process of nutrient enrichment as well as the resulting effects. It is thus a natural process but it is one which can be greatly accelerated by man's activities, in which case it is referred to as *cultural* eutrophication. Human sewage, industrial wastes, and most especially agricultural runoff contribute large quantities of nutrients which can lead to rapid and excessive algal production. When the algae die, their decomposition results in a decrease in the dissolved oxygen content of the water which may result in anaerobic conditions. Further, the dominant species is often the blue-green algae (*Cyanophyta*), whose breakdown products are often toxic and impart bad tastes and odors to the water.

Cultural eutrophication significantly decreases the recreational, municipal, industrial, and agricultural desirability and usability of a body of water. In this chapter we shall be concerned with causes and effects of this form of environmental degradation. We begin by examining the factors which are essential to the growth of algae in a lake or reservoir.

6.1 Light Penetration

As photosynthesis is dependent on light, and the depth to which the light penetrates is dependent on the transparency of the water, a natural approach to the study of a lake is to divide the lake into layers of water according to light penetration. That zone of water (top) in which the light is intense enough to cause the photosynthesis by plants to exceed respiration is called the *euphotic* zone. Below the euphotic zone is the *profundal* zone with an interface between them called the *light compensation level*. These are shown in Figure 6.1.

The light compensation level corresponds to the depth at which light intensity is about 1% of full sunlight, and as is indicated in Figure 6.2, this depth can vary greatly from lake to lake. Lake Tahoe is one of the world's clearest lakes, having a light compensation level of around 90 m.

A very simple test of water transparency can be made by means of a *Secchi disk*, which is a white disk about 20 cm in diameter which is lowered into the water. The depth at which the disk just disappears from view is called the *Secchi disk transparency*. For some well-studied lakes in Wisconsin, this depth corresponds to about the 5% level of light transmittance (Odum 1971). Though this test is crude, its simplicity makes it a commonly used indicator of the lower limit of the photosynthetic zone.

There are many factors which are essential to the production of algae, and the amount of sunlight available is one of the most important. Figure 6.3 shows that the rate of photosynthesis increases with the intensity of light up to some saturation level. Increasing the intensity further has no effect until, at much higher intensities, there is an inhibition effect and the rate of photosynthesis begins to decrease. The figure shows the saturation level to be dependent on the amount of carbon dioxide available, a factor that we will return to later.

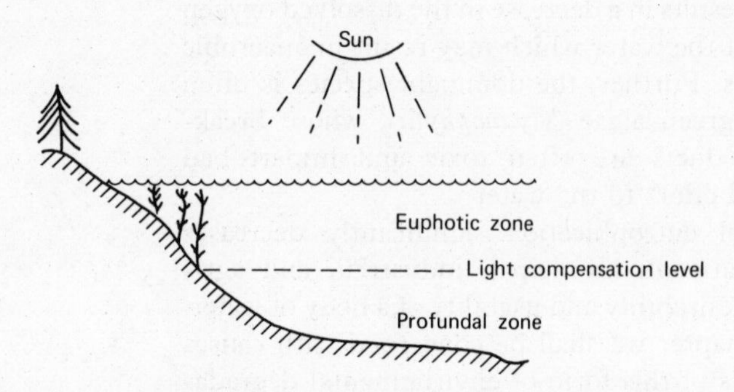

FIGURE 6.1 The division of a lake according to depth of penetration of light. (Odum 1971).

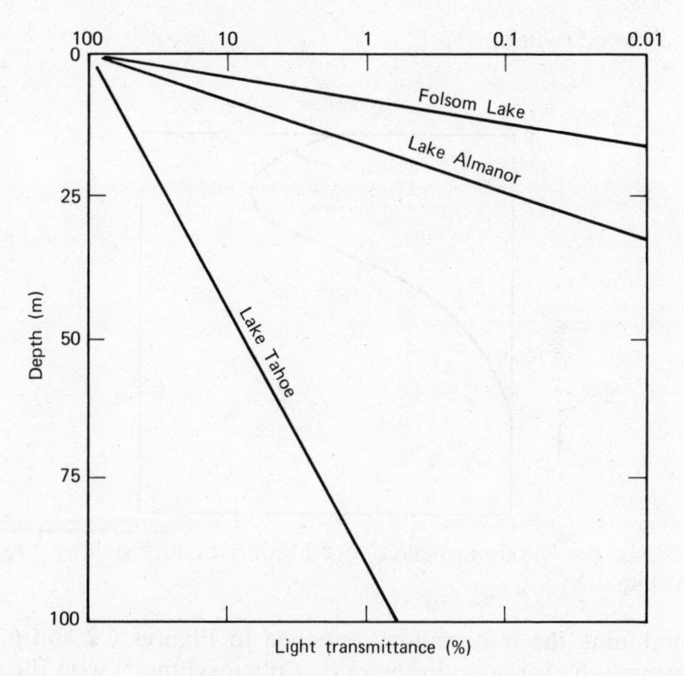

FIGURE 6.2 Transparency of the waters of Lake Tahoe compared to several other lakes. (Leggett and McLaren 1971).

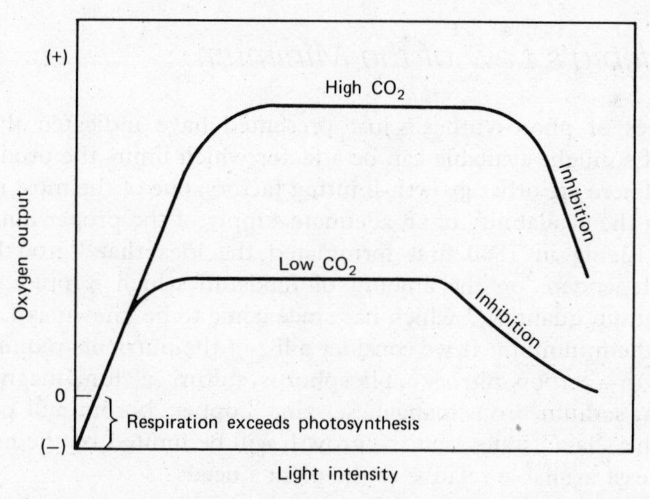

FIGURE 6.3 The rate of photosynthesis is shown here to depend on light intensity and the amount of carbon dioxide available (Fogg 1968).

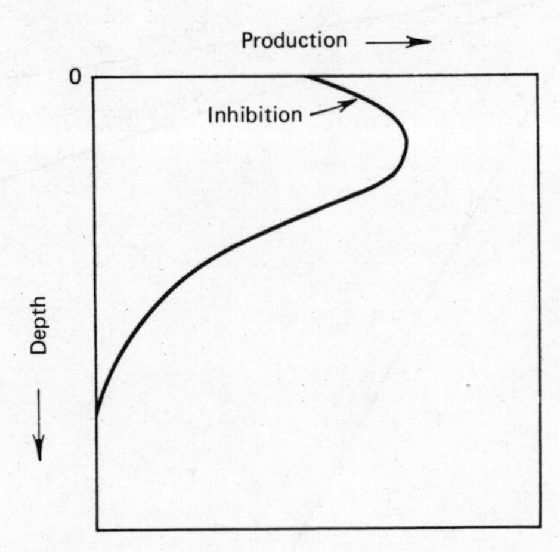

FIGURE 6.4 Maximum algal production occurs a little below the surface.

By combining the information contained in Figures 6.2 and 6.3, we can determine the variation in the rate of photosynthesis with the depth in the water. As shown in Figure 6.4, the intensity of sunlight at the surface may be so great as to be inhibitory so that maximum production may occur at a depth of 1 or 2 m. Below that, photosynthesis becomes light-limited and, so, decreases with depth.

6.2 *Liebig's Law of the Minimum*

The curves of photosynthesis just presented have indicated that the amount of sunlight available can be a factor which limits the production of algae. There are other growth-limiting factors, one of the most important being the availability of an adequate supply of the proper nutrients.

Justus Liebig in 1840 first formulated the idea that "growth of a plant is dependent on the amount of foodstuff which is presented to it in minimum quantity," which has since come to be known as Liebig's "law" of the minimum. If we consider a list of the nutrients required for algal growth—carbon, nitrogen, phosphorus, sulfur, calcium, magnesium, potassium, sodium, iron, manganese, zinc, copper, boron, and perhaps others—this "law" states that the growth will be limited by the nutrient which is least available relative to the plant's needs.

The practical application of this concept says that the rate of eutrophication can be controlled by limiting the availability of any one of the required nutrients. Since it would be extremely difficult to control the sources of the trace elements, efforts are usually directed toward the reduction of the quantitatively more important elements—phosphorus or nitrogen. Liebig's formulation also suggests that if the nutrient which is limiting growth is, say, nitrogen, then efforts to control eutrophication by decreasing the inputs of phosphorus (e.g., by requiring phosphate-free detergents) will not be as effective as efforts directed toward decreasing nitrogen.

It is instructive to work through a stoichiometric analysis of algae to make this point. Suppose an empirical analysis of the chemical composition of some algae yields the following (Stumm and Stumm-Zollinger 1972):

$$\text{algal protoplasm: } C_{106}H_{263}O_{110}N_{16}P$$

Using this formulation for algae and knowing the atomic weights of each element, we can compute the following weight percentages:

C:	106 × 12 =	1272	(35.8%)
H:	263 × 1 =	263	(7.4%)
O:	110 × 16 =	1760	(49.6%)
N:	16 × 14 =	224	(6.3%)
P:	1 × 31 =	31	(0.9%)
		3550	

Thus, for example, to produce 3550 g of this algae requires the presence of at least 224 g of nitrogen *and* at least 31 g of phosphorus. Suppose there is a lot more nitrogen available than the suggested 224 g, but suppose there are less than 31 g of phosphorus. Then phosphorus is the limiting nutrient and the total production of algae will depend on how much phosphorus is available. If there are only 15.5 g of phosphorus then only 1775 g of algae can be produced.

It is more common to express these relationships as concentrations in water. To produce 1 mg/l of this algae requires 0.009 mg/l of phosphorus and 0.063 mg/l of nitrogen. Sawyer (1947) suggests that phosphorus concentrations in excess of 0.015 mg/l and nitrogen concentrations above 0.3 mg/l are sufficient to cause nuisance blooms of algae.

In the above example it was assumed that the limiting nutrient was either nitrogen or phosphorus. It has been proposed (Kuentzel 1969) that carbon dioxide may be limiting in some instances (see Figure 6.3). Since the major source of CO_2 can be from the bacterial decomposition of organic matter, this argument suggests the importance of a low BOD in waste water. Odum (1971) points out that since algal blooms are not a steady-state phenomenon, different nutrients may be limiting during different stages of the bloom. Thus it would be possible for nitrogen, phosphorus, and carbon dioxide to sequentially replace each other in the limiting role.

This small pond, fed by nutrient-rich runoff, has been choked by a thick mat of filamentous algae. (Photograph by W. W. Hill.)

In order to stop cultural eutrophication, attention has been focused on limiting the sources of nitrogen and phosphorus. However, it turns out to be very difficult to achieve control by limiting nitrogen, especially if algae has already become a problem. Eutrophic lakes are often dominated by blue-green algae, which are able to obtain their nitrogen directly from the air. Then when the blue-greens die, their decomposition releases nitrogen in a form which is suitable for other algal species.

Thus attention must be directed to the limitation of phosphorus. Since somewhere around half of the phosphorus in our waters comes from detergents, it may be necessary to impose some sort of special controls on their use—especially around water where phosphorus is the limiting nutrient.

6.3 *Detergents*

By definition, a detergent is anything that cleanses, but it is useful to consider soaps and synthetic detergents (syndets) separately. Ordinary soaps are derived from fats and oils, and do a good job of cleaning in soft water. However, in hard water containing such ions as calcium and magnesium soaps form insoluble precipitates with the ions (the ring around the bathtub), which may be considered objectionable. Synthetic detergents on the other hand do not suffer from this problem. Sales of synthetic detergents have grown from practically zero in 1945 to about 5 billion pounds per year in 1966, accounting for nearly 85% of total detergent sales (Brenner 1969).

The first problem encountered with synthetic detergents was caused by the nondegradability of the basic active ingredient—the surfactant. A surfactant lowers surface tension allowing dirt particles to become linked to the water to be subsequently lifted or floated from the soiled material during the washing process. The original surfactant, alkyl benzene sulfonate (ABS), did not break down easily which led to mountains of foam on sewage treatment plants, and rivers and streams. Groundwater became polluted and in some areas foamy water came out of the tap.

Industry's response to this problem was to change the manufacturing process, substituting a biodegradable surfactant linear alkylate sulfonate (LAS) for the ABS. The conversion was completed in 1965 at an estimated cost of around $150 million, so that all detergents are now biodegradable.

By making detergents biodegradable, a new problem has arisen—namely, the release of large quantities of the nutrient phosphorus during degradation. Phosphates (as sodium tripolyphosphorus, $Na_5P_3O_{10}$) are added to detergents to provide buffering and assist in the suspension, dispersion, and emulsification of soil. When these phosphates are released into streams and lakes, they act as fertilizers to the plants thus speeding eutrophication. In view of the fact that phosphorus is often the limiting nutrient for algal growth, considerable pressure is being exerted to eliminate its use in detergents. However, so far they have not been banned (except in a few localities), largely for the following two reasons. First there are no acceptable substitutes presently available. The sodium salt of nitrilotriacetic acid (NTA) was thought to be an acceptable replacement until data began to accumulate indicating that it might be hazardous to human health [World Health Organization (WHO) 1972].

The second argument contends that since detergents are not the only source of phosphorus, we should be constructing advanced sewage treatment equipment to remove all the phosphorus from sewage effluent. This argument may represent a long-term solution but in the meantime we do not have the required equipment and won't for a long time into the future. It would seem prudent therefore to minimize the use of phosphates in detergents wherever possible, especially near water where phosphorus is the limiting nutrient. To some extent we should even return to the use of just plain soap.

Detergent foam on aeration tanks at the San Jose sewage treatment plant, 1963. Compare with photograph on page 160. (Photograph by B. Wykoff.)

6.4 *Thermal Stratification*

As we have seen, algal blooms are caused by nutrient enrichment. One of the most undesirable aspects of such blooms is the oxygen depletion which results when the algae die and decay. This oxygen depletion is made much more serious by certain physical characteristics of lakes that we shall now consider.

Lakes in the temperate zone tend to have only limited circulation of their waters during certain seasons of the year, which effectively isolates the water near the bottom from the surface waters. To understand this important phenomenon consider the density versus temperature sketch for fresh water given in Figure 6.5. As shown there, water has its maximum density at 4°C. Therefore ice at 0°C is less dense than the water which surrounds it and it floats. Similarly, above 4°C, the fact that water density decreases with temperature indicates that warm water will float on top of colder water. Hence the warm surface water of a lake during summer does not tend to mix with the colder water which is at greater depths. This lack of mixing leads to a layering effect in the water known as *thermal stratification.*

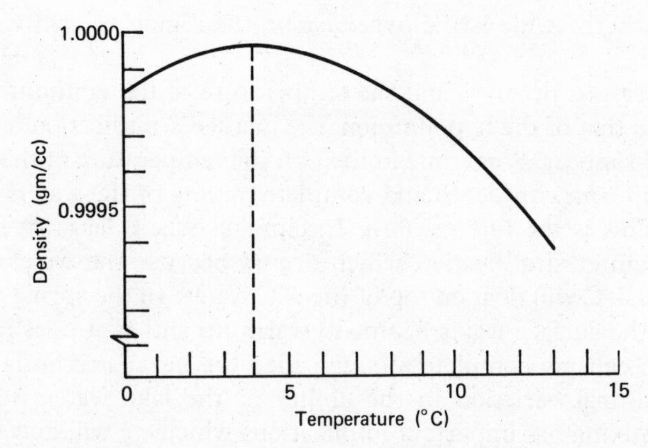

FIGURE 6.5 The density of water is a maximum at 4°C.

Figure 6.6 shows the stratification pattern which typically occurs in a deep lake in the temperate zone during summer. In the upper band, called the *epilimnion*, the water is rather completely mixed by the action of the wind and waves, resulting in an almost uniform temperature distribution. Below that is a transition layer, called the *thermocline*, or metalimnion, wherein the temperature drops rather quickly (by definition, the decrease is greater than or equal to 1°C for each meter of depth).

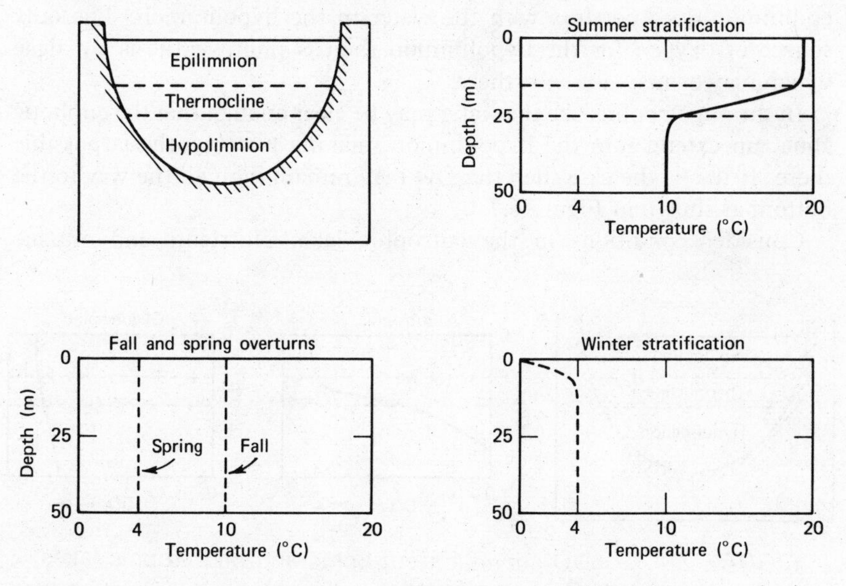

FIGURE 6.6 Characteristic temperature profiles for a typical lake that exhibits both summer and winter stratification.

Below the thermocline is the *hypolimnion*, the region of cold water near the bottom.

As the seasons progress and the temperature of the epilimnion begins to approach that of the hypolimnion, the marked stratification of summer begins to disappear. Sometime in the fall, the temperature of the lake will become uniform with depth and complete mixing of the waters becomes possible. This is the *fall overturn*. In climates where lakes freeze, there will be a winter stratification which occurs because the water which is colder than 4°C will float on top of the 4°C water. In the spring when the ice melts, the surface water begins to warm up and as it does it sinks to the bottom causing complete mixing called the *spring overturn*.

This seasonal variation in the ability of the lake waters to achieve complete mixing has important implications which we will now consider.

6.5 Stratification and Dissolved Oxygen

Dissolved oxygen, one of the most important water quality parameters, is greatly affected by both eutrophication and thermal stratification. Consider the situation in two different thermally stratified lakes—one oligotrophic and one eutrophic (high productivity). In both lakes, the waters of the epilimnion can be expected to have a high concentration of DO since oxygen is made available by reaeration and from algal photosynthesis. However, because of the thermal stratification the oxygen-rich water in the epilimnion does not mix with the water in the hypolimnion. The only source of oxygen for the hypolimnion then is photosynthesis by algae which may or may not exist there.

In the oligotrophic lake the water may be clear enough that the euphotic zone can extend into the hypolimnion, making photosynthesis possible there. If this is the case then the DO may remain high all the way to the bottom as shown in Figure 6.7.

Consider conditions in the eutrophic lake. Nutrients and organic

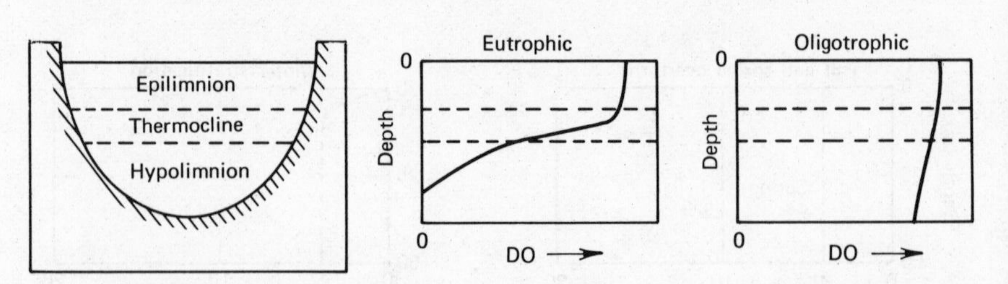

FIGURE 6.7 DO curves for eutrophic and oligotrophic lakes during summer thermal stratification.

matter are washed into the lake from agricultural run-off and urban wastes, causing a thick growth of algae. When the algae die, they can either wash onto the beaches and decay or sink into the hypolimnion. The hypolimnion receives no sunlight so there is no photosynthesis, thus it is completely cut off from all sources of oxygen. The decomposition of the organic debris that rains down from above tends to deplete whatever oxygen that may have existed in the hypolimnion, which frequently leads to anaerobic conditions. The anaerobic decay yields toxic and malodorous metabolic products as described in Section 5.4. Fish are denied access to the hypolimnion by the lack of oxygen, so that species which require the colder water of the hypolimnion may not survive. As lakes eutrophy, the cold-water-, high-oxygen-requiring fish are always the first to be killed.

Those are the conditions in a eutrophic lake during summer stagnation. During the fall and spring overturns, which may last several weeks, the lake's waters become completely mixed. Nutrients from the bottom are distributed throughout the lake and oxygen from the epilimnion becomes mixed with the oxygen-poor hypolimnion. Fish too are able to return to the bottom.

In the winter, the demands for oxygen decrease (lower metabolism of life forms, including decomposers), and the ability of water to hold oxygen increases (see Figure 5.2), thus even though stratification may occur, its effects are not so great. However, if ice forms, there can be a winter fish kill because the oxygen supply from reaeration and photosynthesis may be cut off.

6.6 *Reversible Eutrophication*

In discussing the possibility of reversing eutrophication it is necessary to distinguish between a lake which has aged naturally and one which has been culturally eutrophied. The natural aging of a lake includes the gradual diminution of its volume as sediments build up on the bottom. A culturally eutrophied lake may simply have high productivity without the volume changing appreciably. Thus, while it is not possible to reverse the natural aging of a lake, it may be possible to reverse cultural eutrophication by merely eliminating the source of overburdening nutrients.

It is important, however, to stop the influx of nutrients and organic matter as soon as possible. A highly productive lake quickly builds up a sediment layer which is rich in nutrients so that even if the inputs to the lake are removed, the bottom sediments can yield the necessary fertilization to keep the lake in a eutrophied state for a long period of time.

That it is possible to reverse cultural eutrophication has been beautifully demonstrated at Lake Washington in Seattle. In 1955 the first noticeable bloom of nuisance algae was reported, and by 1957 the hypolimnion was beginning to suffer from oxygen depletion during summer stagnation. Public concern over the deteriorating quality of the lake led to passage of legislation which authorized construction of a sewage diversion project. The actual diversion of sewage (into Puget Sound) began in 1963 and was completed in 1968. As Figure 6.8 indicates, conditions in the lake began to improve almost immediately after the diversion began.

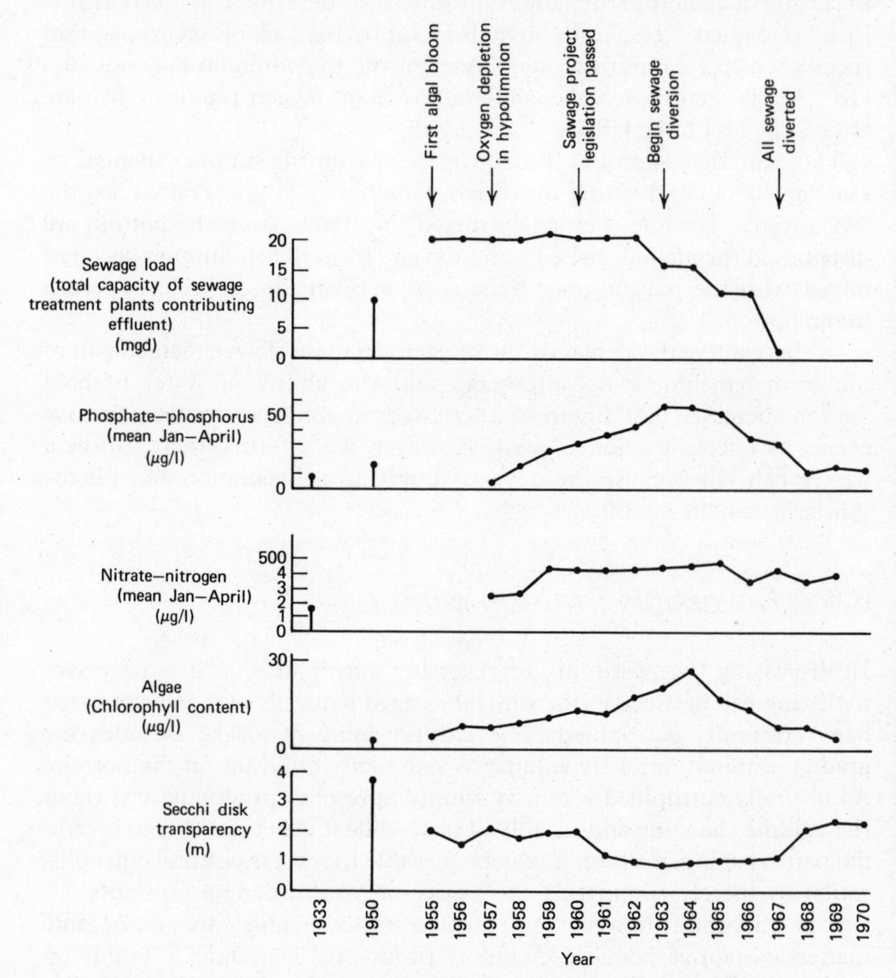

FIGURE 6.8 Reversal of cultural eutrophication in Lake Washington. Sewage diversion project was started in 1963 and completed in 1968 (Edmondson 1971).

By 1969 the concentration of algae was nearly back to its 1950 level and the Secchi disk transparency was $2\frac{1}{2}$ times as great as it was in 1963. The figure also demonstrates that phosphorus was the limiting nutrient since the symptoms of pollution were disappearing while nitrate levels remained at their former levels.

It was possible to achieve impressive results, quickly in Lake Washington because the diversion was made soon after the problem became conspicuous. In other lakes, such as Lake Erie, the problem has been so severe for so long that even if all sources of pollution were immediately removed, the existing nutrients would keep the lake in a eutrophied condition for many years into the future.

6.7 *The Great Lakes*

The story of the Great Lakes provides another good example of the unexpected, undesirable effects which can result from man's activities. The decline of the fishing industry and the deterioration of the general quality of the waters can be attributed to two separate causes: (a) the construction of the Welland Canal, and (b) pollution.

The building of the Welland Canal opened up the Great Lakes to shipping from the Atlantic by providing a bypass around Niagara Falls (Figure 6.9). The unexpected side effect was that it also opened up a path for the sea lamprey to makes its way from the Atlantic. The sea lamprey is a parasite that lives by rasping holes in other fish, sucking their blood and other body fluids, and then leaving them to die. By the 1950's the sea lamprey had killed off nearly all of the lake trout (Figure 6.10) and had then turned to destroying whitefish, chub, blue pike, and suckers.

After a concerted effort using a specific larvicide, the sea lamprey population had been greatly reduced, but a second invader through the canal, the alewife, has since become a problem. The alewife is a small fish which feeds on other fish's eggs and competes with their young for food. The alewife were not a problem until the sea lamprey decimated the larger fish which would have naturally preyed upon them. Without predators, the alewife have become the dominant species. In an effort to control the alewife, various predators are being introduced into the Great Lakes including Pacific coho salmon. Salmon young live in tributaries and by the time they reach the lakes they are large enough to successfully compete with the alewife.

The second major problem in the Great Lakes is pollution, which has been especially serious in Lake Erie. Lake Erie is unique in that it is by far the most shallow of the five (mean depth about 60 feet compared to

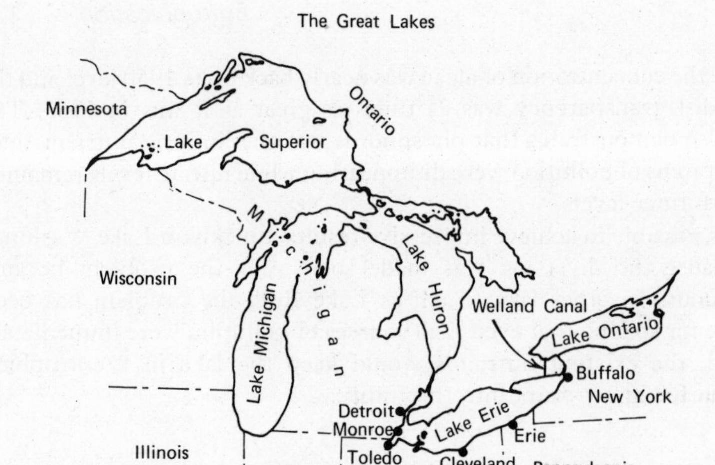

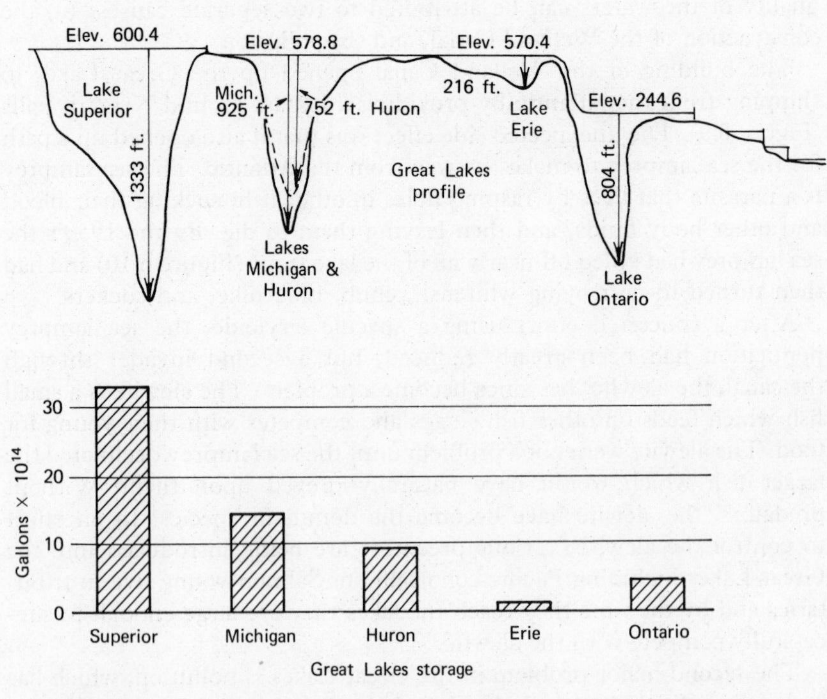

FIGURE 6.9 The Great Lakes. Lake Erie is the most shallow, has the least volume, and is the most polluted. (USDI 1968).

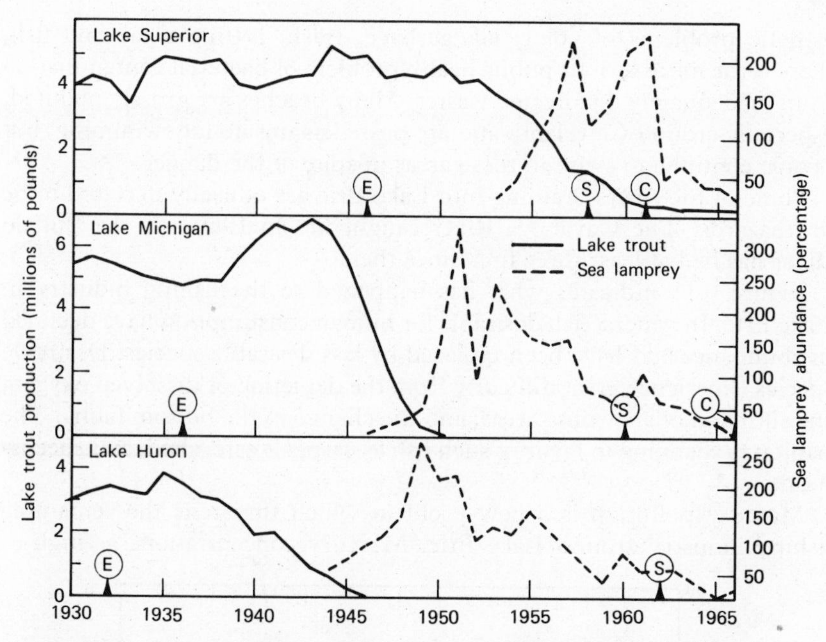

FIGURE 6.10 Production of lake trout and abundance of the sea lamprey in Lakes Superior, Michigan, and Huron. (E) marks the first sea lamprey record, (S) is initiation of chemical control, (C) is completion of the initial series of chemical treatment (Smith 1971).

487 feet for Lake Superior) and has the smallest volume (113 cubic miles compared to 31,820 for Lake Superior) which makes it the least able to dilute the wastes which pour into it daily. It is also the most biologically productive and the most turbid. The lake receives the wastes from such major industrial centers as Detroit, Toledo, Monroe, Cleveland, Erie, and Buffalo as well as receiving much agricultural runoff. Of the 25 million residents living in the communities around the Great Lakes, 11 million are concentrated on Lake Erie.

The western and central positions of the lake display the worst conditions, especially during the summer during periods of thermal stratification. In 1964 it was reported to take only five days of thermal stratification to deplete the dissolved oxygen in the hypolimnion while in 1953 it required 28 (USDI 1968). Algal blooms are common causing extensive problems as they decompose. For example, at Lake Erie State Park, bathers must go to the end of a concrete pier before attempting to enter the water to avoid the foul slimes of decomposing algae. Besides the esthetic problems of color, oil, garbage, trash, rotting algae and fish, there is the more serious public health problem of bacterial contamination from inadequately disinfected wastes. Many beaches are grossly polluted, especially around Cleveland, and are posted as unsafe for swimming, but people continue to swim in these areas in spite of the danger.

Some of the rivers draining into Lake Erie are officially declared to be fire hazards. The Cuyahoga River caught fire in 1969 and the Buffalo River has had at least three fires since then.

Figure 6.11 indicates what has happened to the fishing industry in Lake Erie. In general fish desirable for human consumption have declined in abundance and have been replaced by less desirable species. Desirable species experience great difficulty from the depletion of dissolved oxygen, the silting in of spawning areas, and the change in the bottom fauna. The habitat is changing in favor of such fish as carp, alewife, shad, and sheepshead.

Mercury pollution is a new problem which threatens the remaining fishing industry around Lake Erie. Mercury concentrations as high as

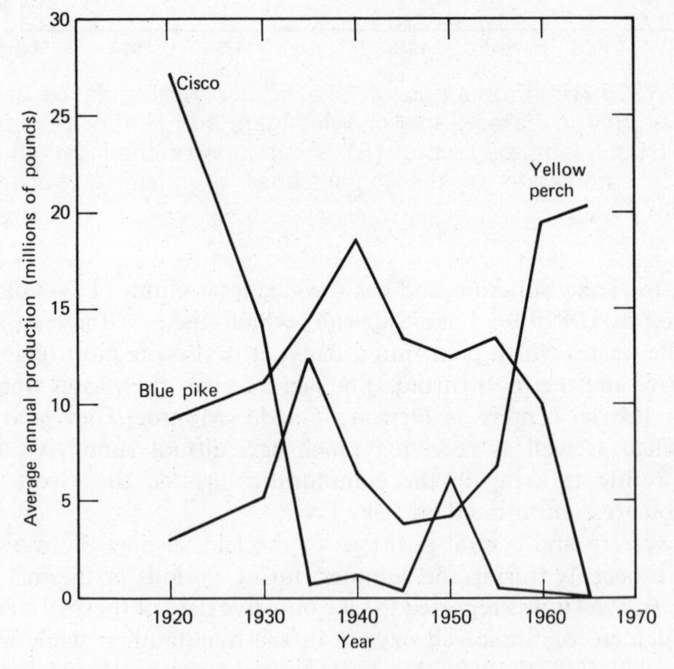

FIGURE 6.11 Average annual production of three major commercial species in Lake Erie. Data points are averages over preceding 5 year periods, except 1920 and 1930 which are over 10-year periods (USDI 1968).

5 ppm were reported in some of the pickerel shipments from Canada in 1970 (USDI 1970). This is ten times the allowable limit of 0.5 ppm set by the Food and Drug Administration, and resulted in a temporary ban on fish caught commercially in Lake Erie, Lake St. Clair, and the St. Clair River.

6.8 *Conclusions*

There exists an interesting conceptual similarity between the pollution of streams (Chapter 5) and of lakes, even though the two ecosystems are quite different (Stumm and Stumm-Zollinger 1972). In both systems, water quality suffers when pollution causes a disturbance in the balance between the production of oxygen (photosynthesis) and the consumption of oxygen (respiration, including decomposition). Biodegradable organics in a stream caüse respiration to exceed photosynthesis with the resulting sag in the dissolved oxygen curve. In a nutrient-rich lake the imbalance is caused by a physical separation of the oxygen production (in the epilimnion) and consumption (in the hypolimnion). The oxygen produced by the algae, near the surface, which is subsequently lost to the atmosphere, is paralleled by a lack of oxygen in the hypolimnion.

The eutrophication of lakes described in this chapter is caused by an unusual form of pollution—nutrient enrichment. A large portion of the excess nutrients which end up in our lakes are a result of our heavy use of fertilizers for agriculture, but there are more subtle sources which are only now beginning to be appreciated. For example, the conversion of detergents into biodegradable products has probably resulted in a significant increase in the rate of eutrophication in some lakes. Also, as we shall see in the next chapter, conventional (primary and secondary) sewage treatment plants have not been designed to remove nutrients so their effluent is a major source of the eutrophication problem. The technology does exist to remove nutrients from sewage plant effluents, but it is costly and it remains to be seen whether we are willing to pay the price.

Bibliography

Beeton, A. M. (1971). Eutrophication of the St. Lawrence Great Lakes. In T. R. Detwyler, ed., *Man's impact on environment.* New York: McGraw-Hill.

Brenner, T. E. (1969). Biodegradable detergents and water pollution. In J. N.

Pitts and R. L. Metcalf, eds., *Advances in environmental sciences,* Vol. 1. New York: Wiley-Interscience.

Colinvaux, P. A. (1973). *Introduction to ecology*. New York: Wiley.

Edmondson, W. T. (1971). Fresh water pollution. In W. W. Murdoch, ed., *Environment, resources, pollution and society*. Stamford, Conn.: Sinaur.

Fogg, G. E. (1968). *Photosynthesis*. New York: American Elsevier.

International Symposium on Eutrophication (1969). *Eutrophication: causes, consequences, correctives.* Washington, D.C.: National Academy of Sciences.

Kuentzel, L. E. (1969). Bacteria, carbon dioxide, and algal blooms. *Jour. Water Pollution Control Fed.,* Oct.

Leggett, J. T., and McLaren, F. R. (1971). Lake Tahoe revisited. *Bull. California Water Pollution Control Assoc.* Jan.

Odum, E. P. (1971). *Fundamentals of ecology*. Philadelphia: Saunders.

Powers, C. F., and Robertson, A. (1966). The aging Great Lakes. *Sci. Am.* Nov.

Sawyer, C. N. (1947). Fertilization of lakes by agricultural and urban drainage. *Jour. New England Water Works Assoc.* 41 (2).

Sawyer, C. N. (1966). Basic concepts of eutrophication. *Jour. Water Pollution Control Fed.* May.

Smith, S. H. (1971). Species succession and fishery exploitation in the Great Lakes. In T. R. Detwyler ed., *Man's impact on the environment*. New York: McGraw-Hill.

Stumm, W., and Stumm-Zollinger, E. (1972). The role of phosphorus in eutrophication. In R. Mitchell ed., *Water pollution microbiology*. New York: Wiley-Interscience.

U.S. Department of the Interior (USDI) (1968). Federal Water Pollution Control Administration. *Proceedings, Progress Evaluation Meeting, Pollution of Lake Erie and its Tributaries.* Washington, D.C. June 4.

USDI (1970). Federal Water Pollution Control Administration. *Conference in the Matter of Pollution of Lake Erie and its Tributaries.* Washington, D.C. June 3, 4.

U.S. Committee on Government Operations (1970). *Phosphates in Detergents and the Eutrophication of America's Waters.* House Report No. 91-1004. Washington, D.C.

World Health Organization (WHO) (1972). *Health hazards of the human environment*. Geneva.

Questions

1. Why is it easier to limit the amount of nitrogen available to algae in an oligotrophic lake than a eutrophic lake?

2. Lake Tahoe appears to be nitrogen-limited while Lake Washington is phosphorus-limited. Around which lake is it more important to control phosphate detergents?

3. Suppose an empirical analysis of some algae yields the following chemical composition:

$$C_{106}H_{181}O_{45}N_{16}P$$

   (a) What is the molecular weight?
   (b) What is the percentage by weight of nitrogen and phosphorus?
   (c) To produce a concentration of this algae equal to 1 mg/l, what concentration of nitrogen and of phosphorus would be required in the water?

   ans. (a) 2428; (b) 9.2%, 1.3%; (c) 0.092 mg/l, 0.013 mg/l.

4. Suppose the density of water was a minimum at 0°C and monotonically increased for all higher temperatures, would you still expect thermal stratification in the summer? In the winter?

   ans. summer—no; winter—yes.

5. Explain the following terms:
   (a) eutrophic
   (b) oligotrophic
   (c) thermocline
   (d) hypolimnion
   (e) epilimnion
   (f) fall overturn
   (g) surfactant
   (h) Secchi disk
   (i) Liebig's law of the minimum
   (j) detergent
   (k) summer stratification

Chapter 7

The Treatment of Water and Wastes

There are many diseases which are associated with the contamination of water supplies by animal or human wastes. They include cholera, typhoid fever, paratyphoid fever, dysentery, tularemia, and infectious hepatitis. It has only been in the last century that these diseases have come under control in the developed world, largely by means of some relatively simple sanitation measures.

For example, the installation of safe water supplies in some 30 rural areas of Japan resulted in a 71.5% reduction in the number of cases of communicable intestinal diseases and a 51.7% decrease in the death rate for infants and children. In the U.S., the death rate from typhoid fever dropped from 20.54 to 0.15 per 100,000 population during the period 1910–1946, largely due to the elimination of unsafe drinking water (WHO 1972).

The basic sanitation techniques which are essential to the control of water-related diseases—the purification of water, and the treatment of sewage—will be described in this chapter. We will see how the emphasis in water-quality control programs has shifted from the area of water-borne diseases, to control of oxygen-consuming wastes, and now to nutrient removal and water reclamation.

7.1 *Biological Health Hazards*

A simple way to categorize water-associated health hazards is according to whether they are communicable. Noncommunicable diseases such as those caused by pesticides, heavy metals, radiation, and nitrates, are treated in other sections of this book. Of the communicable water-associated diseases, those of most interest here are caused by the ingestion of biological agents. There are, however, other modes of transmission including simple water contact, as in the case of schistosomiasis, or insect vectors, as in the case of malaria, which are also important.

The contamination of water by sewage is the principal cause of water-borne diseases. Intestinal discharges of patients or carriers of such diseases contain the biological agents which are responsible for the disease's spread. A carrier may not even be aware that he is infected so it is essential that proper precautions be observed for all human wastes. The disease-causing organisms can be transferred from the excrement of an infected person into the mouths of healthy people by means of contaminated food or water. The spread of such infectious diseases can be controlled both by proper treatment and disposal of human wastes, and by purification of drinking water supplies.

The principal biological agents which are of concern in this context are pathogenic (disease-causing) bacteria, viruses, and parasites. Pathogenic bacteria are the causative agents of the great epidemic diseases—cholera and typhoid—as well as bacillary dysentery, paratyphoid fever, and tularemia. Fortunately, intestinal bacteria that are discharged into natural waterways usually survive for only a matter of days. Figure 7.1 shows the rate at which fecal coliform bacteria (not pathogenic) die away.

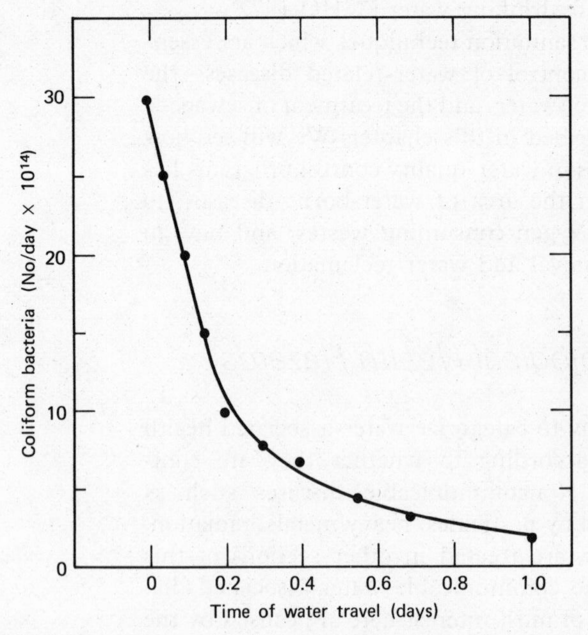

FIGURE 7.1 Pattern of natural purification of coliform bacteria (Kittrell 1969).

These bacteria exhibit an initial rapid decrease that results in about a 90% reduction in 2 days. However, thereafter the rate of decrease is considerably slower. It is interesting to note that this reduction is faster in warm, polluted water than in cold, clean water.

The only viral disease that has been proven to be transmitted through drinking water is infectious hepatitis, although there are others for which transmission through water is theoretically possible. One of the particular difficulties with viruses is that some can survive the normal dosages of chlorine which are applied at water and sewage treatment plants to kill pathogens.

Among the parasites that may be ingested is the protozoa *Entamoeba histolytica,* which causes amebic dysentery. Another parasite is the guinea-worm which causes dracontiasis, a common disease among the rural populations of many developing countries.

7.2 *Urban Water and Sewage Systems*

There are two convenient points to break the chain of events that can result in the spread of water-borne diseases. The first is the water treatment plant where drinking water must be made completely safe before being distributed. The second is the sewage treatment plant where wastes are processed before being released into the receiving water.

As shown in Figure 7.2, the increase in public water supplies and sewer systems in the U.S. is well correlated with the drop in typhoid fever deaths. These improvements are relatively recent, with the major drop in typhoid deaths having occurred in this century. Philadelphia, for example, began filtering its water supply in 1906 and began chlorination in 1913, with the result that the incidence of typhoid fever dropped by 98% during the period 1906–1926 (Benarde 1970).

A schematic version of a city water and sewer system is presented in Figure 7.3, along with some indications of potential trouble spots. Water which may be withdrawn from various sources, including rivers, reservoirs, and wells, is transported to a water treatment plant for purification and perhaps fluoridation (to reduce dental caries). From the treatment plant it is passed to the distribution system of the city.

Waste water is shown being collected from sinks and toilets to be transported through sanitary lines to the sewage treatment plant. Many special precautions must be taken to prevent any kind of cross-connection between the water system carrying potable water and any other system carrying water of doubtful purity. If there are any pathways between the two systems, and if there should be a loss of pressure in the potable water

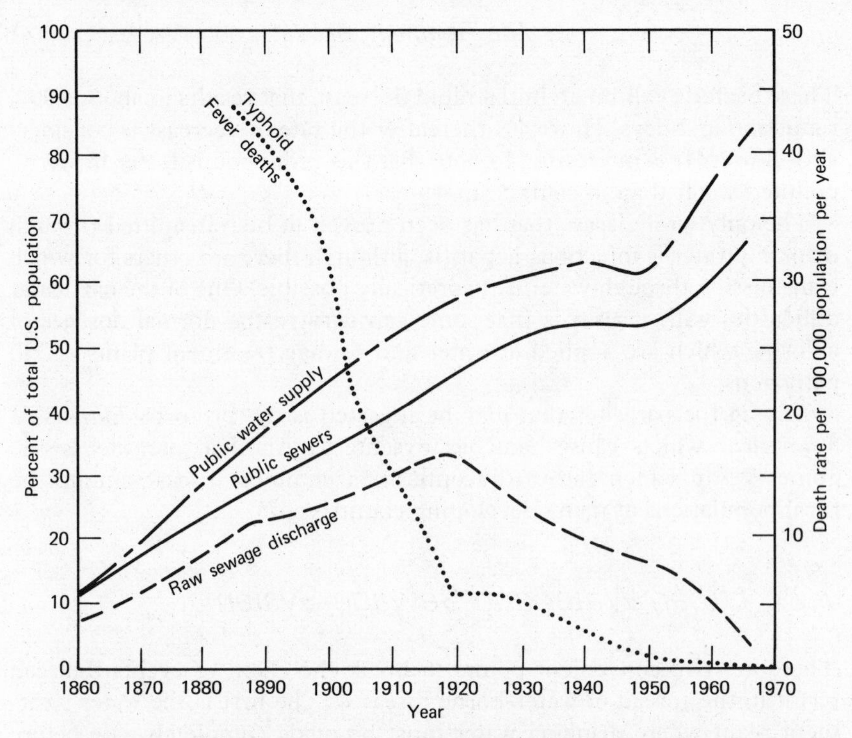

FIGURE 7.2 Growth in the percentage of the U.S. population serviced by public water and sewer systems is correlated with the drop in typhoid fever deaths. (Kruse 1969).

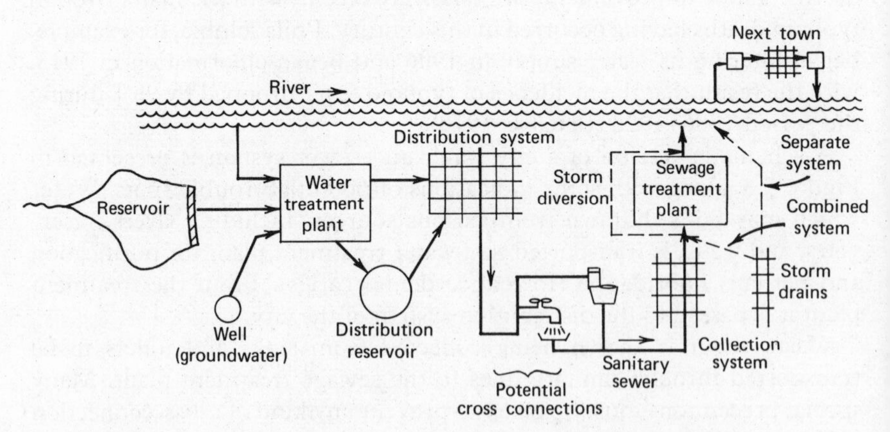

FIGURE 7.3 Municipal water system including sources, treatment plant, distribution system, collection system, sewage plant, and disposal.

system, impure water could be back-siphoned into the drinking water. Cross-connections have been the most frequent source of water-borne disease outbreaks in the past.

Also shown in the system is a storm sewer which collects the runoff from the streets. Early sewer systems in this country simply *combined* the storm runoff with the sanitary lines sending it all to the sewage treatment plant. These combined sewer systems are unsatisfactory during periods of heavy rainfall because the sewage treatment plant can become overloaded, necessitating the diversion of large volumes of untreated waste directly into the receiving water. Newer cities have separate storm sewers and sanitary sewers to avoid the overloading problem. Although these separate systems are much better, they are still not perfect in that the storm runoff is passed, untreated, back into the receiving water, even though it may contain much objectionable material.

The cost of separating existing combined systems would be extremely high, one estimate being $48 billion (American Chemical Society 1969). In some areas it may be possible to avoid separation by temporarily storing storm runoff in large detention basins for treatment at a later time.

Notice in Figure 7.3 that the next town downstream is liable to be withdrawing water that has already been used upstream. At the bottom end of a long river, such as the Mississippi, it is quite likely that a municipal water supply is distributing water that has already passed through another city's sewage system.

Many rural areas do not have any sewage treatment facilities at all, relying mainly on septic-tank systems of the type shown in Figure 4.10. In 1970 some 50 million people in the U.S. relied on these systems for their sanitary services while another 10 million discharged their sewage raw into our waterways. When properly designed and maintained, septic tanks are adequate, but when they go bad they can be a health danger and an esthetic and olfactory nuisance.

7.3 *Water Quality Criteria*

Before going on to discuss techniques of water purification, it is worthwhile to briefly consider how water quality is evaluated. Water quality standards are of course dependent on how the water is to be used; for example, drinking water standards are more stringent than standards applicable to irrigation water.

The Federal Water Pollution Control Administration (FWPCA)* in

*FWPCA was incorporated into the Environmental Protection Agency in 1970.

1969 issued a set of criteria to be used by the states as guidelines in setting their own standards. They used the following water classifications: (a) Recreation and Aesthetics, (b) Public Water Supplies, (c) Fish, Other Aquatic Life, and Wildlife, (d) Agriculture, and (e) Industry. The Public Water Supplies category establishes criteria for the quality of water before being processed at a water treatment plant. The standards which states use to evaluate drinking water quality are based on U.S. Public Health Service (USPHS) standards.

The USPHS drinking water standards (1962) establish limits in four categories: (a) physical characteristics, (b) chemical characteristics, (c) radioactivity, and (d) bacteriological quality. In the category of physical characteristics it is specified that the water shall contain no impurity which would cause offense to the sense of sight, taste, or smell.* The recommended limits for chemical substances are presented in Table 7.1.

Of special interest here is the specification of bacteriological quality. It is very difficult to detect pathogenic microorganisms in a water supply; instead, the water is tested for *any* contamination by human or animal excreta by measuring coliform bacteria. There are literally hundreds of millions of harmless coliform bacteria (principally *Escherichia coli*) per gram of fecal material. In crude sewage there may be hundreds of thousands of fecal coliforms per cubic centimeter, but only a few pathogens. Therefore, the probability of there being any pathogens in a sample, without the accompanying fecal coliforms is essentially zero. On this basis, if a sample of drinking water is found to have no coliform bacteria, then it is reasonable to assume there are no pathogens either, and the water can be considered safe. Statistically, this is a good test, but as can be seen, it does not guarantee water purity.

It should be mentioned that the coliform group comprise not only fecal coliforms but also other coliforms which are principally found in soil and vegetation. One study of the Ohio River indicated fecal coliforms represented 18% of total coliforms (FWPCA 1968). It is possible to distinguish between fecal coliform and the other subgroups but there is no satisfactory way to differentiate between fecal coliforms of human and animal origin.

USPHS drinking water standards for coliform organisms involve considerable detail but roughly they specify that the most probable number (MPN) shall not exceed 1 coliform organism per 100 ml of water. For water-contact recreational activities, many states recommend a limit of 1000 coliform organisms per 100 ml. The coliform measurement is used as a criterion for closing a beach due to pollution.

*The following quantitative limits are set: turbidity, 5 units; color, 15 units; and threshold odor number, 3 units.

TABLE 7.1 USPHS Chemical Standards for Drinking Water (1962)[a]

Substance	Recommended[b] maximum concentration, milligrams per liter	Maximum[c] permissible concentration, milligrams per liter
Alkyl benzene sulfonate (ABS)	0.5	—
Arsenic	0.01	0.05
Barium	—	1.0
Cadmium	—	0.01
Carbon chloroform extract	0.2	—
Chloride	250	—
Chromium	—	0.05
Copper	1.0	—
Cyanide	0.01	0.2
Iron	0.3	—
Lead	—	0.05
Manganese	0.05	—
Nitrate	45.0	—
Phenols	0.001	—
Selenium	—	0.01
Silver	—	0.05
Sulfate	250	—
Total Dissolved Solids	500	—
Zinc	5.0	—

[a]Detailed fluoride specifications have been omitted here.
[b]If exceeded, use more suitable supplies if available.
[c]If exceeded, grounds for rejection of the supply.

7.4 *Water Treatment Fundamentals*

Water treatment plants are designed to bring raw water up to drinking water quality. A typical plant might include the following sequence of steps (Figure 7.4): mixing, coagulation, settling, filtration, and chlorination. Basically the idea is to coagulate the suspended particles which cause turbidity, taste, odor, and color, so that they can be removed by settling and filtration.

In the mixer, a coagulant such as alum, $Al_2(SO_4)_3 \cdot 18H_2O$ is added to the raw water and rapidly mixed. The coagulant enables colloidal particles to stick together when contact is made, thus forming a floc nucleus. It is essential at this stage to obtain rapid and uniform dispersion of the coagulant to assure complete reaction.

In the flocculation basin, gentle and prolonged agitation enables the submicroscopic coagulated particles to assemble into large, plainly visible, agglomerates. These particles are large enough to settle at a rapid rate, or be removed from suspension by filtration.

From the flocculator, the water is passed into a settling basin where it may typically be held for from 2 to 4 hours. Here the large floc particles are allowed to settle under the influence of gravity, whereupon they are collected as sludge and disposed of. The effluent from the settling basin then goes to the filtration unit.

One of the most widely used filtration units is called a rapid-sand filter, which consists of a layer of carefully sieved sand, 24–30 inches thick, on top of a 12–18 inch bed of graded gravels. The pore openings between grains of sand are often greater than the size of the floc particles that are to be removed; so much of the filtration is accomplished by means other than simple straining. Adsorption, continued flocculation, and sedimentation in the pore spaces are also important removal mechanisms. When the filter becomes clogged with particles, the inlet valve is closed and the filter is cleaned by backwashing for 3 or 4 minutes. During the coagulation, settling, and filtration, practically all of the suspended solids, most of the color, and about 98% of the bacteria are removed. For safety, the effluent must be disinfected, usually by chlorination.

Chlorination, and perhaps fluoridation, are the final steps in water treatment before storage and distribution. Chlorine is particularly effective against pathogenic bacteria but its ability to destroy amoeba and viruses is questionable. In Delhi, India, in 1955, there was a dramatic example of the inability of chlorination to destroy the virus responsible for infectious hepatitis. A reversal in the flow of the Jamuna River caused raw sewage which was normally deposited downstream to be taken up by the water treatment plant. The chloride concentrations were immediately increased which apparently was effective in preventing any significant increase in

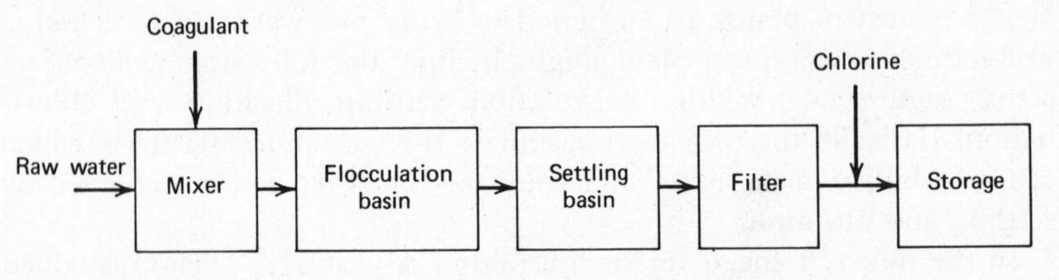

FIGURE 7.4 Flow diagram for a water treatment plant.

typhoid or other enteric bacterial diseases, but it did not prevent the spread of some 35,000 cases of infectious hepatitis.

7.5 *Desalination*

Table 7.1 indicates that the Public Health Service recommends 500 ppm as the maximum concentration of total dissolved solids (TDS) in drinking water. This concentration is regularly exceeded in many communities in the United States; in fact, over 3 million people in the U.S. receive water with a total dissolved solids concentration exceeding 1000 ppm (American Water Works Association 1971).

It has also been pointed out that as water evaporates, the salts are left behind so that many of man's uses of water (e.g., irrigation) have the effect of increasing the salt content (measured as total dissolved solids) of the return water. Conventional water treatment plants remove suspended and colloidal particles but are not effective in reducing the concentration of dissolved solids. There is, therefore, increasing interest being shown in desalination techniques not only for "brackish" water (roughly 1,000–10,000 ppm TDS) but also for seawater conversion (35,000 ppm TDS).

Desalination technology exists but is not commonly used except under special circumstances (e.g. on board ships or on islands). One factor which has been influential in restricting the proliferation of vast desalination projects is the requirement for large amounts of energy. From a relatively simple thermodynamic analysis (e.g., Spiegler 1962, or Harte and Socolow 1971), it can be shown that the theoretical minimum amount of energy required to desalt seawater, at 25°C, by any technique, is 2.65 kilowatt-hour (kWh) per 1000 U.S. gallons. An actual desalination plant would require considerably more energy than this theoretical minimum. Assuming a plant efficiency of 5% (an average based on Table 10.1 in Spiegler 1962) suggests that a value of 53 kWh per 1000 gallons is more realistic. To supply the U.S. average per capita water usage of 1600 gallons per day for all uses (including industry and agriculture) would therefore require close to 85 kWh of energy per person per day. For comparison, this is about one-third of the total amount of energy, per capita, consumed in the U.S., and is over three times the per capita consumption of electrical energy.

Though these amounts of energy are very large, it may be possible in the future to construct combined nuclear power plants and desalination facilities, with much of the energy for desalination coming from power plant waste heat. This makes desalination much more feasible than the above energy calculation might imply.

There are a number of techniques for desalinating water, including distillation, freezing, reverse osmosis, ion exchange, and electrodialysis. The particular technique which is most appropriate for a given area depends on such factors as the volume of water to be recovered, the salt concentration, and sources of energy available. We shall briefly describe two of the most frequently used processes: distillation and electrodialysis.

Distillation techniques are based on the fact that salts do not evaporate with water and hence if water is caused to vaporize and then to condense, the condensate will be pure water. Most of the world's seawater desalination uses some variation of this technique, which is schematically illustrated in Figure 7.5. Notice the incoming cold salt water is used to condense the steam. This serves the second purpose of warming the salt water so that less energy is required in the boiler, thus increasing the process efficiency. The efficiency is increased even more by connecting many stages, similar to those shown, in series, causing the incoming salt water temperature to gradually be raised as it passes from stage to stage.

The largest distillation plant in the U.S. is in Key West, Florida, and produces 2.6 million gallons per day (mgd). At current rates of municipal consumption, this could supply the needs of about 20,000 people. The cost of seawater desalting is currently about $1 per 1000 gallons, but as Figure 7.6 indicates, as plant sizes increase in the future the cost is projected to decrease rapidly. For comparison, in 1965 the average cost to the consumer for water supplied by municipal systems was 29.2 cents per 1000 gallons (U.S. Water Resources Council 1968).

For brackish waters, a popular technique for desalination is *electrodialysis*. In electrodialysis the ions forming the salt are pulled out of solution by an electric field. Selective membranes, some of which pass only positively charged ions and some of which pass only negatively charged ions, make possible the separation of salt water from fresh, as shown in Figure 7.7. The two types of membranes are alternately spaced,

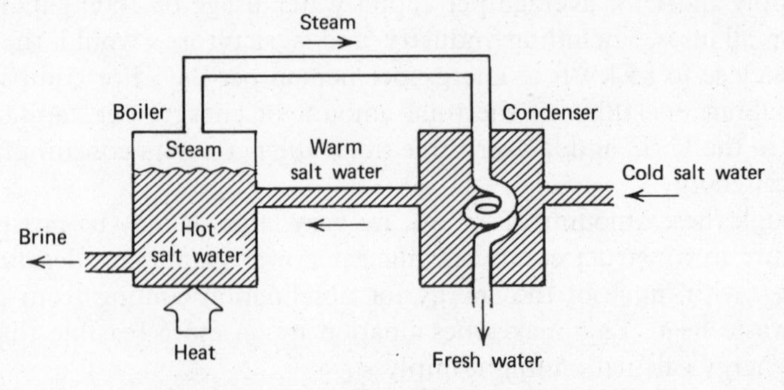

FIGURE 7.5 Single-stage distillation.

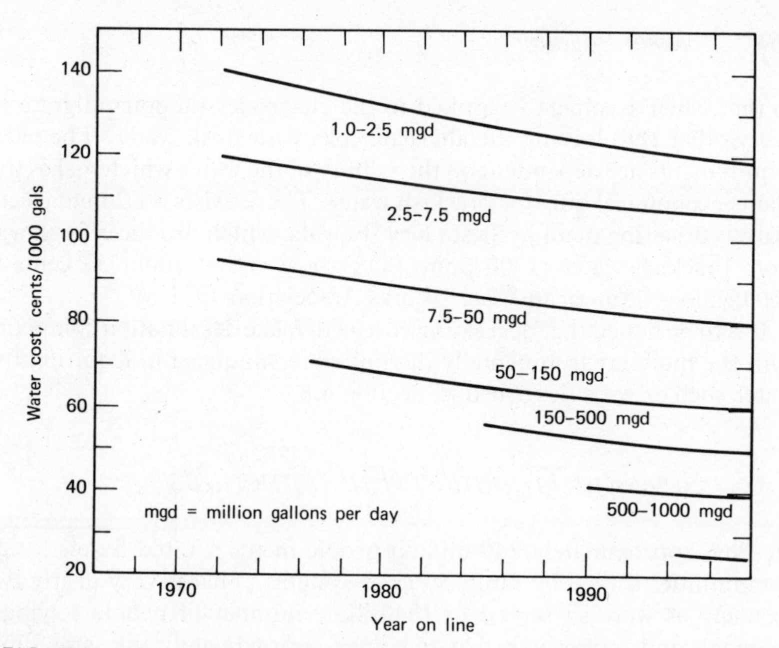

FIGURE 7.6 Seawater desalting costs, distillation technology, for a range of plant sizes (USDI 1973).

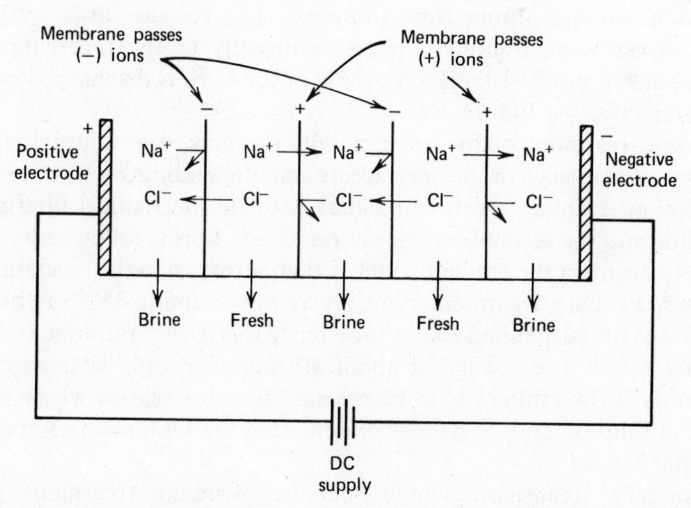

FIGURE 7.7 Electrodialysis method of desalination. Selective membranes and electric field create alternate cells of fresh water and brine.

so that when a voltage is applied to the electrodes the ions migrate into every other cell, leaving the alternate cells with fresh water. The energy requirements are dependent on the salinity of the water which makes these plants economical only for brackish water. There exists a 1.2 mgd electrodialysis desalting plant in Siesta Key, Florida, which produces fresh water from brackish water (1300 ppm TDS) at a cost of about 33 cents per 1000 gallons (American Water Works Association 1971).

It is to be hoped that decreasing costs will make desalination competitive with the more environmentally disruptive techniques for acquiring fresh water such as were described in Section 4.6.

7.6 Sewage Treatment Fundamentals

In 1968 approximately 140 million people in the United States lived in communities served by sanitary sewer systems. That is very nearly twice as many as were so served in 1940. The number of people relying on cesspools and septic tanks has remained approximately the same during that period, being roughly 60 million. Our concern here is with the treatment and disposal of wastes collected in these sewered communities.

Municipal sewage is about 99.9% water and only about 0.03–0.06% solids. Typical values for BOD range from 75 to 276 mg/l and average flow rates in the U.S. are around 135 gallons per capita per day. The composition of the sewage, of course, varies from city to city, being largely dependent on the inputs from industry. The sewage flows by gravity either to a sewage treatment plant or directly to the receiving water. The sewage of nearly 10 million people in the U.S. is discharged raw into our waterways (see Figure 7.8).

Sewage treatment plants are generally designated as supplying either primary, secondary, or tertiary treatment depending on the degree of purification. Primary treatment plants provide mechanical filtering and screening and the settlable solids are removed. This is followed by chlorination to disinfect the effluent which is then returned to the receiving body of water. Primary treatment typically removes about 35% of the BOD and 60% of the suspended solids. Obviously this is very minimal treatment but this is how the sewage of about 40 million people is treated. Very frequently if the effluent is to be released into the ocean, where there is plenty of dilution and no danger of depleting the DO, primary treatment is all that is used.

Secondary treatment, which includes primary treatment (minus chlorination) as the first step, provides conditions for the biological oxidation of the organic wastes, much the same as would occur in nature.

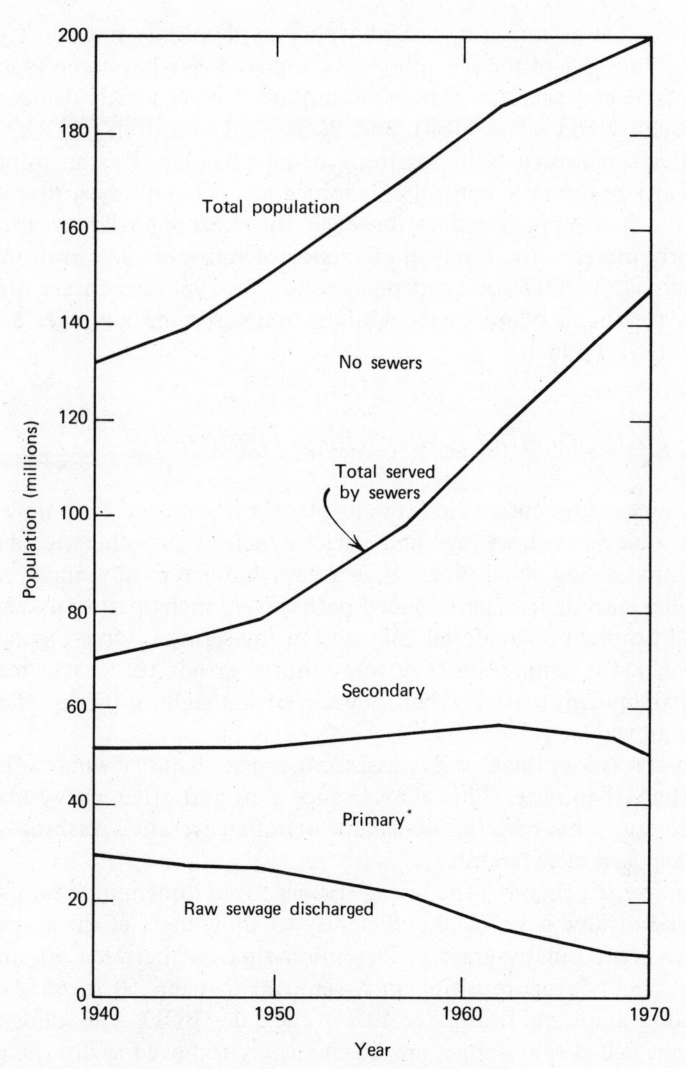

FIGURE 7.8 Methods of sewage disposal in the United States. (USDI 1968b; EPA 1972).

Secondary treatment removes about 90% of the BOD and 90% of the suspended solids. Most communities have this degree of treatment. The major complaint about secondary treatment though, is that while it does a good job of removing materials which would directly cause a drop in the DO of the receiving waters, it does a poor job of removing the nutrients such as nitrogen and phosphorus. Typically only 50% of the nitrogen and 30% of the phosphorus is removed. We have seen that these nutrients encourage the growth of nuisance algae which decrease the water quality and whose death and decay lead to a drop in DO.

Tertiary treatment is increasingly being recognized as an important part of any program to control eutrophication. Though most designs are still in the development and testing stage, the effect of tertiary treatment is to greatly increase the removal efficiency of nutrients and also improve the removal of BOD and suspended solids. In 1968 there were only ten tertiary treatment plants in the United States serving just over 325,000 people (USDI 1968b).

7.7 *Primary and Secondary Treatment*

When sewage first enters a treatment plant it is screened to remove large floating objects such as rags and sticks which might otherwise damage the pumps or clog small pipes. Screens vary, but typically might consist of parallel steel or iron bars spaced perhaps 1/2 inch apart. To avoid the disposal problem for material collected in the screens, some plants use a device called a comminuter. A comminuter grinds the coarse material into small enough particles that they can be left right in the sewage flow, to be removed later.

After screening, the sewage passes into a grit chamber where it is held for perhaps 1 minute. This allows sand, grit, and other heavy material to settle out. This material is usually nonoffensive after washing and is often disposed of as landfill.

From the grit chamber the sewage passes to a sedimentation tank where the speed of flow is reduced sufficiently to allow most of the suspended solids to settle out by gravity. Detention times of between 90 and 150 minutes are typical, resulting in a removal of from 50 to 65% of the suspended solids and from 25 to 40% of the 5 day BOD. The solids which settle out, called raw sludge, are mechanically removed as are the grease and scum which float to the top. A simplified cross section of a rectangular sedimentation tank is shown in Figure 7.9; circular tanks are also common. As shown, an endless conveyor scrapes the floating material into a scum trough while it also pushes the settled solids into a sludge hopper.

If this is a primary treatment plant, the effluent at this point is chlorinated to destroy disease-causing bacteria and help control odors. Then it is released.

Most modern sewage treatment plants follow primary treatment by biological, or secondary, treatment. Primary treatment removes the solids

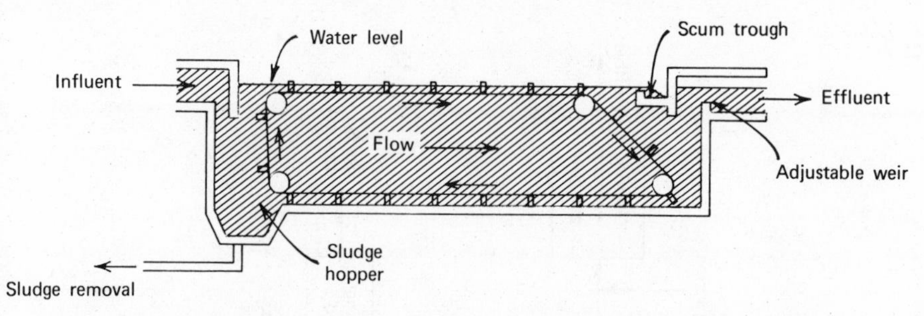

FIGURE 7.9 Rectangular sedimentation tank.

which settle easily, while secondary treatment is effective in removing most of the remaining organic matter by biological processes. Secondary treatment is very similar in concept to the processes of decomposition which occur in nature. Organic wastes are consumed by bacteria under controlled conditions, so that most of the BOD is removed in the treatment plant rather than in the receiving water.

Figure 7.10 shows a flow diagram for a secondary treatment plant which uses the activated-sludge process. After primary treatment (screen, grit chamber, primary settling), the effluent passes into an aeration tank where the organic matter is brought into contact with sludge which is heavily laden with bacteria. To maintain aerobic conditions, air is pumped into the tank and the mixture is kept thoroughly agitated. The bacteria convert a portion of the organic waste into stabilized, low-energy compounds such as nitrates, sulfates, and carbon dioxide, but mostly new bacterial cells are synthesized.

After about 6 hours of aeration, the sewage (now referred to as the mixed liquor) passes on to a secondary settling tank where the solids (mostly bacterial masses) are separated from the liquid by subsidence. A portion of this activated sludge is recycled back to the aeration tank to maintain the proper level of biological activity there. The remainder of the sludge is removed for processing and disposal. The effluent from the secondary settling tank is then chlorinated and released.

Many biological treatment plants use a trickling filter instead of an aeration tank. The liquid effluent from the primary sedimentation tank is sprinkled over a bed of rocks which are covered by a layer of biological slime. The rock bed may be anywhere from about 3 to 8 feet in depth, with enough openings between rocks to allow air to easily circulate. As the waste water passes through the filter, the bacteria adsorb and consume the organic matter in much the same manner as occurs in the activated-sludge process. The biological community attached to the rocks is quite complex, consisting not only of various kinds of bacteria, but also fungi, algae,

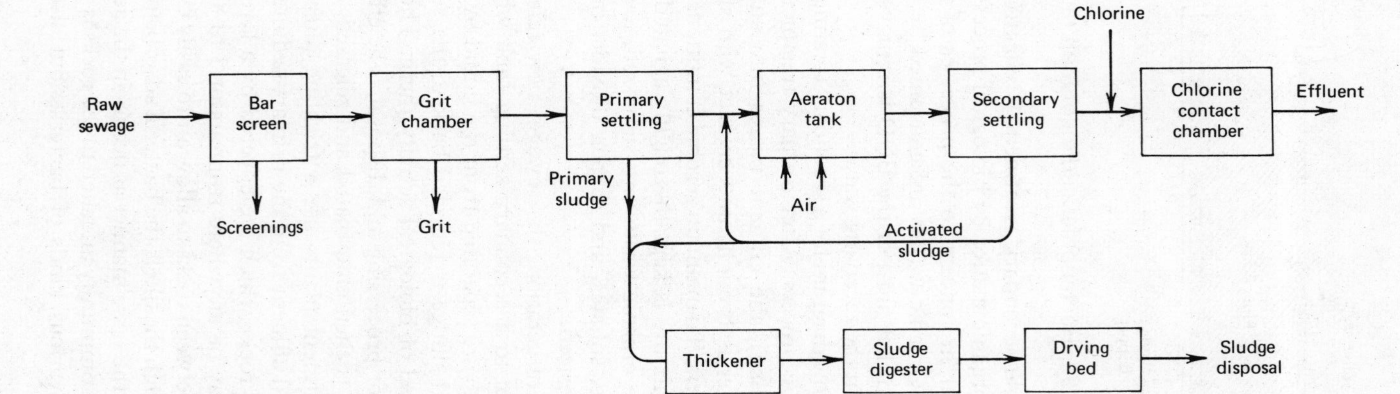

FIGURE 7.10 Flow diagram for an activated-sludge treatment plant.

protozoa, worms, insect larvae, and snails. Periodically this biological slime is washed off of individual rocks by hydraulic action, to be later removed in the secondary settling tank.

The photograph on page 159 shows a typical circular trickling filter in which the liquid sewage is spread by means of rotating distribution arms. The effluent is collected at the base of the filter and passed on to the secondary settling tank.

Sludge which has been collected from the settling tanks is concentrated and sent to an anaerobic digester. Anaerobic decomposition in the digester is slow but has the advantage that only a small percentage of the wastes are converted into new bacterial cells. Most of it is converted into methane gas and carbon dioxide. The remaining solids are well stabilized and can be dried and used for landfill or fertilizer. About 65–70% of sewage gas is methane (CH_4) which can be captured and used as a fuel for running blowers and boilers, pumping sewage, and generating electricity for the sewage plant. On a daily per capita basis, a secondary sewage plant produces about 1.0 cubic foot of sewage gas, with a heating valve of about 600 BTU (Metcalf and Eddy 1972). This amount of energy is about equal to the energy consumed by a 60 watt lightbulb burning for 3 hours.

There are other ways to treat sewage besides those mentioned here. Some communities use what are called stabilization or oxidation ponds, which are large ponds about 3 feet deep with a surface area of about 1

Secondary treatment plants may use trickling filters such as this one.

Activated-sludge process aeration tanks.

acre for every 1000 people to be served. Oxygen is supplied to the ponds either by mechanical aeration or through the photosynthetic activity of algae. The decomposition by bacteria is essentially the same as occurs in the activated-sludge process. Oxidation ponds may be used alone, in which case they may be as effective as secondary treatment, or as an addition to other waste treatment processes. At Santee, California, the effluent from a secondary treatment plant is kept in such a lagoon for 30 days. After chlorination, the lagoon water is allowed to trickle down through sandy soil into a lake which has such high quality water that it is used for swimming, boating, and fishing.

7.8 *Tertiary Treatment*

It has only been relatively recently that the processing of sewage beyond the secondary level has received much attention. The interest has been generated by the need to control eutrophication and the desire to process waste water to a level at which it can be reused.

Secondary treatment plants do a poor job of removing nutrients. Total nitrogen removal of from 25 to 55%, and total phosphorus removal of from 10 to 30% are typical (McCarty 1970). As was discussed in Chapter 6, nutrient stimulation can lead to excessive growths of algae and other aquatic plants. There are so many techniques available for increasing the nutrient removal efficiencies of waste water treatment plants (see, for example, Metcalf and Eddy 1972 Table 14.2), that it is convenient to classify them as being physical, chemical, or biological processes. Physical processes include filtration, distillation, and reverse osmosis; chemical processes include electrodialysis (Section 7.5), chemical precipitation, carbon adsorption, ammonia stripping, and ion exchange; biological processes include the harvesting of algae grown on the nutrients, bacterial assimilation, and bacterial nitrification and denitrification.

Obviously it would be impossible to discuss all of these techniques in one short section so we shall instead describe the techniques which are in use at one of the nation's most advanced wastewater treatment plants—the 7.5 mgd South Tahoe facility in California. Effluent from the activated-sludge secondary treatment portion of the plant is passed through the sequence of steps diagrammed in Figure 7.11.

The advanced treatment begins with the addition of lime which acts as a coagulant (Section 7.4). The flocculated, high-pH water flows to the chemical clarifier where suspended matter and most of the phosphates are settled out as a lime sludge. The effluent from the clarifier passes to the ammonia stripping tower for nitrogen removal. Ammonium ions in waste water exist in equilibrium with ammonia and hydrogen ions according to the following equation:

$$NH_4^+ \rightleftharpoons NH_3 + H^+ \tag{7-1}$$

At the high-pH levels created by the lime, the equilibrium is shifted far to the right so that the ammonia is virtually all present as a dissolved gas rather than as ammonium ion in solution. When the water is agitated in the presence of large amounts of air, the ammonia is liberated to the atmosphere. This happens in the 50 foot high, forced ventilation stripping tower, resulting in a removal of from 50 to 98% of the nitrogen.

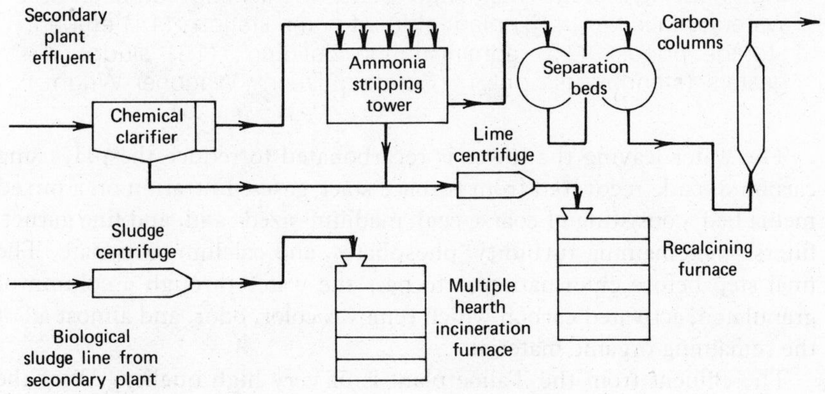

FIGURE 7.11 South Tahoe advanced treatment flow diagram.

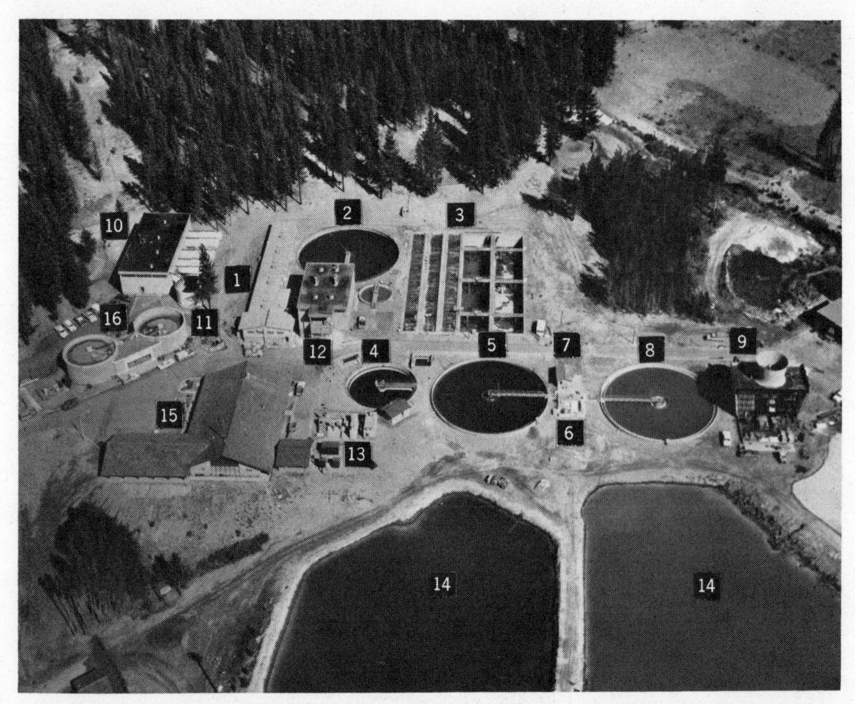

This wastewater purification system at South Lake Tahoe is one of the most advanced in the world: (1) headworks and primary sedimentation basin; (2) primary sedimentation basin; (3) activated sludge aeration tanks; (4) and (5) secondary clarifier; (6) chemical flocculation basin (start of advanced treatment process); (7) return sludge pump station; (8) chemical clarifier; (9) nitrogen removal tower; (10) separation beds and carbon columns; (11) backwash decant tank; (12) solids disposal building, housing sludge incinerator, lime recalcining furnace, dewatering facilities; (13) plant effluent pump station; (14) effluent storage ponds; (15) administration building; (16) sludge digesters (standby use only). (Courtesy Denny-Wagoner-Wright.)

The water leaving the tower is recarbonated to reduce the pH, using carbon dioxide recovered from furnace stack gases. Filtration on a mixed media bed, consisting of coarse coal, medium-sized sand, and fine garnet, filters out remaining turbidity, phosphates, and calcium carbonate. The final step before chlorination is to pass the water through a column of granulated, activated carbon which removes color, odor, and almost all of the remaining organic material.

The effluent from the Tahoe plant is of very high quality. All of the

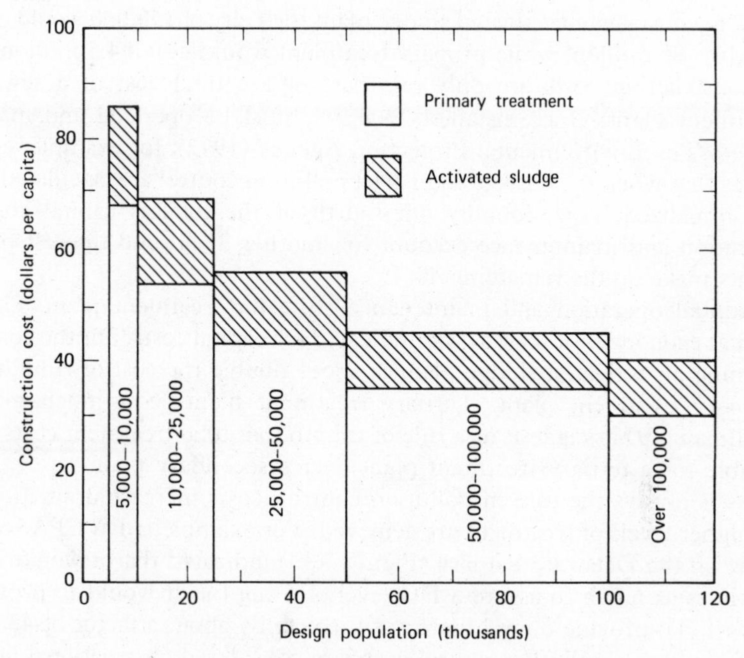

FIGURE 7.12 Construction costs per capita.

suspended solids, 99.8% of the BOD, 94% of the phosphorus, and from 50 to 98% of the nitrogen have been removed. The median coliform count is less than 2.2 per 100 ml (Culp and Moyer 1969). By law, all waste water must be removed from Tahoe basin, so the effluent is pumped over a ridge and released into a specially constructed reservoir at Indian Creek, 27 miles away.

7.9 *The Cost of Water Pollution Control*

It is important to at least briefly indicate the order-of-magnitude costs which are involved in water pollution control. Figure 7.12 indicates the economies of scale that can be achieved when large sewage treatment plants are built. For example, construction costs for an activated-sludge secondary treatment plant serving 5,000–10,000 people are about $86 per capita, while the same facilities designed for a population of over 100,000 would cost only $40 per capita. It is also indicated that the cost of constructing an activated-sludge plant is only about 30% more than the cost of simple primary treatment. For example, at the rates indicated in Figure

7.13, to construct an activated sludge plant for a city of 150,000 would cost roughly \$6 million, while primary treatment would cost \$4.5 million.

Construction costs are only one part of the total cost of a sewage treatment plant. Once installed, facilities must be operated and maintained. The Environmental Protection Agency (1972), for example, estimates that when the costs of industrial pollution control are calculated on an annualized basis, roughly one-fourth of the total is capital costs; operation and maintenance account for another 35 %; and interest payments make up the remaining 40 %.

Annual operation and maintenance costs for treatment plants show similar economies of scale as was indicated for capital costs, but the cost of running an activated-sludge plant is about double the cost of running a primary treatment plant. Tertiary treatment plants cost much more. Wollman (1971) suggests as a rule of thumb that total treatment costs are double for a tertiary treatment plant over a secondary plant.

As is always the case in pollution control, costs increase dramatically as higher levels of treatment are achieved. For example, a FWCPA study done on the Delaware Estuary (Figure 7.13) indicated that it would cost $1\frac{1}{2}$ times as much to assure a DO level of 5 mg/l as it would to provide 4 mg/l. To provide 6 mg/l increases the cost by about a factor of 4.

When water pollution control costs are considered on a national level,

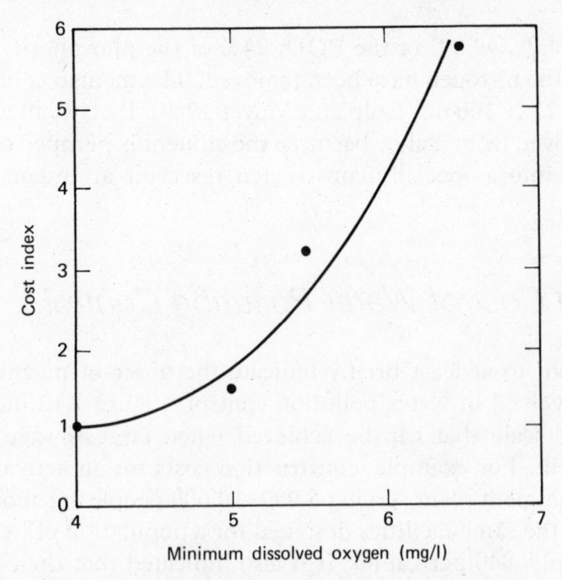

FIGURE 7.13 Cost index for various levels of dissolved oxygen control for the Delaware estuary. (EPA, 1972).

TABLE 7.2 The Costs of Water Pollution Control for the Years *1971–1980*, Billions of 1971 Dollars[a]

	Cumulative requirements, 1971–1980		
Sector	Capital investment	Operating costs	Cash flow
Public:			
Federal	1.2	2.7	3.9
State and local:			
Treatment systems	18.9	23.6	42.5
Combined sewers	(17–56)	n.a.	(17–56)
Private:			
Manufacturing	11.9	14.2	26.1
Utilities	4.5	4.2	8.7
Feed lots	1.9	1.8	3.7
Construction sediment	0.9	0.1	1.0
Vessels	0.9	0.5	1.4
Total	40.2	47.1	87.3

[a]Source: CEQ 1972.

the necessary expenditures to achieve a reasonable level of water quality can be staggering. Table 7.2 summarizes some estimates reported by the Council on Environmental Quality in 1972. The data indicate the expenditures which would be required in the 1970's to bring water quality up to standards which were already law in mid-1972. The capital investment in treatment plants and the interceptor sewers, pumping stations, and outfalls associated with the plants would cost \$18.9 billion. Operating the plants would require another \$23.6 billion, for a total of \$42.5 billion. Expenditures required by the private sector, mostly manufacturing, are another \$40.9 billion. Total costs for the 10 year period are estimated at \$87.3 billion—excluding an estimated \$17–56 billion more which could be necessary if combined sewer systems were to be updated.

Congressional awareness of the water pollution problem resulted in passage of the Federal Water Pollution Control Act Ammendments of 1972, which some consider to be the most comprehensive and expensive environmental legislation in the nation's history.* The bill contains authorizations which total almost \$24.7 billion for fiscal years 1972–1975,

*Congress cleared the bill on October 4, 1972, President Nixon vetoed it October 17, and Congress overrode the veto on October 18.

including $18 billion in contract authority for federal grants to the states for waste treatment facilities construction. It also requires municipal plants to provide secondary treatment by 1977.

The bill sets as a national *goal* the elimination of all pollutant discharges into U.S. waters by 1985, and as an interim goal, achievement of water quality safe for fish, shellfish, wildlife and recreation by 1983. By July 1, 1977, all industries are required to use the "best practicable" technology for treatment of any discharges, and by July 1, 1983, they are required to install the "best available technology economically achievable."

7.10 *Conclusions*

It is interesting to note the historical shift in emphasis that has occurred in the water quality control programs of the United States. Initially, attention was focused on the need to control the dreadful outbreaks of water-borne diseases such as typhoid and cholera. The relatively simple techniques of filtration and chlorination of drinking water can be credited with saving many thousands of lives. It would in fact be possible to eliminate the majority of human disease in the world through the single step of supplying proper sanitation to all people.

Water-borne diseases however, are no longer the principal water pollution concern in this country. It will, of course, always be vitally important to diligently monitor our sanitation systems for pathogenic organisms, but after relative control was achieved, attention did shift to the control of oxygen-consuming wastes. Sewage treatment plants were designed to reduce the BOD load placed on receiving waters, and when properly operated, secondary treatment plants are quite satisfactory in this regard. Unfortunately, too many municipalities are still using inadequate primary treatment facilities.

More recently, water pollution problems have become more complex. Now it is necessary in some cases to be able to control the release of nutrients to slow the eutrophication of our lakes. Industry must also learn to do a better job of controlling the release of its many obnoxious effluents—mercury, cadmium, acid mine wastes, pesticides, heat, etc. The cost of the necessary programs is high, but not at all unreasonable in view of the danger involved in continuing to treat these problems lightly.

Bibliography

American Chemical Society (1969). *Cleaning our environment, the chemical basis for action.* Washington, D.C.

American Water Works Association (1971). *Water quality and treatment, a handbook of public water supplies.* New York: McGraw-Hill.

Benarde, M. A. (1970). *Our precarious habitat.* New York: W. W. Norton.

Council on Environmental Quality (1972). *Environmental Quality.* 3rd Annual Report, Washington, D.C. Aug.

Culp, R. L., and Moyer, H. E. (1969). Wastewater reclamation and export at South Tahoe. *Civil Eng.–ASCE* 39 (6): 38–42.

Federal Water Pollution Control Administration (FWPCA) (1968). *Water Quality Criteria.* Washington, D.C. April 1.

Goldman, M. I. (1972). *Ecology and economics: Controlling pollution in the 70s.* Englewood Cliffs, N.J.: Prentice-Hall.

Harte, J., and Socolow, R. H. (1971). *Patient earth.* New York: Holt, Rinehart, and Winston.

Kittrell, F. W. (1969). *A practical guide to water quality studies of streams.* U.S. Department of the Interior. Federal Water Pollution Control Administration. Washington, D.C.

Kruse, C. W. (1969). Our nation's water: Its pollution control and management. In J. N. Pitts and R. L. Metcalf eds., *Advances in environmental sciences.* New York: Wiley-Interscience.

McCarty, P. L. (1970). Phosphorus and nitrogen removal by biological systems. U.C. Berkeley, 2nd Annual Sanitary Engineering Research Laboratory Workshop, Wastewater Reclamation and Reuse. Tahoe City, Cal. June 26.

Metcalf and Eddy, Inc. (1972). *Wastewater engineering.* New York: McGraw-Hill.

Salvato, J. A., Jr. (1972). *Environmental engineering and sanitation.* 2d ed. New York: Wiley-Interscience.

Spiegler, K. S. (1962). *Salt-water purification.* New York: Wiley.

U.S. Environmental Protection Agency (EPA) (1972). *The Economics of Clean Water, Summary.* Washington, D.C.

U.S. Department of Health, Education, and Welfare (HEW) (1962). *Public Health Service Drinking Water Standards.* PHS Pub. 956. Washington, D.C.

U.S. Department of the Interior (USDI) (1968a). Federal Water Pollution Control Administration. *The Cost of Clean Water, Vol. 1, Summary Report.* Washington, D.C. Jan. 10.

USDI (1968b). Federal Water Quality Administration. *Municipal Waste Facilities in the United States.* Washington, D.C.

USDI (1973). *1972–1973 Saline Water Conversion Summary Report*. Office of Saline Water, Washington, D.C.

U.S. Water Resources Council (1968). *The Nation's Water Resources*. Washington, D.C.

Wollman, N., and Bonem, G. (1971). *The outlook for water—quality, quantity, and national growth:* Resources for the Future. Baltimore: Johns Hopkins Press.

World Health Organization (WHO) (1972). *Health hazards of the human environment*. Geneva.

Questions

1. Draw three oxygen sag curves corresponding to sewage which is discharged into a stream:
   (a) raw
   (b) after primary treatment
   (c) after secondary treatment

2. Explain the rationale for using the coliform count as an indicator of biological purity.

3. Why is secondary sewage treatment often not satisfactory?

4. Using the information in this chapter, what would be the approximate construction cost of a primary treatment plant for a town of 40,000 people? For secondary treatment?
   ans. $1.7 million; $2.2 million

5. Explain the mechanism of transfer of typhoid fever from a carrier to a well person.

6. Identify the terms:
   (a) electrodialysis
   (b) flocculation
   (c) cross connections
   (d) ammonia stripping
   (e) pathogenic bacteria
   (f) combined sewer system
   (g) sludge
   (h) activated sludge

Part III

Air Pollution

Chapter 8

Introduction to Air Pollution

Man is, of course, totally dependent upon air for survival. While an average adult male daily consumes about 3 pounds of food and about 5 pounds of water, his requirement for air is much greater—about 32 pounds per day. Moreover, if cut off from all three, the lack of air would be the first factor to become critical.

What we call air is a mixture of gases consisting of roughly 78% nitrogen, 21% oxygen, and 1% argon. Table 8.1 indicates the relative concentrations of the various components which make up the normal, dry atmosphere at sea level. The units are parts per million (ppm) by volume.* Thus, for example, 315 ppm of carbon dioxide means that in 1 million cubic feet of air there will be about 315 cubic feet of CO_2.

Strictly speaking, air pollution may be caused not only by man's activities but also by various natural occurrences such as forest fires and volcanic eruptions. Such sources, however, are normally a minor part of the total air pollution problem and since they are beyond our control anyway, our attention will be directed only toward man-made pollution.

8.1 *General Considerations*

The rise of air pollution as a problem can be directly related to two principal factors: (a) man's exploitation of energy; and, (b) the increased concentration of his numbers in cities. The total quantity of pollution is principally determined by the amount of energy which is at our control. The first air pollution was probably caused by the burning of wood for heat and cooking, but by the fourteenth century the smoke and gases released from the burning of coal had become the major problem. The first attempts at control were already being

*Recall that "ppm" referred to a concentration by weight in the section on water pollution.

TABLE 8.1 Concentrations of Gases Comprising Normal Dry Air at Sea Level (Stern 1968)

Gas	Concentration, ppm
Nitrogen	780,900
Oxygen	209,400
Argon	9,300
Carbon dioxide	315
Neon	18
Helium	5.2
Methane	1.0–1.2
Krypton	1.0
Nitrous oxide	0.5
Hydrogen	0.5
Xenon	0.08
Nitrogen dioxide	0.02
Ozone	0.01–0.04

made as early as 1307 when a commission was appointed:

> To enquire of all who burnt sea coal in the City (of London) or parts adjoining and to punish them for the first offence with great fines and upon the second offence to demolish their furnaces (Holland 1972).

The burning of coal has continued ever since to be a principal source of air pollution, although in some areas where cleaner fuels are now used, the automobile has become the dominant contributor.

The deleterious effects of air pollution are caused not so much by the quantity of emissions as by their concentration in the air. The tendency for man to concentrate the sources of pollution into small geographical areas can burden the local air resources with more pollution than can be easily diluted and dispersed. In the United States, more than 50% of all the emissions are released over less than $1\frac{1}{2}$% of the nation's land area. As was indicated in Figure 2.9, this tendency toward urbanization is predicted to continue to grow. In 1950, the urban population of the developed regions of the world was about equal to the rural population; it is predicted to increase to about 80% of the total population by the year 2000. In the United States, more than one-fourth of the population lives in and around the ten biggest cities.

What happens to the thousands of tons of pollutants emitted per day over a typical large city? If there were no self-purification mechanisms, the air would quickly become lethal, but fortunately, there are a wide variety of removal processes. Particles can become incorporated into cloud or fog droplets, or can be washed out when it rains. Small particles can grow to larger size by coagulation and can then settle out by gravitation. Some pollutants are removed when they come into contact with buildings or plants. Others take part in chemical reactions and are transformed into new substances. Most important of all, in terms of reducing their damaging effects, is the dilution and the dispersion of the pollutants. For the air to remain clean over a city it is important that the pollutants be able to mix with plenty of air and then be blown away. Of course as the urban sprawl begins to join cities together, it becomes more and more difficult for there to be an "away."

The amount of dilution possible is dependent on vertical mixing ability, local geographical conditions, and wind velocities. Dilution is greatly diminished when the vertical movement of pollutants is restricted by an *inversion layer,* which is a layer of air in which the temperature of the atmosphere increases with altitude rather than decreasing as it normally does. Inversion layers will be discussed in the next chapter, but for now it is accurate to say that they act as a lid on the atmosphere, trapping pollutants below them.

When cities are located in valleys, the horizontal movement of pollutants can be restricted by hills or mountains. Should an inversion layer occur, pollutants can be trapped above the city and their concentration can increase to dangerous levels.

8.2 *Emission Sources*

Two of the most important questions regarding air pollution emissions are: (a) where is the pollution coming from; and (b) is any progress being made in our efforts to decrease those emissions. The answers to these questions will be in terms of quantities of specific pollutants. There are five major primary pollutants: hydrocarbons (HC), carbon monoxide (CO), nitrogen oxides (NO_x), sulfur oxides (SO_x), and particulates. Later in this chapter each of these pollutants will be examined in some detail but it is worthwhile to very briefly comment on each of them now.

Hydrocarbons are substances whose molecules contain only hydrogen and carbon atoms. They are emitted mainly as a result of the partial combustion of fossil fuels (complete combustion would yield simply carbon dioxide and water).

Carbon monoxide is a colorless, odorless, tasteless gas which can cause dizziness, unconsciousness, or even death by lessening the ability of blood to carry oxygen. It results from the incomplete combustion of hydrocarbons and its main source is the automobile.

Nitrogen oxides, mainly nitric oxide (NO) and nitrogen dioxide (NO_2), are formed when nitrogen and oxygen from the air are combined under high-temperature conditions. Thus they are characteristic of any high-temperature combustion process such as occurs in an automobile engine or a fossil-fueled electric power plant.

Sulfur oxides, mostly sulfur dioxide (SO_2) with some sulfur trioxide (SO_3), are emitted when fossil fuels containing sulfur impurities are burned. They are especially dangerous in combination with particulates.

Particulates is a loose category which includes a wide range of solid or liquid particles which are typically emitted during combustion or from the grinding of materials. Some of the deleterious properties of particulates are caused by their chemical composition, while others are merely a result of their size.

Table 8.2 shows estimates of the emissions of these pollutants, by weight, for 1970. Transportation, which is almost totally motor vehicles, is seen to be the largest source of CO, hydrocarbons, and NO_x. Stationary fuel combustion emissions come from power plants, industry, commercial, and residential sources, with power plants being the single largest contributor. This category accounts for most of the SO_x and a sizeable portion of the particulates. The industrial processes category includes all noncombustion operations such as occur in the manufacture of such diverse products as petroleum products, steel, plastics, and cement. Solid waste disposal sources include the emissions which result from the common practices of burning municipal wastes in incinerators or in the open at the dump itself, as well as such operations as the burning of sawdust and bark in large "wigwam" burners at lumber mills.

TABLE 8.2 Estimated Emissions of Air Pollutants by Weight, U.S., 1970 (CEQ 1972) in Millions of Tons per Year

Source	CO	Particulates	SO_x	HC	NO_x
Transportation	111.0	0.7	1.0	19.5	11.7
Fuel combustion in stationary sources	0.8	6.8	26.5	0.6	10.0
Industrial processes	11.4	13.1	6.0	5.5	0.2
Solid waste disposal	7.2	1.4	0.1	2.0	0.4
Miscellaneous	16.8	3.4	0.3	7.1	0.4
Total	147.2	25.4	33.9	34.7	22.7
Percent change 1969–1970	−4.5	−7.4	0	0	+4.5

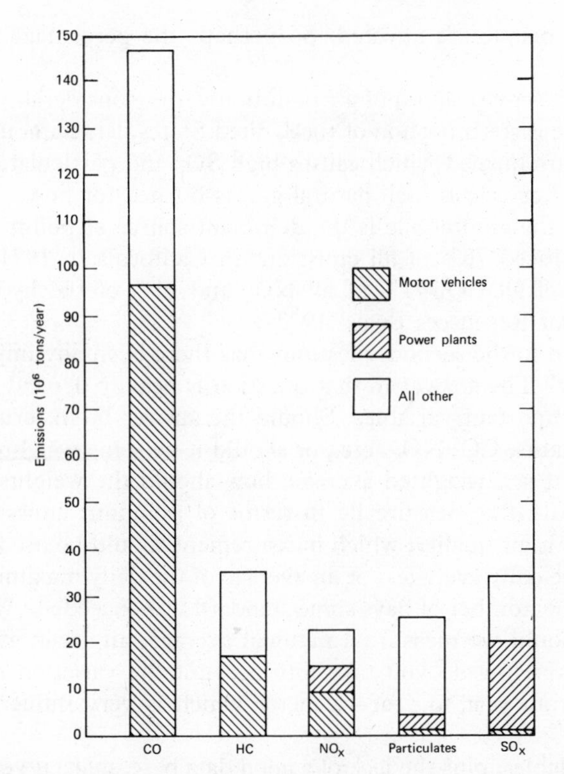

FIGURE 8.1 The motor vehicle and power plant contribution to total emissions in the U.S., 1970 (EPA 1973).

Figure 8.1 shows the contribution to air pollution made by the two single largest sources: motor vehicles and electricity-generating power plants. To try to estimate the total contribution to air pollution made by motor vehicles is an ambiguous exercise. If we simply compare the total tons of pollutants emitted by cars in 1970 to the total emissions from all sources, we arrive at a figure of 47 %. This is probably not a fair comparison because it does not take into account the differing severity of health effects associated with the individual pollutants. Carbon monoxide is, for example, much less dangerous, pound for pound, than SO_x, but it is the largest contributor and hence tends to dominate this crude calculation. On a health basis, then , it could be argued that cars are less than 47 % of the problem (see, for example, Caretto and Sawyer 1970). On the other hand, motor vehicle emissions are at ground level where they can do the most damage, while many other sources emit from tall smokestacks, located away from population centers, which allows some dilution to take place before people are affected. So perhaps the percentage should be increased.

Further, the composition of air pollution varies considerably from city to city. In the eastern portion of the United States, large amounts of coal and fuel oil are burned which causes high SO_x and particulate levels. In California, a very clean fuel, natural gas, is burned for power and space heating, and the automobile is the dominant source of pollution. Motor vehicles produced 76% of all emissions in California in 1971. This included 92% of all CO, 71% of all NO_x and 61% of the hydrocarbons (California Air Resources Board 1972).

Let us turn to the second question—has there been any improvement in air quality? The answer to that question is heavily dependent upon a number of important variables. Should the answer be in terms of individual pollutants, CO, NO_x, etc., or should it be some weighted average of them? If it is a weighted average, how should the weights be determined? Should the measure be in terms of pollutant emissions or air quality? If it is air quality, which measurement should be used—a yearly average of the daily averages; or an average of the daily maximum values, or perhaps the number of days some standard was exceeded? What about location? Should we measure a national average, air basin averages, or city averages? How do you take into account the variation in weather conditions from year to year—a factor which is very influential in air quality?

These variables, plus the lack of a good data base, make it very difficult to make any general statements yet about overall trends. Figure 8.2 shows the emissions of the five primary pollutants over the past 30 years, based on some rather sketchy data. As might be expected emissions have increased along with population and industrialization. Serious control efforts have only recently been undertaken so there is little, if any, improvement to be noted in the graphs. However, there is considerable reason for optimism as the much tougher emission regulations which are already law, begin to take effect (see Chapter 10).

8.3 *Photochemical Smog*

The nature of air pollution is dependent upon the types of pollutants emitted in a given region as well as upon the geographical and meteorological conditions. It is convenient to make a distinction between a type of pollution which consists mostly of a combination of sulfur oxides and particulates—sometimes called *London* smog—and *photochemical* smog

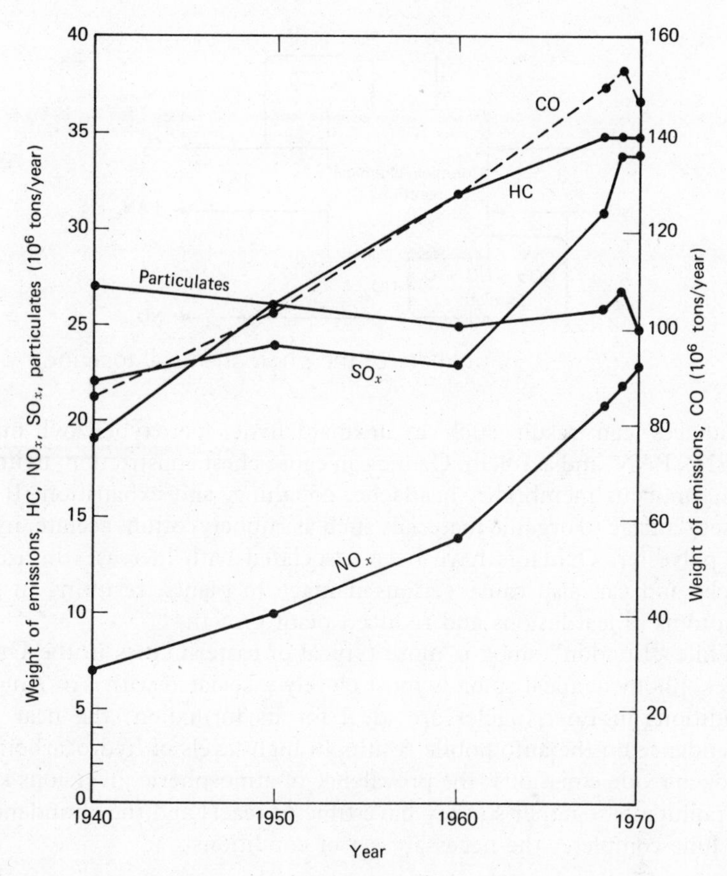

FIGURE 8.2 Emissions of the five major pollutants (CEQ 1972).

which is the result of a chemical reaction between hydrocarbons, oxides of nitrogen, and sunlight.

The primary pollutants involved in photochemical smog are nitric oxide (NO) and hydrocarbons. When these primary pollutants are together in the presence of sunlight, a partially understood complex series of reactions takes place which results in various harmful *secondary* pollutants, including nitrogen dioxide (NO_2), ozone (O_3), and peroxyacetyl nitrate ("PAN," $CH_3CO_3NO_2$). Ozone and PAN are usually referred to as *photochemical oxidants.* The reactions which take place during the formation of photochemical smog are summarized in Figure 8.3.

One of the effects of photochemical air pollution is eye irritation. Ozone itself is not an eye irritant, but when it reacts with hydrocarbons, irritating

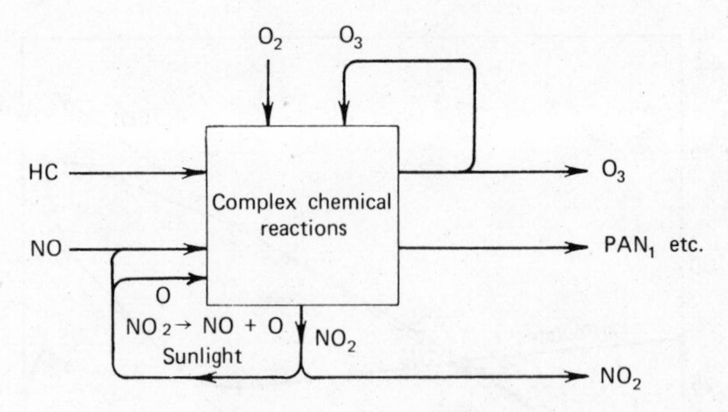

FIGURE 8.3 Summary of the photochemical reaction.

substances can result such as formaldehyde, peroxybenzoyl nitrate (PBzN), PAN, and acrolein. Ozone can cause chest constriction, irritation of the mucous membrane, headache, coughing, and exhaustion. It also causes damage to organic materials such as rubber, cotton, acetate, nylon, and polyester. Oxidants have been associated with increases in asthma attacks and can also cause serious damage to plants, resulting in such symptoms as leaf lesions and reduced plant growth.

While "London" smog is more typical of eastern cities in the United States, photochemical smog is most closely associated with Los Angeles. Conditions in Los Angeles are ideal for its formation: the near total dependence on the automobile results in high levels of hydrocarbon and nitrogen oxide emissions; the prevalence of atmospheric inversions keeps the pollutants together so they have time to react; and the abundance of sunshine completes the necessary set of conditions.

8.4 *Respiratory Diseases*

Air pollution can affect the human body through contact with the skin and eyes, or it can be brought into the body by way of the respiratory system. It is this latter avenue of attack that is the most damaging to health. The respiratory diseases of most importance in the study of the effects of air pollution are bronchitis, emphysema, asthma, and lung cancer.

First consider Figure 8.4(*a*), which shows the major anatomical features of the human respiratory system. It is convenient to divide the system into three parts; the nasopharyngeal structure, the tracheobronchial system, and the pulmonary structure where oxygen and carbon dioxide are exchanged. Large particles which enter the respiratory system can be trapped by hairs and the lining in the nose, to be driven out by a cough or

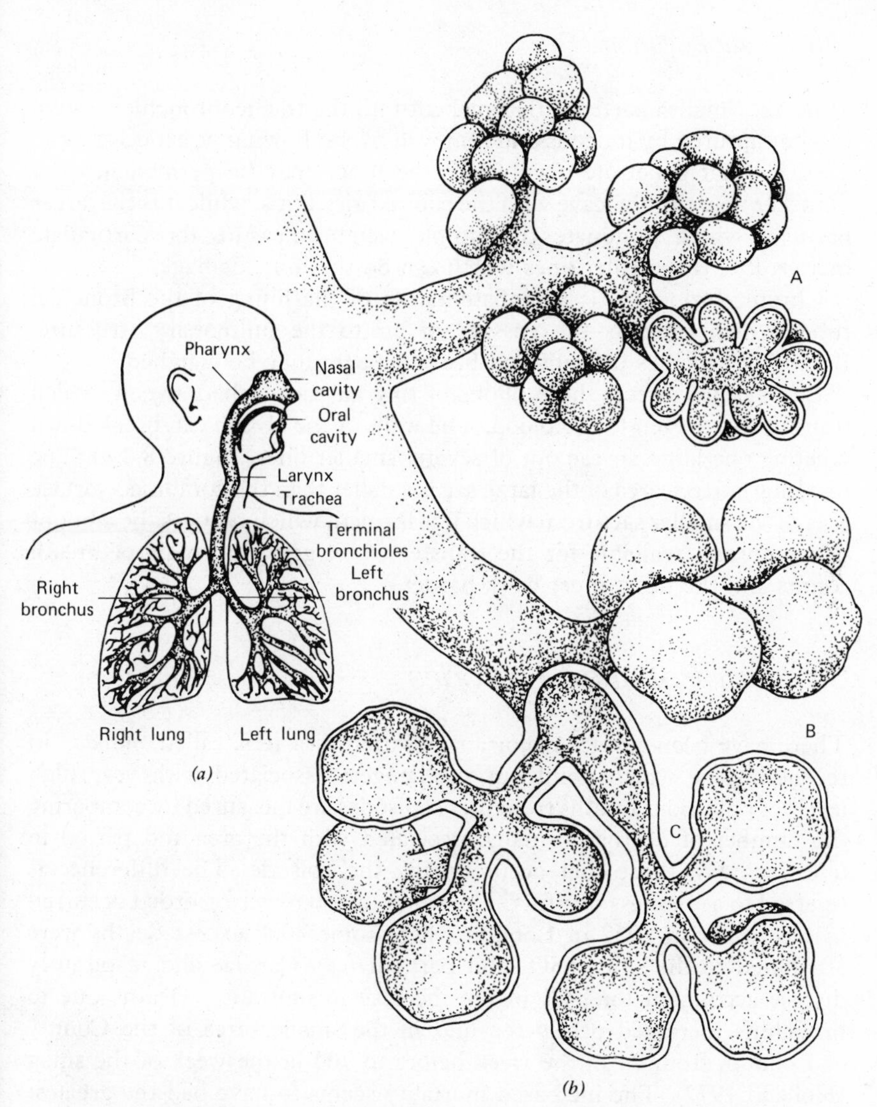

FIGURE 8.4 (*a*) Showing the major anatomical features of the respiratory system (HEW 1969a). (*b*) Bronchitis-Emphysema is a chronic lung disease that is apparently aggravated by air pollution. In the normal lung the air passes through the bronchial tubes to enter millions of alveoli (A), tiny cells in which the oxygen is transferred to the blood. In a diseased lung the walls of many of the alveoli break down (B), causing a reduction in the amount of membrane available to carry out the oxygen transfer. At the same time there is a narrowing of the smallest branches of the bronchial tree (C), further restricting air exchange. From McDermott, W., "Air Pollution and Public Health." Copyright © 1961 by Scientific American, Inc. All rights reserved.

a sneeze. Smaller particles that make it into the tracheobronchial system can be captured by mucous and removed by swallowing or expectorating. Particles that penetrate deeply into the lungs may be permanently retained or may simply leave with the expired air. Thus, while it is the larger particles which dominate the simple weight measure for particulate matter, it is the smaller ones which can do the most damage.

Chronic bronchitis is an inflammation of the lining of the bronchial tubes which restricts the passage of air to the pulmonary structure. Breathing becomes difficult and heavy phlegm may be coughed.

Emphysema affects the millions of tiny air sacs, called alveoli, which transfer the oxygen to the blood. The walls of the alveoli can break down creating one large air sac out of several smaller ones, Figure 8.4(*b*). The resulting surface area of the large sac is smaller than the combined surface area of the smaller sacs from which it is formed, which reduces the amount of membrane available for the transfer of oxygen. Shortness of breath results and the heart must work harder.

8.5 *Air Pollution Episodes*

There have been several major air pollution disasters, called *episodes*, in recent history which point out the dangers associated with very high levels of pollution. Deaths from these episodes are measured by comparing the number of deaths normally associated with the area and period in question with those that occur during the episode. The difference is referred to as "excess deaths." The worst disaster ever recorded occurred in December of 1952 in London where some 4000 excess deaths were attributed to the smog (SO_x and particulates). Cardiac and respiratory disease accounted for 84% of the increase in mortality. Deaths due to bronchitis increased nearly ten-fold in the smaller area of the County of London, from 74 in the week before to 704 in the week of the smog (Holland 1972). The increased mortality seems to have had the greatest affect on people who were already suffering from chronic respiratory or cardiac diseases. Figure 8.5 shows the rise in SO_2 and smoke levels and the corresponding elevated death rate.

The first recorded air pollution episode in the United States occurred in Donora, Pennsylvania, during October, 1948. Donora was then a town of 14,000 having among its industries a large steel mill, a sulfuric acid plant, and a large zinc production plant. Within a 4 day period, 20 deaths and more than 5900 illnesses were attributed to the smog. This is the highest per capita death rate ever recorded for an air pollution episode.

Table 8.3 summarizes some of the history of air pollution episodes from all over the world. Larsen (1970) has tried to relate elevated death

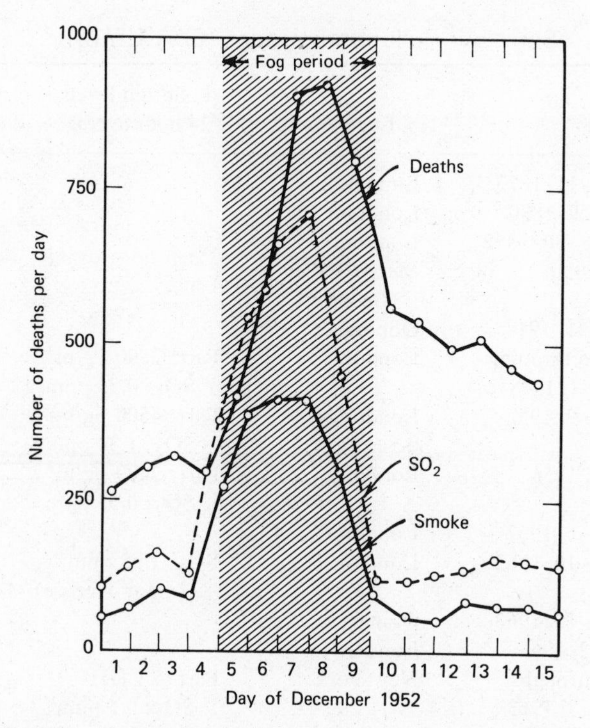

FIGURE 8.5 Deaths and air pollution in London County during December 1952 (Frenkiel 1972).

rates to atmospheric concentrations of SO_2 and particulates. These two pollutants usually occur together and seem to act synergistically—that is, their effect together is greater than the sum of their effects taken separately. Figure 8.6 suggests that excess deaths may be related to the product of their concentrations.

The following list summarizes some of the major characteristics of air pollution episodes (EPA 1971a).

1. Stagnant air produced by low wind speed and temperature inversion.
2. As concentrations of smoke, SO_2, particulates, and other pollutants increase—coughing, eye irritation, and sickness increase.
3. Deaths increase as pollutant levels reach peaks.
4. Excess deaths increase with increasing age.
5. Death and illness occur in all age groups.
6. Deaths are generally caused by respiratory or heart problems.
7. The various impacts on health are rapid and are due to a combination of several pollutants.
8. The episode lasts 2–7 days.

TABLE 8.3 Some Air Pollution Episodes (EPA 1971a)

Date	Location	Pollution levels, 24 hour average	Excess deaths
December 9–11, 1873	London		650
January 26–29, 1880	London		1,176
December 28–30, 1892	London		779
December, 1930	Meuse Valley (Belgium)		60–80
October 27–31, 1948	Donora		20
November 26 through December 1, 1948	London	Part: 2800 $\mu g/m^3$ SO_2: 0.75 ppm	700–800
December 5–9, 1952	London	Part: 4500 $\mu g/m^3$ SO_2: 1.34 ppm	4,000
January 3–6, 1956	London	Part: 2400 $\mu g/m^3$ SO_2: 0.55 ppm	1,000
December 2–5, 1957	London		200–250
December 5–10, 1962	London	SO_2: 1.98 ppm (1 hour average)	700
December 7–10, 1962	Osaka		60
January 7–22, 1963	London		700
January 29 through February 12, 1963	New York	Part: 7 COH[a] SO_2: 0.5 ppm	200–400
February 27 through March 10, 1964	New York		168

[a]COH is Coefficient of Haze—a measurement of the quantity of dust and smoke in a theoretical 1000 linear feet of air; often used for particulates instead of micrograms per cubic meter because the measurement is quicker and easier. There is no simple correlation between the two kinds of measurements.

8.6 *Chronic Health Effects*

The evidence relating high levels of air pollution to excess mortalities is relatively easy to obtain and is clearly of great importance. It is on the other hand quite difficult to obtain conclusive data concering the effects of continued exposure to low levels of air pollution. There are two general approaches to such studies: epidemiological and toxicological. Epidemiological studies are statistical analyses of the effects of air pollution on human populations under natural conditions. Such studies are extremely important, but due to the multiplicity of unknown factors it is not possible to prove cause-and-effect relations. For example, in Los Angeles, epi-

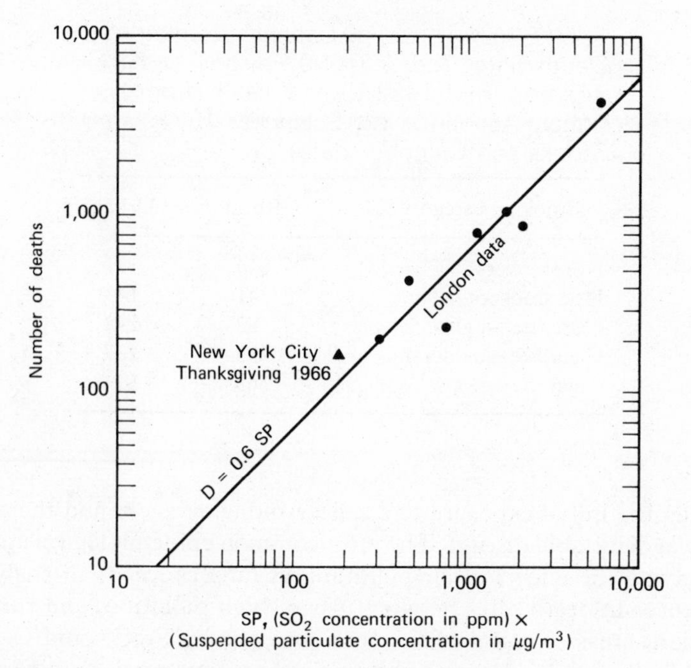

FIGURE 8.6 Number of deaths in London or New York air pollution episodes as a function of the product of sulfur dioxide and suspended particulate concentrations (Larsen 1970).

sodes of high air pollution tend to occur during periods of high temperature. Increased mortality could be due to either factor alone, or to both acting together.

Toxicological studies are performed in the laboratory under controlled conditions. Such variables as individual pollutant concentrations, exposure duration, temperature, etc., can all be manipulated by the experimenter. These experiments permit conclusions as to causation but their relevance to the natural setting is sometimes questionable. The two types of studies are complementary. We shall briefly consider epidemiological evidence here and in the later sections on individual pollutants some toxicological data will be mentioned.

Let us first consider the link between lung cancer and air pollution. Lung cancer deaths have increased enormously in the United States in recent years, so that by 1969 it was the greatest single cause of cancer deaths in men, killing about 50,000 men and 10,000 women. Cause-and-effect studies are complicated by the long interval—perhaps 30 years—

TABLE 8.4 Lung Cancer Standardized Death Rates (per 100,000 Persons, age 14–74) from mid-1952 to mid-1954 (from National Academy of Sciences 1972 after Stocks and Campbell data)

Smoking category	Rural	Urban
Nonsmokers	14	131
Pipe smokers	41	143
Cigarette—light	87	297
Cigarette—moderate	183	287
Cigarette—heavy	363	394

between the initial exposure to a cancer-inducing agent and the appearance of a detectable disease. The most common epidemiological approach to studying the effects of air pollution on lung cancer is to analyze the differences in death rates between urban (high pollution) and rural (low pollution) areas. Typical of the studies is one by Stocks and Campbell comparing rural residents to those in urban Liverpool, summarized in Table 8.4 [National Academy of Sciences (NAS) 1972]. These data indicate a strong correlation between smoking and lung cancer, but they also indicate that there is definitely an urban factor involved. In all smoking categories, but most especially for those who smoke very little if at all, the increase in lung cancer deaths associated with urban living is marked. Urban dwellers as a whole seem to have approximately twice as high an incidence of lung cancer as those living in rural areas. Further the rate of incidence seems to increase with duration of residence in the urban environment (NAS 1972, Haenszel data).

While it is definitely possible to state that the urban environment contributes to lung cancer, it is not proven that the difference is caused by air pollution. However, certain substances commonly found in polluted air, namely polycyclic organic matter, are known to cause cancer of the lung and other organs in experimental animals.

Epidemiological data for chronic bronchitis and emphysema are quite similar to those reported for lung cancer. Bronchitis-emphysema rates are higher for smokers, urban dwellers, and men than for nonsmokers, rural dwellers, and women. There is considerable evidence to indicate that air pollution definitely aggravates bronchitis-emphysema and some studies indicate that it may also be a causative agent.

8.7 *Carbon Monoxide*

Carbon monoxide constitutes the single largest pollutant in the urban atmosphere. Natural sources create a background concentration of about 1 ppm but this is insignificant compared to normal urban levels. The principal cause of CO emissions is the incomplete combustion of hydrocarbons which may result when any of the following four variables are not kept sufficiently high: (a) oxygen supply, (b) flame temperature; (c) gas residence time at high temperature, and (d) combustion chamber turbulence.

Since CO levels do not continuously increase, there must be some natural processes which account for its removal from the atmosphere. Not enough is known to enable a complete description of the removal process, but it apparently involves the conversion of CO to CO_2.

As was demonstrated in Figure 8.1 the principal source of CO emissions in the United States is motor vehicles (66 %). Hourly atmospheric concentrations of CO often reflect city driving patterns. Peaks occur on weekdays during the morning and late afternoon rush hours, while on weekends the peaks are absent. Emissions from motor vehicles are highly dependent on average driving speed—the higher the speed the lower the emissions—so that good urban traffic-flow planning can be a significant help in reducing emissions.

At levels of CO which are liable to occur in urban air, there seems to be no effect on vegetation and associated microorganisms, but there are effects on humans. The effects on humans result from the fact that CO readily reacts with hemoglobin in the blood to form carboxyhemoglobin (COHb). Some of the hemoglobin, instead of picking up oxygen in the lungs, picks up CO so that the amount of oxygen carried by the blood is reduced. The reduction in oxygen can cause headache, dizziness, or even death. To maintain a reasonable level of oxygen requires that the heart work harder and so cardiovascular difficulties are typical as CO concentrations increase.

The amount of COHb formed is dependent upon the CO concentration, the length of time exposed, and the rate of breathing. Also the distinction should be made between smokers and nonsmokers since smokers have a rather high background level of COHb (about 5%) compared to nonsmokers (about 0.5%).

As Figure 8.7 indicates, the level of COHb rises with exposure time to a steady-state value which is dependent on the concentration of CO.*

*See Question 2 at the end of the chapter.

Studies have indicated that adverse health effects can be noted at COHb levels as low as 2.5% (students were less able to distinguish the duration of a 1000 hertz tone signal) (HEW 1970a). A 2.5% COHb level can be reached during an 8 hour exposure to a 15 ppm concentration of CO; 30% of the days in Chicago and 10% of the days in Los Angeles have 8 hour exposures exceeding 15 ppm CO (HEW 1970a, 1962–1967 data). The impaired time interval discrimination, and changes in visual acuity and other psychomotor responses mentioned in Figure 8.7 are thought to be factors which could increase the auto accident rate under conditions of high CO concentration, but more research needs to be done to determine if this is true.

8.8 *Oxides of Nitrogen*

The two oxides of nitrogen that are important in the study of air pollution are nitric oxide (NO) and nitrogen dioxide (NO_2). Under high-temperature conditions (above about 2000° F), such as occur during combustion,

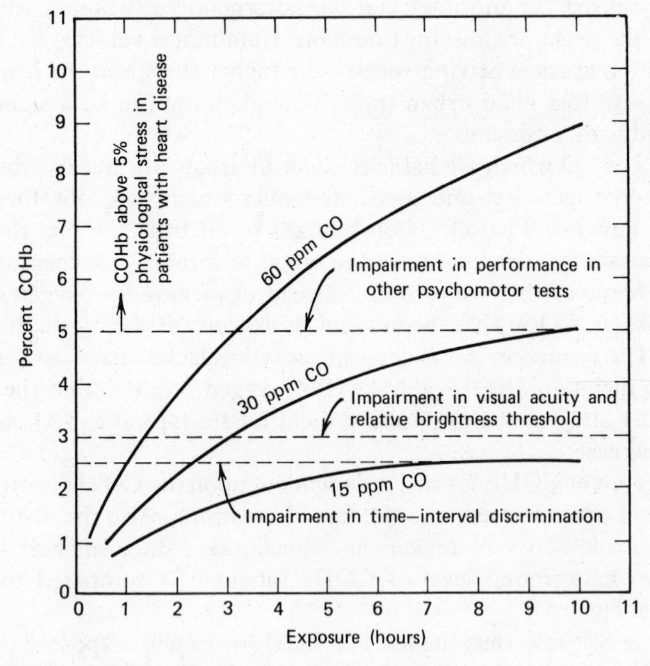

FIGURE 8.7 COHb increases with concentration of CO and exposure time. Adverse health effects are noted above 2.5% COHb. Curves are approximate for male nonsmokers engaged in sedentary activity.

atmospheric nitrogen and oxygen combine to form mostly NO with a small amount of NO_2. Nitric oxide has no known adverse health effects at concentrations found in the atmosphere but it does convert to NO_2, and NO_2 is corrosive and may be physiologically irritating and toxic. Nitrogen dioxide reduces the brightness and contrast of distant objects and is responsible for the yellow to reddish-brown color of smog. The creation of NO_2 in the presence of light and hydrocarbons is a part of the complex set of reactions that cause photochemical smog. The NO that is emitted converts to NO_2 and the NO_2, in turn, is influential in the formation of O_3.

Motor vehicles contributed 9.1 million tons of NO_x in 1970, which is 40 % of the total emissions (EPA 1973). That so much comes from motor vehicles is especially unfortunate since that means a large dose of NO enters the atmosphere in the early morning (along with hydrocarbons) allowing all day for the sunlight to convert it to photochemical smog. Power plants (4.7 million tons) and industrial combustion of fuels (4.5 million tons) are the other major source categories.

Nitrogen oxides have been determined to cause various textile dyes to fade and can cause cotton and nylon fibers to deteriorate (runs in nylons). Corrosion of nickel-brass wire springs in telephone company equipment and other types of corrosion have been associated with typical atmospheric NO_x concentrations.

Nitrogen dioxide can cause damage to plants. For example, 0.5 ppm NO_2 applied for 35 days results in leaf drop and chlorosis in citrus and 0.25 ppm or less for 8 months causes reduced yield in navel oranges (EPA 1971b).

In an important epidemiological experiment in Chattanooga, NO_2 was associated with an increase in acute respiratory disease at commonly occurring urban concentrations. The frequency of acute bronchitis increased among infants and school children when the range of 24 hour NO_2 concentrations, measured over a 6 month period, was between 0.063 and 0.083 ppm. Yearly average NO_2 concentrations exceed the Chattanooga health effect related value of 0.06 ppm in 54% of American cities with populations between 50,000 and 500,000, and 85% of cities with populations over 500,000 (EPA 1971b).

8.9 *Hydrocarbons*

Hydrocarbons are compounds made up of just the two atoms carbon and hydrogen, such as methane (CH_4), ethane (C_2H_6), and ethylene (C_2H_4). Natural sources of hydrocarbons are mostly biological in nature accounting for some 3×10^8 tons of methane and 4.4×10^8 tons of volatile terpenes and isoprenes per year in the world. Nonurban air naturally contains from about 1.0 to 1.5 ppm methane and less than 0.1 ppm each of other hydrocarbons. The methane is virtually inert and so it is customary to subtract it from the total hydrocarbon measurement. These natural sources can cause a natural smog or haze such as occurs in the Great Smoky Mountains.

Hydrocarbon emissions originate principally from two types of processes—inefficient combustion and evaporation. Of the 35 million tons of emissions in 1970, 16.7 million tons were emitted by motor vehicles (48 %), 5.5 million tons resulted from industrial processes (15.7 %), and 3.1 million tons came from organic solvent evaporation (8.9 %) (EPA 1973).

Although urban concentrations of hydrocarbons have not been shown to cause any adverse health effects directly, they are essential in the formation of photochemical smog which can be deleterious to health. It has been determined that 0.3 ppm of nonmethane hydrocarbons during the 3 hour period from 6:00 to 9:00 A.M. can be expected to cause an average 1 hour photochemical oxidant concentration of 0.1 ppm about 2–4 hours later—a level which has been determined to have adverse health effects. Federal air quality standards are therefore based not on the effects of hydrocarbons but on the level of oxidant which can be expected to result.

The only hydrocarbon produced from combustion sources which is known to have a direct effect on plants at atmospheric concentrations is ethylene. Ethylene is also important because it causes the formation of the eye irritant formaldehyde in the photochemical reaction.

8.10 *Sulfur Oxides*

Fossil fuels generally contain appreciable quantities of sulfur (roughly 0.5–6.0 % in coal), either in the form of inorganic sulfides or as organic sulfur. When the fuels are burned, the sulfur is released, mostly as sulfur dioxide (SO_2) with a much smaller percentage of sulfur trioxide (SO_3).

Sulfur dioxide can be detected in air by taste at concentrations above 0.3 ppm and in concentrations above 3 ppm it has a pungent, irritating odor. It reacts with water to form sulfurous acid H_2SO_3, while SO_3 reacts immediately with water to form droplets of sulfuric acid (H_2SO_4). Thus SO_3 is normally not found in the atmosphere, being already converted to sulfuric acid. About 5–20% of the particulate matter in urban air consists of sulfuric acid and other sulfates (HEW 1969b).

As Table 8.2 indicates, over three-fourths of the SO_x emissions are the result of fuel combustion in stationary sources, with over 80% of that coming from the combustion of coal. The SO_x problem mainly affects the industrial northeastern states, although, due to rapid depletion of cleaner fuels such as natural gas, it is likely to spread across the country soon. Even with the heavy emphasis on nuclear power plants in the future, the use of coal-burning power plants is going to continue to grow.

Sulfur dioxide is responsible for the rapid corrosion of materials. Mild steel panels exposed to 0.12 ppm SO_2 for one year have experienced weight losses, due to corrosion, of about 16% (see Figure 8.8). The average annual SO_2 concentration in Chicago is about this value, 0.12 ppm, although it is much less in most other cities. Other examples of effects on materials include a one-third reduction in the life of overhead power line hardware and guy wires and the necessity to use more expensive, less corrodible metals, such as gold in electrical contacts. Leather loses its strength, building materials are discolored and deteriorate, statuary and other works of art are damaged. Sulfur dioxide may also cause leaf injury to plants and can cause supression of growth and yield.

With regard to health effects of sulfur oxides, for the most part they are related to irritation of the respiratory system. Those people most affected are individuals with chronic pulmonary disease or cardiac disorders as well as very young or old individuals. These effects are enhanced by a factor of 3 or 4 when particulate matter is present along with SO_2. Thus

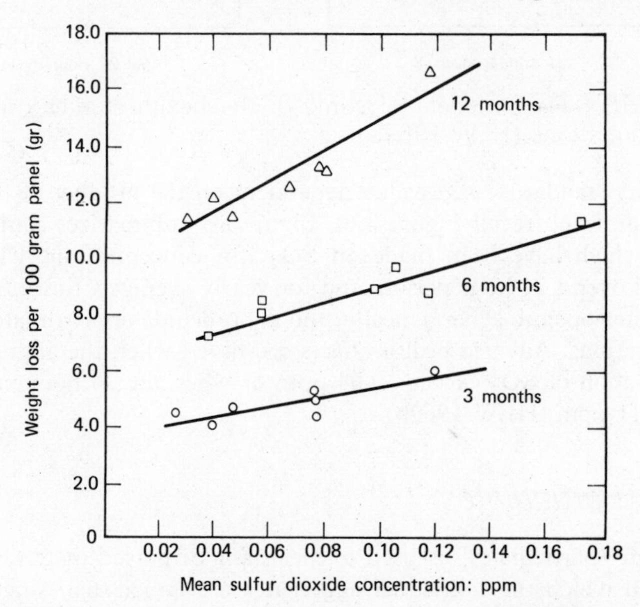

FIGURE 8.8 Corrosion of mild steel due to sulfur dioxide exposure for 3, 6, and 12 months (HEW 1969b).

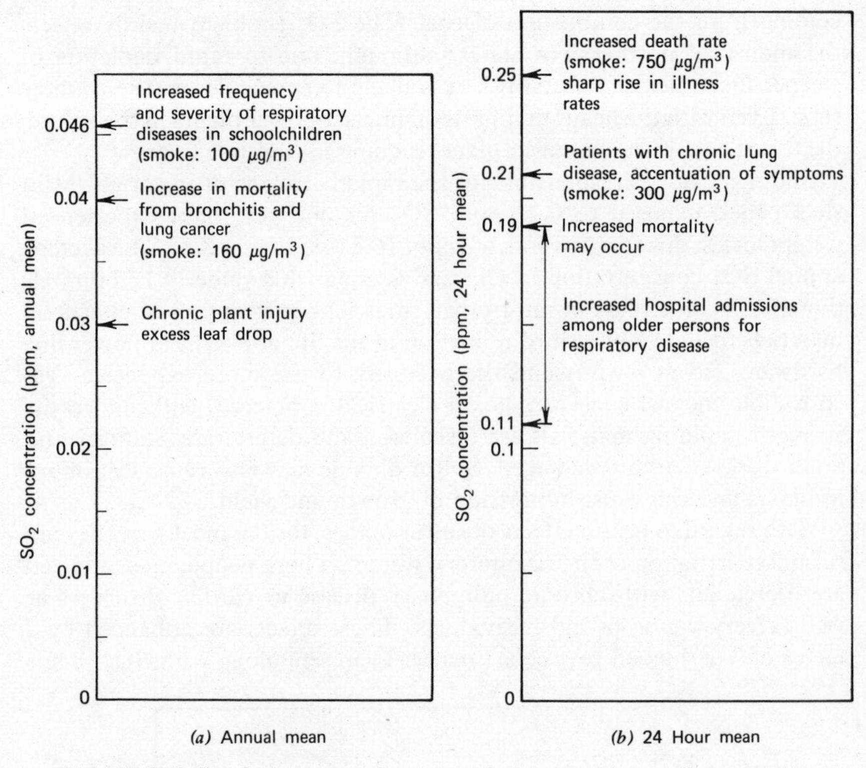

FIGURE 8.9 Summary of some of the health studies on SO_2 concentrations (HEW 1969b).

air quality standards use a value dependent on the product of SO_2 and particulate level (recall Figure 8.6). Figure 8.9 summarizes some of the studies which have been made on SO_2, for concentrations which are averaged over a 24 hour period, and for yearly averages. Since SO_2 and particulates operate synergistically, the data include approximate smoke concentrations. Adverse health effects are noted when the annual mean concentration of SO_2 exceeds 0.04 ppm or when the 24 hour mean exceeds 0.11 ppm (HEW 1969b).

8.11 *Particulates*

The term "particulates" is used to mean any dispersed matter, solid or liquid, in which the individual aggregates are larger than single small molecules (about 0.0002 μ in diameter), but smaller than about 500 μ

(where 1μ is 1 micron, which is equal to 10^{-6} meter). Particulates include aerosols, fumes, dust, mist, and soot.* The particles of most interest lie between 0.1 μ and 10 μ in diameter, which is roughly the size of bacteria (the unaided human eye has a resolving power of about 100 μ). Particles smaller than 0.1 μ undergo random (Brownian) motion and through coagulation generally grow to sizes larger than 0.1 μ, and particles larger than 10 μ settle quickly as dust. A 10 μ particle has a settling velocity of about 7 inches per minute. Particles smaller than about 1 μ result largely from the condensation of vaporized materials after combustion, while particles larger than about 10 μ result largely from mechanical processes such as grinding and erosion.

It should be noted that there are many difficulties associated with generalizing the properties of particles. Some properties are functions of size and some are dependent on the nature of the particles themselves (e.g. their toxicity) but most measurements take neither into account and merely report the overall weight per unit volume (micrograms per cubic meter). This measure is crude and tends to overemphasize the importance of large particles. As was pointed out in section 8.4, it is the smaller particles which pass deeply into the lungs and therefore do the most damage to health.

We shall see in the next chapter some of the effects particulates may be having on global and local weather conditions. By blocking the sun they may be contributing to a drop in the earth's temperature and by providing condensation nuclei they contribute to increased fog and rain in cities. Notice it is possible for the following positive feedback mechanism to aggrevate the visibility problem: As visibility drops, more lights are turned on, which causes more power plant pollution, thus reducing visibility further.

Particulate concentrations correspond quite well with visibility reductions, as expressed by the following approximate relationship:

$$r \cong \frac{750}{c} \quad \text{where} \quad \begin{aligned} r &= \text{visible range (miles)} \\ c &= \text{particulate concentration} \\ &\quad \text{(microgram per cubic meter)} \end{aligned}$$

*Some definitions: *Aerosol*, particle of solid or liquid matter that can remain suspended in the air because of its small size (generally under 1 μ); *Dust*, solid particulate matter; *Fume*, solid particles under 1 μ in diameter, formed as vapors condense or as chemical reactions take place; *Mist*, liquid particles up to 100 μ in diameter; *Soot*, very finely divided carbon particles clustered together in long chains.

Thus for example, the California air quality standard of 100 $\mu g/m^3$ 24 hour mean would, by this relation, correspond to a visibility of about 7.5 miles. Visibilities much below this tend to slow operations at airports and make conditions hazardous for flying.

High concentrations of particulate matter and sulfur dioxide have been implicated in every severe air pollution episode. Epidemiological studies have shown a good correlation between death rates for the respiratory diseases asthma, bronchitis, and emphysema, and average particulate levels in the area of residence. Adverse health effects have been noted when the annual geometric mean level of particulate matter exceeds 80 $\mu g/m^3$.

As has been mentioned, some particulates are especially dangerous because of their toxic nature. Such a pollutant is *lead*, which has a particularly interesting history. The wealthy class of Rome, in the second century B.C., was particularly afflicted with sterility, child mortality, and permanent mental damage. It has been suggested that these symptoms may have been caused by lead leached from wine and food utensils, and thus the fall of Rome may have, in some part, been due to lead poisoning (Gilfillan 1965).

Even today, lead poisoning from ingestion is not uncommon. The illicit manufacture of moonshine whiskey, using stills made from old lead-coated automobile radiators, is the principal recognized source of excess lead in dietary consumer products (EPA 1971c). Another particularly sad example of lead poisoning results from the tendency of some children in city slums to eat chips of lead-based paint which peel off the walls of deteriorating housing. The all-too-frequent result is mental retardation, cerebral palsy, recurrent seizures, chronic kidney disease, and even death.

Lead is also an atmospheric pollutant. It is added to gasoline as an antiknock agent and the resulting exhaust emissions account for about 97% of atmospheric lead. Most of the particles have such small size (0.5 μ or less) that they are easily distributed over large distances. Fortunately this source of lead may be reduced in the near future, not only because it is being recognized as a dangerous air pollutant, but also because leaded fuels are incompatible with catalytic converters. These catalytic devices are seriously being considered by American automobile manufacturers as pollution control devices (see Chapter 10).

8.12 *Hazardous Air Pollutants*

Some air pollutants are particularly dangerous to health and may be placed in a separate "hazardous" category.

Asbestos is an especially dangerous air pollutant. The inhalation of asbestos fibers has been related to a number of human diseases, including an otherwise rare form of lung cancer called mesothelioma. Asbestos workers have been particularly affected, but evidence now indicates that asbestos fibers can be found in ambient air and in the lungs of nonoccupationally exposed persons. Not enough is known to evaluate the seriousness of these findings. Asbestos enters the atmosphere from a wide variety of sources, including asbestos mining and milling operations and the manufacture and disposal of asbestos-containing products. Asbestos-containing insulation is sprayed onto the girders of skyscrapers during construction. Asbestos is emitted as brake linings wear down and as asbestos-asphalt roadways are used.

Beryllium is an extremely dangerous and toxic pollutant which is known to cause several forms of lung disease. High exposures can rapidly lead to death. Low-level exposures can result in chronic berylliosis which may not develop until after months or years of exposure. Emissions result from manufacturing processes which involve the grinding, burning, or cutting of beryllium. Another source is rocket-motor firings, since it is used in their fuel.

Mercury is a pollutant that has received much attention lately, especially since it was discovered that inorganic mercury can be converted to an organic, toxic form, methyl mercury, by certain microorganisms and then concentrated in the food (see Section 5.8). Mercury is also a hazardous atmospheric pollutant. As such it exists in the form of elemental mercury vapor, a form which is known to produce detrimental effects in the central nervous system. Major sources of atmospheric mercury include paint, coal-fired power plants, mercury processing plants, and primary nonferrous smelters.

8.13 *The Clean Air Act*

The Clean Air Act was passed by Congress in December of 1963 with important, strengthening amendments added in 1967 and 1970. One of the provisions of the amended Act required the Federal Government to publish a set of *Air Quality Criteria* documents describing in detail the effects of different concentrations of air pollutants. Using these reports, the states have been required to set air quality standards which are to be set at levels which will assure that the public is not exposed to any significant risk of health damage, regardless of whether these standards can be achieved with present technology and control methods.

The Federal Government has also established national ambient (outdoor) air quality standards which are listed in Table 8.5. National Pri-

TABLE 8.5 National Ambient Air Quality Standards, Not to Be Exceeded More than Once a Year

Pollutant	Averaging time[a]	Primary standard[b]	Secondary standard[b]
Particulate matter	Annual geometric mean	75 $\mu g/m^3$	60 $\mu g/m^3$
	24 hours	260 $\mu g/m^3$	150 $\mu g/m^3$
Sulfur dioxide	Annual arithmetic mean	0.03 ppm	0.02 ppm
	24 hours	0.14 ppm	0.10 ppm
	3 hours	—	0.5 ppm
Carbon monoxide	8 hours	9 ppm	same as primary
	1 hour	35 ppm	
Photochemical oxidants	1 hour	0.08 ppm	same as primary
Hydrocarbons (corrected for methane)	3 hours (6–9 A.M.)	0.24 ppm	same as primary
Nitrogen dioxide	Annual arithmetic mean	0.05 ppm	same as primary

[a]The arithmetic mean of n samples $x_1, x_2, \ldots, x_n$ is given by: $(x_1 + x_2 + \cdots + x_n)/n$. The geometric mean is given by $\sqrt[n]{x_1 \cdot x_2 \cdots x_n}$.

[b]Conversion factors for pollution units based on volume (parts per million) to units based on weight (milligrams per cubic meter) are given below:

CO 1 ppm = 1.15 mg/m^3	HC 1 ppm = 0.655 mg/m^3 (as methane)
O_3 1 ppm = 1.96 mg/m^3	SO_2 1 ppm = 2.86 mg/m^3
PAN 1 ppm = 5.4 mg/m^3	NO_2 1 ppm = 1.886 mg/m^3

mary Standards are set, "with an adequate margin of safety," to protect the public health. National Secondary Standards, which are more severe, are set at levels which protect the public welfare. As part of the Clean Air Act, states have been required to submit "implementation plans" to the EPA describing programs which will bring pollution levels below the Federal Primary Standards within 3 years of the program's approval. Including an allowable 2-year extension, the standards are therefore supposed to be met by 1977. Elements in these plans may include emission limitations for motor vehicles and stationary sources; measures to reduce traffic such as gas rationing, parking restrictions, and expansion of mass transit facilities; closing or relocation of polluters; emission charges, taxes, or other economic incentives; etc. This portion of the law, if actually enforced, would require severe life-style changes in our most polluted cities. Los Angeles, for example, would practically have to eliminate the use of automobiles on city streets by 1977. Federal Secondary standards need only be met within a "reasonable time."

The Clean Air Act also empowers the EPA to take emergency action when pollution reaches levels that present "imminent and substantial endangerment" to human health. Table 8.6 indicates those levels of pollution that the EPA has determined could cause "significant harm" to the health of persons. In mid-November 1971, Birmingham, Alabama endured a week-long period of air stagnation which drove particulate levels to 771 $\mu g/m^3$ (24 hour average). Local industry was asked by the EPA to curtail operations but most either failed to comply or took only token actions. This led the EPA to obtain a federal injunction against 23 Birmingham companies which forced them to shut down until the crisis passed (BAAPCD 1971).

To prevent air pollution levels from reaching the danger point, states have been required to design emergency episode procedures for their most heavily polluted regions. These contingency plans should be modeled after the 3-stage program suggested by the EPA. The three stages are called *alert*, *warning*, and *emergency* and air quality standards are specified for each level. During the alert stage, open burning is banned and industrial pollution sources are directed to cut pollution in accordance with

TABLE 8.6 Air Pollution Levels that Could Cause "Significant Harm" to the Health of Persons[a]

Pollutant	Averaging time	Concentration
Sulfur dioxide	24 hours	1.0 ppm
Particulates	24 hours	8 COHs[b]
Product of SO_2 and COH[b]	24 hours	1.5
Carbon monoxide	8 hours	50 ppm
	4 hours	75 ppm
	1 hour	125 ppm
Oxidant	4 hours	0.4 ppm
	2 hours	0.6 ppm
	1 hour	0.7 ppm
Nitrogen dioxide	24 hours	0.5 ppm
	1 hour	2.0 ppm

[a]Established by EPA, October 19, 1971.
[b]See footnote, Table 8.3, p. 182.

prearranged emission reduction schedules. The public is asked to voluntarily reduce unnecessary driving and cut back use of electricity (power plants pollute).

The warning stage is similar to alert but further restrictions in emissions are required. The emergency stage is triggered when levels of pollutants pose imminent and substantial danger to health. Use of motor vehicles is banned except in emergencies. Schools, libraries, and government offices are ordered to close. Nonessential office, retail, wholesale, and commercial operations are directed to close. Manufacturing plants are ordered to put maximum pollution abatement procedures into effect and to stop operations if possible.

One of the most important sections of the Clean Air Act sets deadlines for controlling major emissions from automobiles and will be discussed in Chapter 10.

8.14 *Conclusions*

Two dominant trends which have become most clear since the industrial revolution are the growth in man's use of energy and his increasing concentration in cities. Invariably as man congregates in cities the technology required to sustain life becomes more complicated and more pervasive. The consistent result of these activities has been higher and higher levels of pollution. The point is being reached in some big cities where the resource which limits further development is not space, or materials, or water—but air.

As we have seen, air pollution can be damaging to health at levels which are typical in most urban centers today. An unfortunate shift in weather conditions can raise concentrations to levels which are lethal. The damage which is being inflicted as a result of daily living in moderately polluted air is essentially unknown.

It seems there is reason for both optimism and pessimism concerning the near future. Total emissions from automobiles shall drop considerably if the emission standards specified in the Clean Air Act can be met. Photochemical smog problems should therefore ease some. A most serious problem is emerging now in the form of an "energy crisis" (Part IV). Our cleaner burning fuels are rapidly being depleted which is going to force higher consumption of coal. The burning of coal in power plants releases tremendous quantities of particulates and sulfur oxides—the most objectionable pollutants.

Bibliography

Bay Area Air Pollution Control District (1971). "Birmingham Air Pollution Episode." *Air Currents*, San Francisco, Nov.

California Air Resources Board (1972). *Air Pollution in California, 1971, Annual Report*. Sacramento, Jan.

[illegible] J. N. and Sawyer, R. F. (1970). Air pollution sources reevaluated. *Environ. Sci. Tech.* June: 453–455.

[illegible], [illegible] (1972). *Air pollution and public health*. New York: Dryden.

Council on Environmental Quality (CEQ) (1972). *Environmental Quality, 3rd Annual Report*. Washington, D.C. Aug.

Frenkiel, F. N. (1972). Atmospheric pollution in growing communities. In W. E. Britton, et al. eds. Boulder, Colo.: Colorado Associated Univer. Press.

Gilfillan, S. C. (1965). Lead poisoning and the fall of rome. *Jour. Occup. Med.* 7:53–60.

Holland, W. W. ed. (1972). *Air pollution and respiratory disease*. Westport, Conn.: Technomic.

Larsen, R. I. (1970). Relating air pollutant effects to concentration and control. *Jour. Air Pollution Control Assoc.* 20 (4).

McDermott, W. (1961). Air pollution and public health. *Sci. Am.*, Oct.

National Academy of Sciences (NAS) (1972). *Particulate polycylic organic matter*. Committee on Biologic Effects of Atmospheric Pollutants. Washington, D.C.

Purdom, P. W. ed. (1971). *Environmental health*. New York: Academic Press.

Stern, A. C. (1968). *Air pollution*. (3 volumes) New York: Academic Press.

U.S. Environmental Protection Agency (EPA) (1971a). Air Pollution Control Office. *Guide for Air Pollution Episode Avoidance*. AP-76. Washington, D.C.

U.S. EPA (1971b). Air Pollution Control Office. *Air Quality Criteria for Nitrogen Oxides*. AP-84. Washington, D.C. Jan.

U.S. EPA (1971c). Air Pollution Control Office. *Environmental Lead and Public Health*. AP-90. Research. Triangle Park, N.C. March

U.S. EPA (1971d). Air Pollution Control Office. *Background Information—Proposed National Emission Standards for Hazardous Air Pollutants: Asbestos Beryllium Mercury*. APTD-0753. North Carolina. Dec.

U.S. EPA (1971e). Air Pollution Control Office. *Air Pollution Episodes, A Citizen's Handbook*. Dec.

U.S. EPA (1973). Office of Air and Water Programs *Nationwide Air Pollutant Emission Trends 1940–1970*. AP-115. North Carolina, Jan.

U.S. Department of Health, Education, and Welfare (HEW) (1969a). National Air Pollution Control Administration. *Air Quality Criteria for Particulate Matter*. AP-49. Washington, D.C. Jan.

U.S. HEW (1969b). National Air Pollution Control Administration. *Air Quality Criteria for Sulfur Oxides*. AP-50. Washington, D.C. Jan.

U.S. HEW (1970a). National Air Pollution Control Administration. *Air Quality Criteria for Carbon Monoxide*. AP-62. Washington, D.C. March.

U.S. HEW (1970b). National Air Pollution Control Administration. *Air Quality Criteria for Photochemical Oxidants*. AP-63. Washington, D.C. March.

U.S. HEW (1970c). National Air Pollution Control Administration. *Air Quality Criteria for Hydrocarbons*. AP-64. Washington, D.C. March.

U.S. HEW (1970d). National Air Pollution Control Administration. *Nationwide Inventory of Air Pollutant Emissions 1968*. AP-73. Raleigh, N.C. Aug.

Questions

1. The incidence of lung cancer in urban areas is about double that of rural areas. What factors besides air pollution could be causing this difference?

2. The following formula can be used to estimate the equilibrium value of COHb attained after continuous exposure to a constant concentration of CO:

$$\text{COHb}\ (\%) = 0.5 + 0.16 \times \text{CO}\ (\text{ppm})$$

   Estimate the percentage COHb which would result from continuous exposure to 10 mg/m^3 of CO.

   ans. 1.9%

3. What measurement would you propose as the best indicator of whether air pollution is getting better or worse?

4. What does it mean to say sulfur oxides and particulates act synergistically?

5. Estimate the visibility that must have occurred during the 1971 Birmingham, Alabama episode.

   ans. 1 mile

6. What is the arithmetic mean of the numbers 1, 2, 3, 4? What is the geometric mean?

   ans. 2.5, 2.2

7. For your own region find out what provisions are included in the implementation plans to bring ambient air quality within Federal Primary Standards.

Chapter 9

Air Pollution Meteorology

Even in our most polluted cities, there are frequent periods when the air is quite clear, interspersed with periods of poor air quality. These periodic fluctuations are not caused by gross changes in the emissions of the local polluters but rather are a function of the variations in meteorological conditions.

It is therefore important to consider the role of meteorology in the accentuation of air pollution problems. We begin by examining the atmospheric conditions required for good vertical mixing of pollutants. As was mentioned in the last chapter, this mixing and dilution is greatly limited when temperature inversions are present, so we shall look into the various mechanisms which lead to the formation of inversion layers.

Then we shall reverse the point of view, and see how atmospheric pollution may be affecting the world's climate. And, on a smaller scale, we shall see that air pollution does affect the climate of cities.

9.1 *The Atmosphere*

In Figure 9.1 an approximate temperature profile for the earth's atmosphere is given. The lower portion of the atmosphere is called the *troposphere,* and is characterized by a decrease in temperature with altitude. The rate of temperature drop is called the *lapse* rate, about which we will have more to say later. In general, the atmosphere is not heated by the sun's rays, but rather is heated from below by the warm earth. Thus, it is quite reasonable for temperature to decrease with increasing altitude. About 80% of the mass of the atmosphere is contained in the troposphere which typically extends up to about 40,000 feet. The upper limit of the troposphere is called the tropopause.

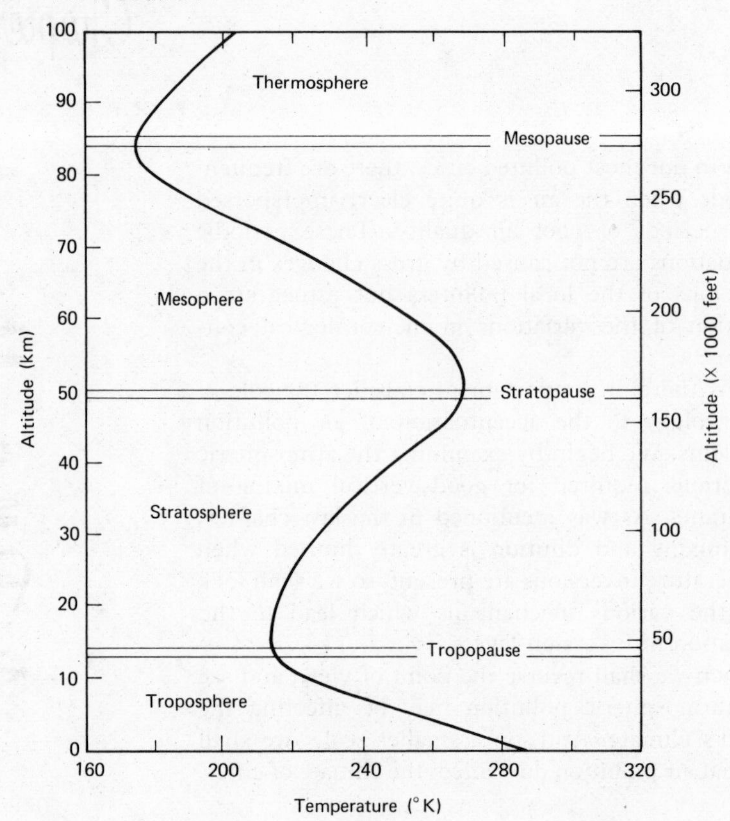

FIGURE 9.1 Temperature profile for the atmosphere (note °C = °K − 273).

Above the tropopause is a very stable layer of air called the *stratosphere,* the temperature of which is constant or increasing with altitude. In the stratosphere, ozone absorbs UV radiation from the sun and decomposes into oxygen atoms and molecular oxygen. When these particles recombine to form ozone again, energy is released in the form of heat. The net effect of the absorption of UV energy is the liberation of heat, hence the characteristic temperature increase. The stratosphere extends upward to an altitude of about 30 miles to the stratopause.

Above the stratopause, the temperature again decreases with altitude in a region called the *mesosphere.* Then the mesopause marks the separation between the mesosphere and another layer of atmosphere called the *thermosphere* wherein temperature increases rapidly with height due to the absorption of solar energy by atomic oxygen. Within the thermosphere is a relatively dense band of charged particles, called the *ionosphere* which is extremely important to worldwide communications because of its ability to reflect radio waves back to earth.

Changes in atmospheric conditions can make the difference between clean and heavily polluted air. These photographs were taken only a few days apart. (Courtesy Peninsula Conservation Center.)

9.2 Atmospheric Stability

The rate of change of atmospheric temperature with increasing altitude is of crucial importance to the ease with which pollutants can be diluted by vertical mixing with cleaner air.

Let us imagine an insulated balloon which we move up or down in the atmosphere. The insulation does not allow any heat to enter or leave the balloon. If we move the balloon up, it will expand due to the decrease in atmospheric pressure. As a gas expands, it temperature drops, so the temperature of the air inside the balloon will decrease. The decrease in temperature inside the balloon with increasing altitude is called the *adiabatic* (no heat transfer) *lapse rate,* and is equal to about -1°C per 100 meters, or -5.4°F per 1000 feet.

When the temperature of the atmosphere decreases faster than the adiabatic lapse rate (superadiabatic) the atmosphere is unstable and the air from one altitude eagerly mixes with the air from other altitudes. From the point of view of air pollution, this is very desirable since pollutants will be rapidly dispersed throughout the atmosphere. The situation is exactly analogous to the turnover period in a stratified lake. Cold air descends towards the earth, is warmed, and rises again.

If the temperature decreases less rapidly with altitude than the adiabatic lapse rate, then the air is stable. This means a parcel of air from one altitude has forces acting on it which inhibit the vertical movement of that parcel. Under these conditions there is very little movement of air from one altitude to another so any pollutants in the air will likewise be restricted. With no vertical mixing, pollution concentrations can build very rapidly.

The extreme case, which is called inversion, occurs when temperature increases with altitude, forming a very stable atmosphere. These cases are illustrated in Figure 9.2. As this figure indicates, the adiabatic lapse forms a boundary between atmospheric stability and instability. Let us briefly justify that statement.

One version of the perfect gas law states

$$P = \rho R T \tag{9-1}$$

where

P = pressure
ρ = density
R = gas constant
T = absolute temperature

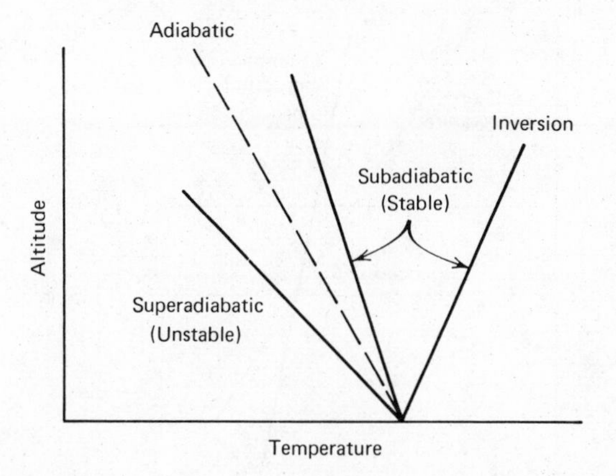

FIGURE 9.2 The adiabatic lapse rate is the boundary between stable and unstable air.

If we again imagine the insulated balloon suspended in the atmosphere, we can write, for the air inside the balloon,

$$P_b = \rho_b R T_b \tag{9-2}$$

and for the air surrounding the balloon,

$$P_a = \rho_a R T_a \tag{9-3}$$

Neglecting any effect caused by the tension in the balloon, we can say that these pressures are equal, and hence can equate (9-2) and (9-3) to give

$$\rho_b = \rho_a \left(\frac{T_a}{T_b}\right) \tag{9-4}$$

Let us demonstrate the case of stability for a subadiabatic lapse rate. If the atmosphere is stable, the balloon will remain at a constant altitude. Trying to move it up will generate forces which will push it back down, and trying to push it down will generate forces which will raise it back up.

We start with the balloon at position 1 in Figure 9.3, having the same internal temperature and pressure as the surrounding atmosphere. Raise the balloon slightly to position 2. The decrease in pressure on the balloon means it will expand and the temperature inside will drop according to the adiabatic lapse line in the figure. Now since $T_b < T_a$, we know from equation (9-4) that $\rho_b > \rho_a$. That is, the density of the air inside the balloon is greater than the surrounding air, so the balloon sinks back to its original position.

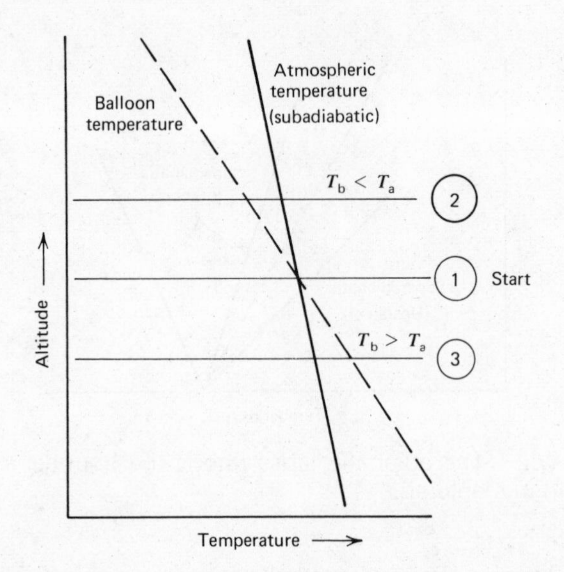

FIGURE 9.3 Raising the balloon from position 1 to position 2 causes the density inside to be greater than the density outside so the balloon returns to 1. Lowering it to position 3 causes the density of the balloon's air to be less than atmospheric density so it again returns to 1. Hence this is a stable atmosphere.

Now move the balloon down slightly to position 3. As the figure indicates, $T_b > T_a$ so, from equation (9-4) $\rho_b < \rho_a$. The air in the balloon becomes less dense than the surrounding atmosphere and so buoyancy forces return the balloon back to position 1. Therefore, the balloon tends to remain at position 1 and the atmosphere is stable. It is simple enough to reverse the arguments and show instability for superadiabatic lapse rates.

It is interesting to observe the effects of the lapse rate on smokestack plumes, as shown in Figure 9.4. When the atmosphere is very unstable, the result is a looping plume which causes rapid dispersal of pollutants. A fanning plume is caused by thermal inversion. Vertical movement of pollutants is severely limited and the plume spreads out horizontally. The third case shown is called fumigation and results when there is an inversion above the plume and instability below. When this occurs, pollutants are brought down to the ground thereby creating abnormally high concentrations.

Emissions from this copper smelter are trapped by the clearly evident inversion layer. (Photograph by J. Randolph.)

9.3 *Hadley Cells and Subsidence Inversion*

There are several ways that inversion layers can be formed. One of the most common ways is for air to be heated by compression as it descends in a high pressure system. If, as it descends, its temperature rises above that of the air below, it creates what is known as a *subsidence* inversion.

There is a simple model which helps explain the prevalence of subsidence inversions at latitudes which are typical of parts of the United States. We will start by taking a very simplistic view of the atmospheric circulation around an idealized globe. If we consider a smooth earth, not rotating, with heat arriving uniformly around the equator, we might expect the air around the equator to rise because it is being heated. As it rises, it expands and cools, and at the tropopause some of it heads north and some south. We might expect at the same time for the air above the cold poles to be sinking. This suggests the simple atmospheric circulation pattern shown in Figure 9.5.

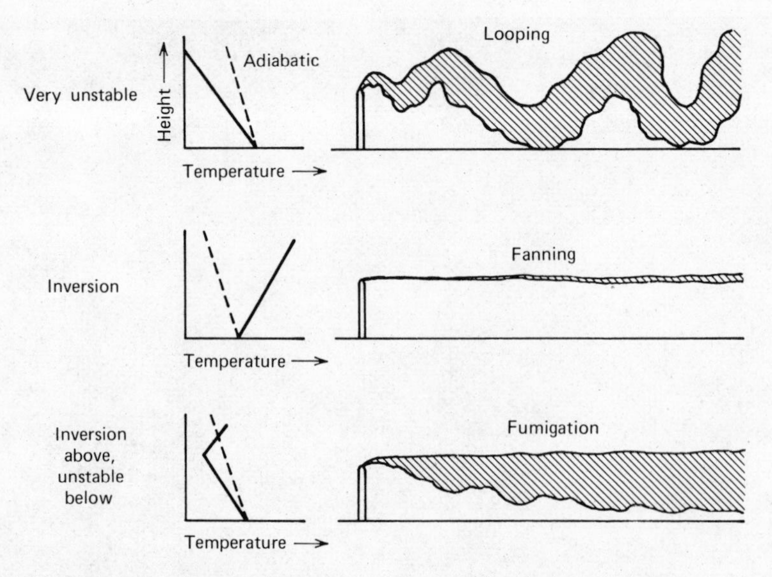

FIGURE 9.4 Atmospheric stability and plume behavior. The broken line is the adiabatic lapse rate, the solid line the actual lapse rate.

The air is shown moving in "Hadley cells" named after an eighteenth century English meteorologist named George Hadley. In actuality the circulation pattern involves more cells than shown in Figure 9.5. The air which rises at the equator cools and descends at a latitude of about 30° as shown in Figure 9.6. Cells also develop between 30° and 60° and between 60° and 90° as shown.

The pattern shown in Figure 9.6 is extremely influential in determining the climate in various parts of the world and is also important to us in

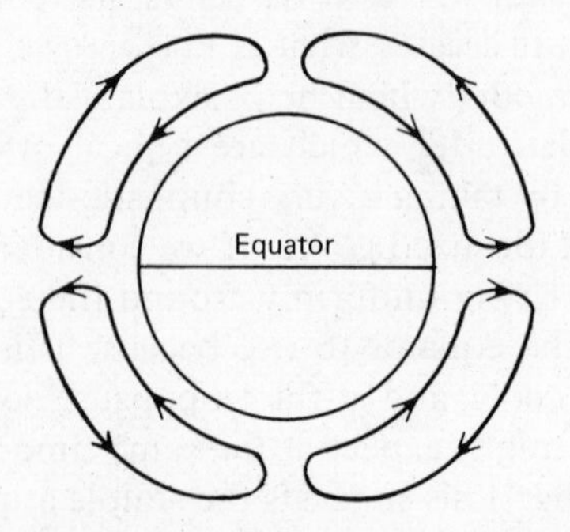

FIGURE 9.5 One possible circulation pattern for a smooth, nonrotating earth, heated at the equator.

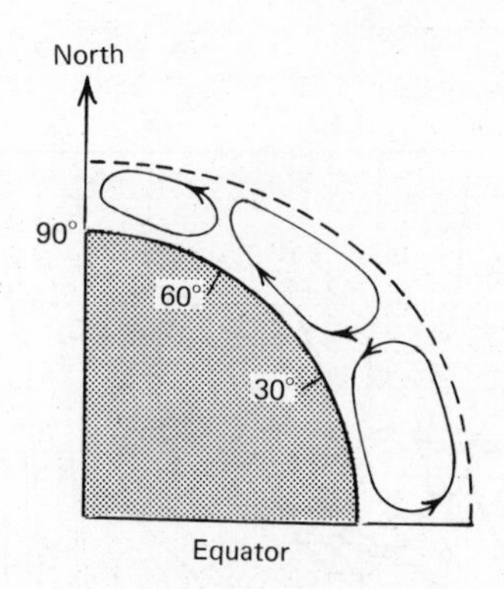

FIGURE 9.6 Hadley cells in one quadrant in the northern hemisphere.

this chapter because it helps provide the explanation for commonly occurring subsidence inversions at midlatitudes.

When air is heated, it rises and as it rises, it expands and cools. The cooler the air is, the less able it is to hold moisture and the result is rain or fog. So there is lots of rain near the equator and near 60° latitude (north or south). The air which, in our model, is descending at 30° is being compressed and heated. The resulting high-pressure area has associated with it clear skies and little rain. The descending air, heated by compression, may become warmer than the air below it, thus leading to subsidence inversion. Figure 9.7 shows the circulation patterns and subsidence inversions associated with the first cells.

Figure 9.7 is an idealized version of the circulation which would take place if the sun were to circle a smooth earth at the equator (corresponding to the equinoxes, March 21 and September 23). The system shown in the

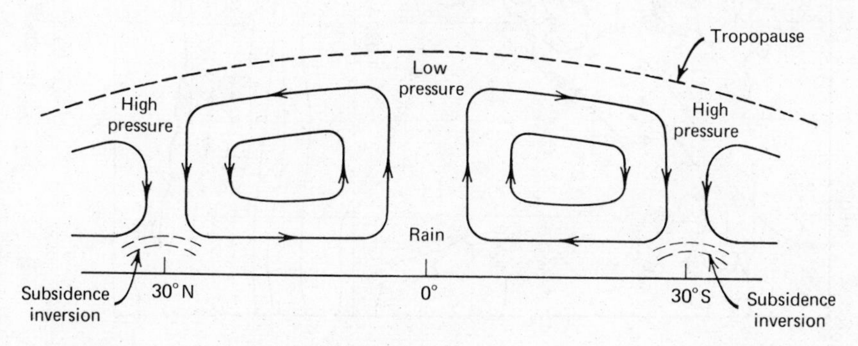

FIGURE 9.7 Creation of subsidence inversions by Hadley cells (Petterssen 1969).

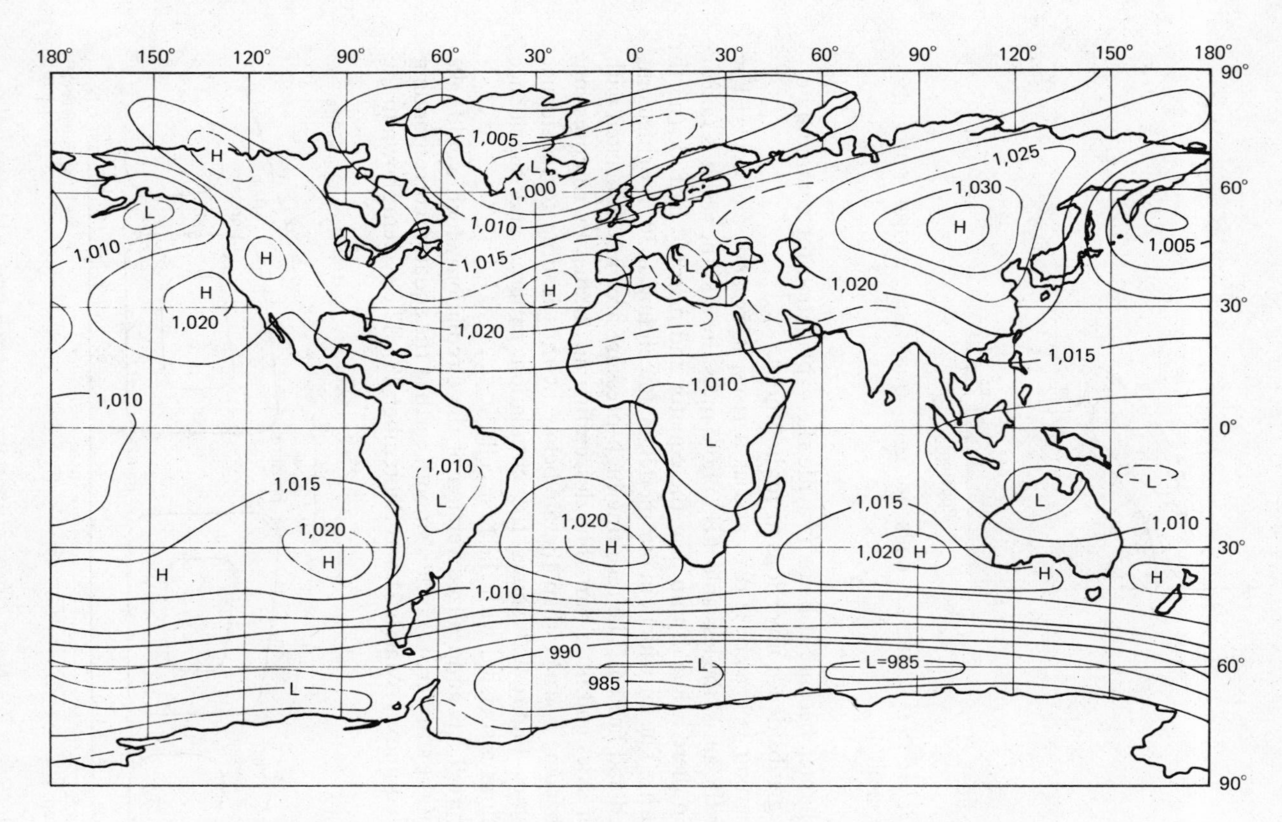

FIGURE 9.8 Normal sea-level pressure map of the world, fall equinox (Petterssen 1969).

figure shifts north during the summer and south during the winter. In spite of the crudeness of this model, it does help to explain the atmospheric sealevel pressure map of the world shown in Figure 9.8. The bands of high pressure areas around 30°N and 30°S are clearly evident as are the low pressure areas around the equator. Notice the high-pressure area off the coast of California which brings the warm, dry air which is responsible for California's good weather. Unfortunately, it also causes subsidence inversions which contribute to the area's air pollution problems.

9.4 *Other Causes of Inversions*

Besides the subsidence inversions associated with high-pressure systems, there are several other important mechanisms which create inversions. *Radiation* inversions occur in the morning at the surface of the earth, following a cold, clear night. The surface of the earth cools at night by radiating energy toward space. If it is a cloudy night, some of the energy is reflected back to the earth and there is less cooling. However, on a clear night the earth cools rapidly and the air directly above the cold ground cools below the temperature of the upper air, creating an inversion. As the sun warms the earth during the day, the inversion disappears. Figure 9.9 shows a temperature profile at various times of the day, demonstrating the appearance and disappearance of the inversion layer. As the sun warms the earth which in turn warms the lower atmosphere, a layer where mixing of the atmosphere occurs is formed as shown in Figure 9.9*b*. Pollutants which had been suspended aloft in the stable atmosphere which existed before dawn, are suddenly free to move throughout the mixing layer and many are brought rapidly down to earth resulting in momentarily high concentrations. This process is known as *fumigation*. Recall the effects of fumigation on a smokestack plume, as were shown in Figure 9.4.

The nighttime inversion shown in Figure 9.9*a* is of crucial importance in the far north during the long polar night. Such inversions can persist for several weeks and in such places as Fairbanks, Alaska, can contribute greatly to the air pollution problem.

Radiation inversion can occur near the surface of the earth at the same time that a subsidence inversion exists at a higher altitude as suggested by Figure 9.10.

Another cause of inversions is cold air sliding under warmer air. For example, consider a coastal city in the summer. The land warms up during the day and heats the air above it. The warm air rises drawing in cool air from the sea, which results in an inversion. As the air moves further in-

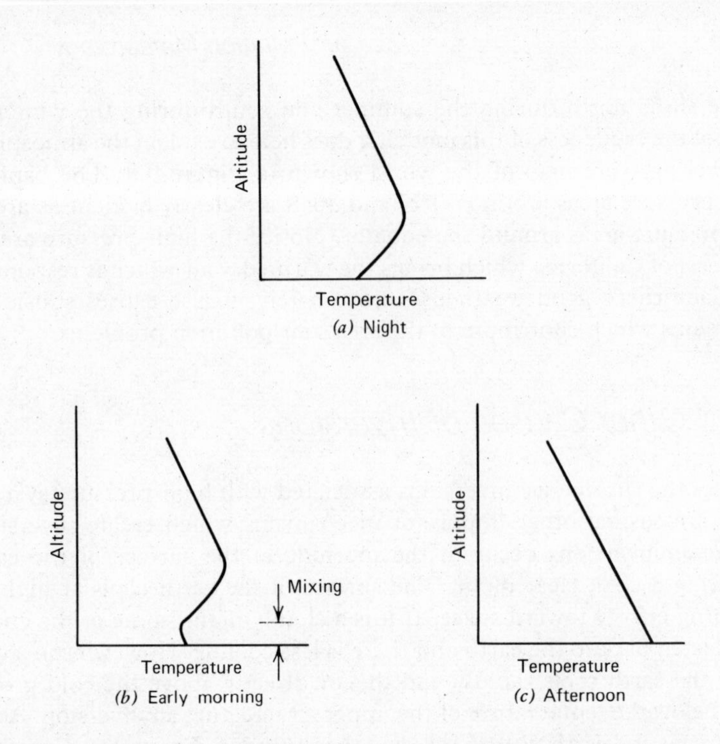

FIGURE 9.9 As the day progresses the morning radiation inversion disappears. In (*b*) the mixing layer formed causes fumigation.

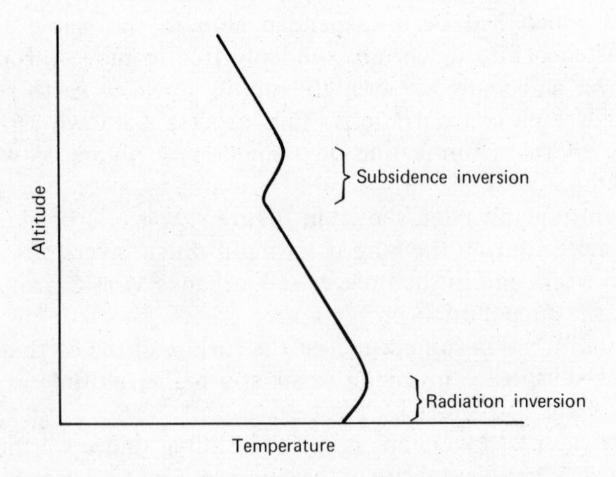

FIGURE 9.10 Temperature profile through two inversion layers—subsidence and radiation.

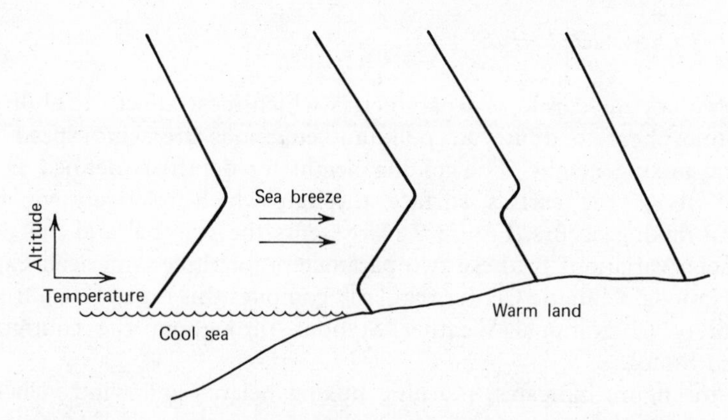

FIGURE 9.11 Decay of an inversion as cool air moves inland during the day.

land, the inversion is degraded and finally disappears as shown in Figure 9.11. At night the direction of the breeze is reversed, and a radiation inversion may form over the land.

Another local geographical condition which can contribute to the formation of inversions is suggested in Figure 9.12. When hillsides cool at night, the cold dense air which forms flows down the sides and collects in the valleys. Light breezes of warm air can pass over the valley and not disturb the stagnant, cold air below. The effect can be maintained well into the day, at least until the hills no longer block the direct rays of the sun. When industrial towns are built in such valleys, the air pollution problem can be particularly severe since horizontal air motion may be blocked by the hills and vertical dilution can be restricted by the inversion. These factors were influential in the air pollution disasters in the Meuse Valley in Belgium, and at Donora, Pennsylvania.

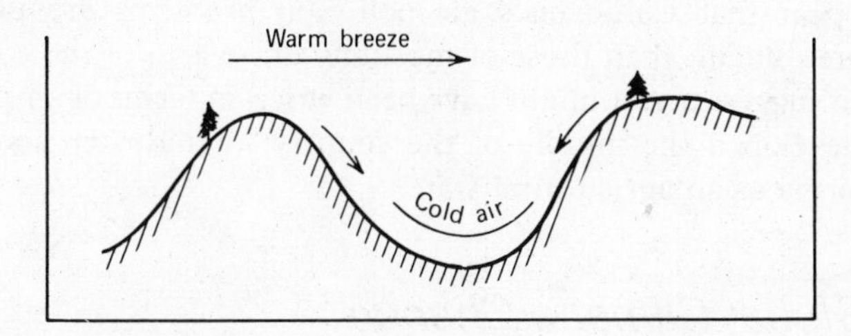

FIGURE 9.12 The manner in which an inversion can form in a valley. Cold air flows down the sides of the hills and settles in the valley. If the breeze is strong enough or the sun warm enough, the inversion disappears.

9.5 *Episode-Days*

The two key meteorological parameters which most affect the ability of the atmosphere to dilute air pollutant emissions are wind speed and vertical mixing height. The mixing height (or depth) is defined as the height above the earth's surface through which relatively vigorous vertical mixing occurs. Figure 9.13 presents the seasonal and morning/afternoon variations in these two parameters for three American Cities. The National Climatic Center regularly compiles this type of data from a system of 62 National Weather Stations throughout the contiguous United States.

As the figure indicates, morning mixing heights and wind velocities are consistently lower than the corresponding afternoon values. Stations situated near large bodies of water typically have less daily temperature variation and so exhibit less change in mixing heights, than do inland stations.

High concentrations of air pollution in an area are most likely to occur after several days of low mixing height, low wind speed, and no precipitation. If this combination of circumstances occurs very frequently in an area, then it can be said that the area has a high potential for air pollution episodes. Holzworth (1972) has analyzed the air pollution potential across the nation in terms of episode-days. An episode-day occurs when the mixing height is less than 1500 m, the wind speed is less than 4 m/sec (about 9 mph), and there is no significant precipitation. Also these conditions must persist for at least 2 days.

Figure 9.14 summarizes the air pollution potential in terms of the total number of episode-days which occurred over a 5 year period. As can be seen, the highest number of episode-days in the nation, occur around Southern California. There are almost none in the midwest, and rather few in the east except for a region around Appalachia. From this figure it would appear that California's air pollution problems are potentially much more difficult than those of most any other area of the country. It would also suggest that it might have been wiser, in terms of air pollution, to have developed the middle of the country for industry and to have left California as an agricultural state.

9.6 *Global Climatic Change*

So far in this chapter we have discussed some of the influences meteorology has on air pollution and now we will reverse the point of view and see how air pollution can affect climate.

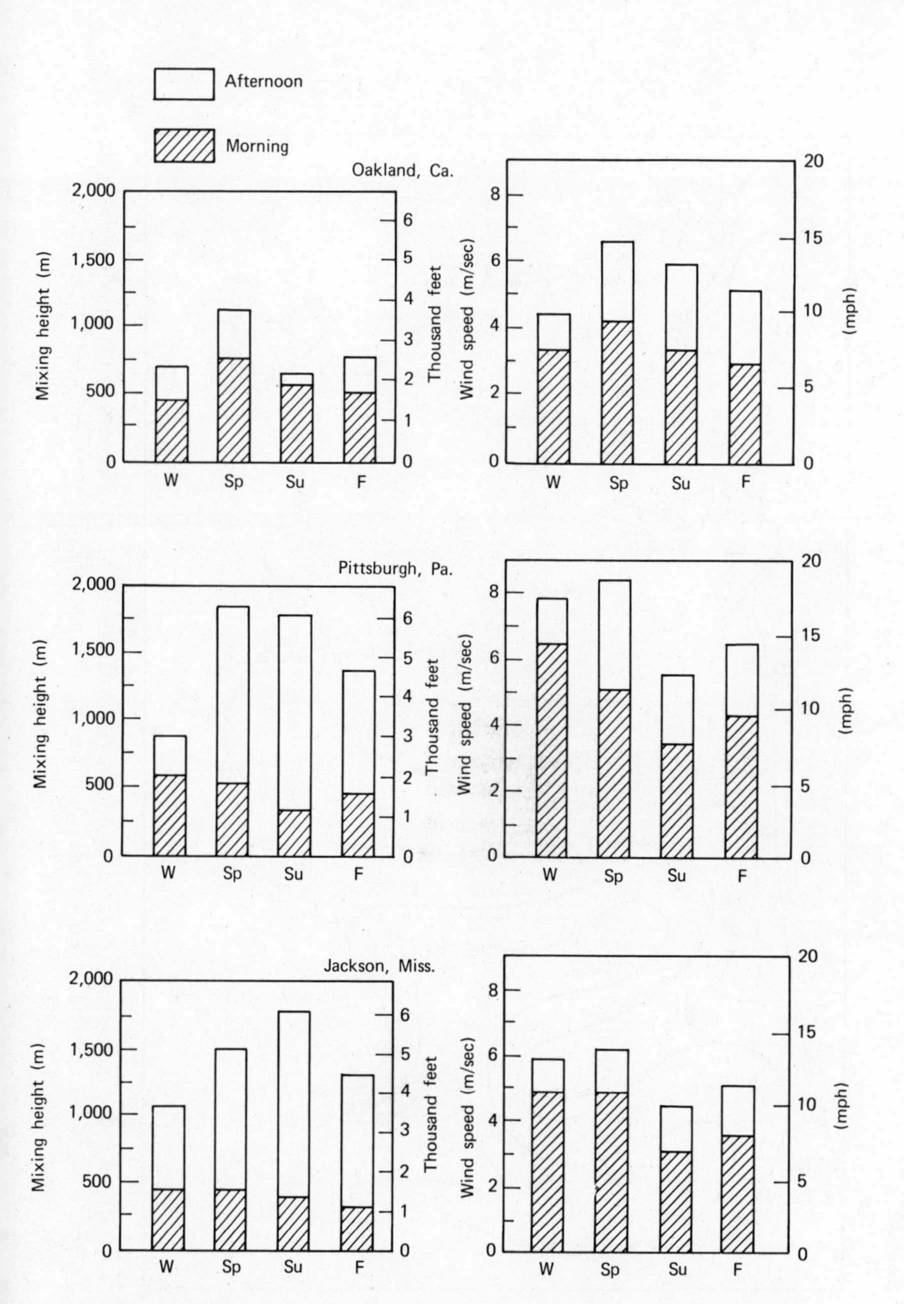

FIGURE 9.13 Seasonal and morning/afternoon variations in mean mixing height and wind speed for three cities (Holzworth 1972).

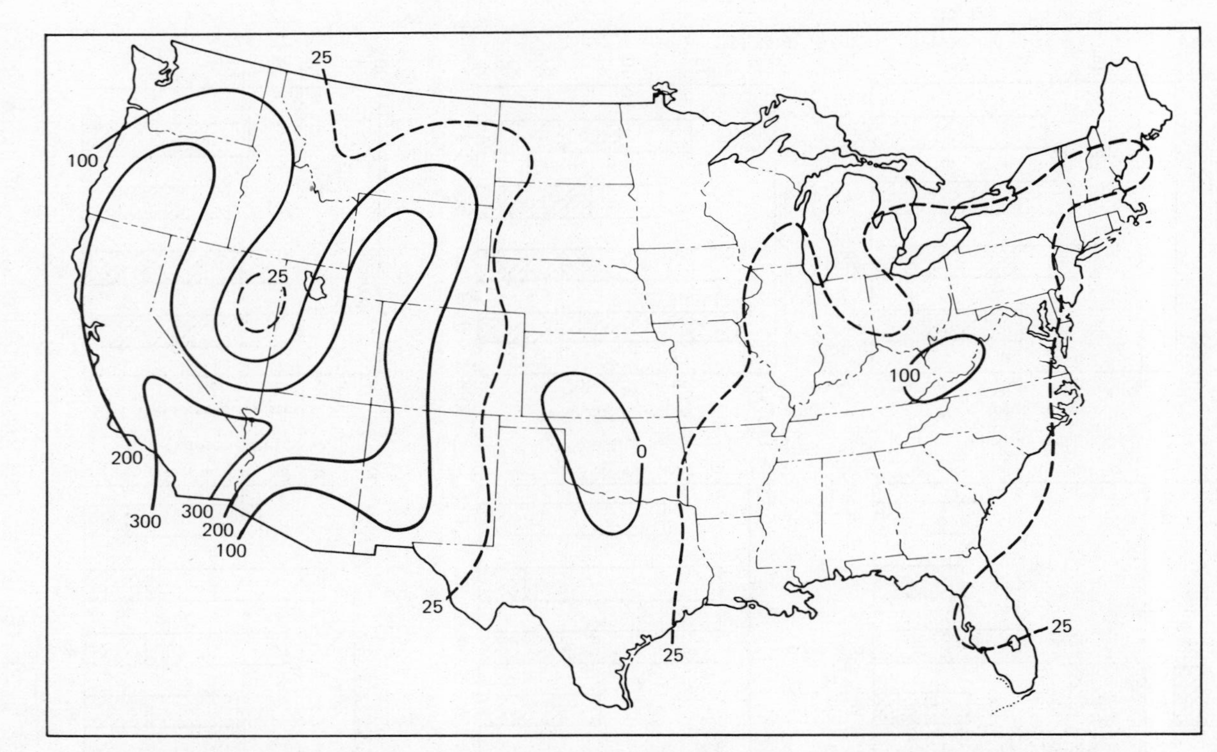

FIGURE 9.14 Isopleths of total number of episode-days in 5 years with mixing heights ≤1500 m, wind speeds ≤4.0 m/sec, and no significant precipitation—for episodes lasting at least *2* days. (Holzworth 1972.)

Most air pollution problems are side effects of combustion, e.g., SO_x from sulfur impurities in coal and oil, NO_x from air reacting at high temperature, HC and CO from incomplete combustion, lead from the burning of gasoline. However, even if we were to achieve complete combustion of an impurity-free fuel, we would still produce two reaction products, water and carbon dioxide, which are usually not considered pollutants:

$$\text{fuel} + O_2 \rightarrow CO_2 + H_2O + \text{energy} \tag{9-5}$$

Equation (9-5) describes not only the complete combustion of fossil fuels but also applies to respiration by animals and plants. In fact, about 80% of the input CO_2 to the atmosphere comes from respiration, the other 20% comes mostly from the burning of fuels. There are at least two natural "sinks" for CO_2: (a) water, be it rain, lakes or oceans, absorbs CO_2, and (b) plants use it for photosynthesis. However, the natural balance of production and absorption seems to have been offset by combustion and the concentration of CO_2 in the atmosphere is increasing. In 1880 the concentration was about 284 ppm, at present it is about 330 ppm, and by the year 2000 it may increase to about 379 ppm [Study of Critical Environmental Problems (SCEP) 1970]. The concentration of CO_2 is increasing at an average rate of about 0.2% per year.

It is thought that the buildup of CO_2 may influence the overall heat balance of the earth, causing a global temperature rise by a process known as the "greenhouse effect." In a greenhouse, the energy from the sun easily passes through the glass, heating the interior. The characteristic frequencies of this incoming energy are much higher than those of the energy which is being reradiated from the heated interior. The warm interior radiates infrared radiation (IR) but the IR does not easily pass through the glass so the energy is effectively captured. This effect plus the reduction of wind keeps the interior of a greenhouse warmer than its surroundings. The CO_2 in the atmosphere, along with water vapor, acts like the glass of the greenhouse, passing the incoming solar energy but absorbing much of the earth's reradiated IR energy. As the concentration of CO_2 increases, the balance between incoming energy and outgoing energy is disturbed. More energy is captured than is reradiated to space, so the earth's temperature should rise.

A detailed study by Manabe and Strickler (1964) predicted that a 10% increase in CO_2 concentration would produce a 0.3°C increase in the world's mean surface temperature. It might only take a rise of a few degrees to begin an irreversible melting of the earth's ice caps.

Figure 9.15 shows the trend in CO_2 concentration and world temperature change over about the last 80 years. As can be seen, the temperature

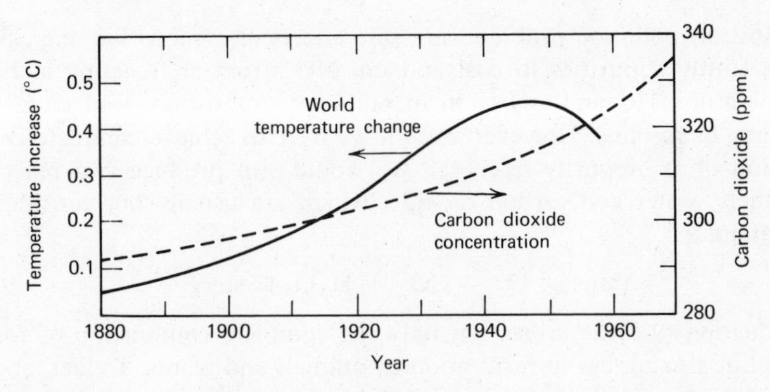

FIGURE 9.15 Trends in mean annual world temperature and carbon dioxide concentration in the atmosphere (Lovelock 1971).

was increasing, in agreement with the greenhouse theory, until about 1945, and since then it has been decreasing. Evidently there is some other mechanism which is becoming more important than the greenhouse effect and which is causing the temperature to drop.

The prevailing theory ascribes the decrease in temperature to the increase in atmospheric turbidity. The suggestion is that the increasing number of particles in the atmosphere are causing more of the incoming solar radiation to be reflected back into space before it can reach the earth's surface. That is, the effective albedo (or reflectivity) of the planet is increasing. An increase of only 1% in albedo, from perhaps 37 to 38%, would lower the mean temperature of the earth by about 1.7°C (Bryson 1971).

Data reported in *Air Quality Criteria for Particulate Matter* (HEW 1969) indicate that in about 60 years the turbidity over Washington, D.C., has increased by 57 %, and by 70 % over Davos, Switzerland. Total solar radiation received by cities is generally 15–20 % less than is received in rural environments. Global turbidity is growing even faster than the growth of combustion which suggests that there may be some secondary effects which are also important. For example, fuel is burned in farm machinery creating turbidity from the exhaust, but in addition to that, the dust that is blown off of plowed fields probably contributes even more to the turbidity.

Lovelock (1971) states that if the turbidity continues to increase at the present rate of about 30% per decade, then it is possible that within just a few decades the average temperature of the Northern Hemisphere will approach that of the last ice age.

In addition to the greenhouse effect and the increase in turbidity as factors probably affecting the world's climate, there is another factor which may exert some influence, and that is sunspot activity. Sunspot activity has been increasing since around 1900. Since sunspots would increase the solar flux, their contribution would be to raise the global temperature.

It is important to note that the reasons given here for the fluctuations in temperature are largely speculative and are far from being scientifically confirmed.

9.7 The Effect of Cities

Cities affect climate in several ways. The uneven surface presented by buildings reduces the wind; the concentration of air pollutants provides condensation nuclei thereby increasing fog and rain; and the extra heat associated with the city creates certain atmospheric thermal effects. A study by Landsberg (1962) compared the climate of cities with their rural environs and the results are summarized in Table 9.1. It can be seen from the table that in the city there is more fog (100% more in the winter), more cloudiness (5–10%), more rain (5–10%), and more dust particles (10 times more).

There is lower humidity in the city, probably due to the quick runoff of water from hard city surfaces. The annual mean temperature in the city is higher than the rural surroundings, often by much more than indicated in the table. The "heat island" effect is due to the greater heat-holding capacity of city surfaces, the reduced evaporation which results from the increased runoff, and to the heat generated by combustion processes in the city itself.

The heat island creates air circulation patterns as shown in Figure 9.16. The rising air cools and is thus less able to hold moisture. Com-

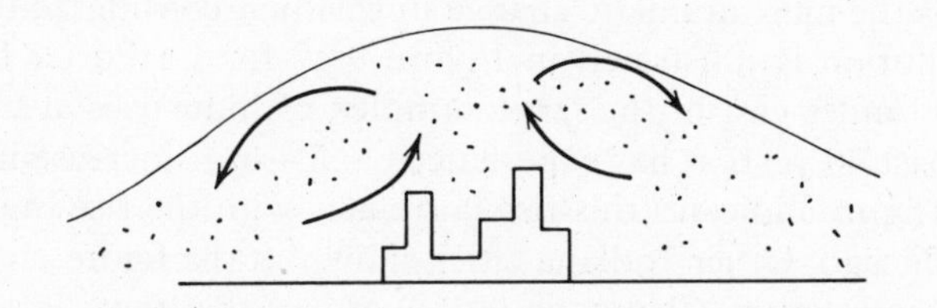

FIGURE 9.16 The heat from the city causes the air to rise and then settle over the cooler environs. Dust moves with the air.

TABLE 9.1 Climatic Changes Produced by Cities (Landsberg 1962)

Element	Comparison with rural environs
Temperature	
Annual mean	1.0–1.5°F higher
Winter minima	2.0–3.0°F higher
Relative humidity	
Annual mean	6% lower
Winter	2% lower
Summer	8% lower
Dust particles	10 times more
Cloudiness	
Clouds	5–10% more
Fog, winter	100% more
Fog, summer	30% more
Radiation	
Total on horizontal surface	15–20% less
Ultraviolet, winter	30% less
Ultraviolet, summer	5% less
Wind speed	
Annual mean	20–30% lower
Extreme gusts	10–20% lower
Calms	5–20% more
Precipitation	
Amounts	5–10% more
Days with < 0.2 inch	10% more

bined with the increased number of dust particles which act as condensation nuclei, the city has more clouds, fog, and rain.

Perhaps the most dramatic change in weather, considered to be caused by air pollution is illustrated in Figure 9.17 for La Porte, Indiana. La Porte is 30 miles east of the large complex of industries at Chicago and over the past 40 years it has experienced a 30–40 % increase in precipitation. The figure indicates this rise correlates with the rise in smoke-haze days in Chicago. Other Indiana cities shown in the figure do not exhibit this same correlation. Changnon (1968) argues that there is a cause-and-effect relationship between the industrial activities around Chicago and the change in weather at La Porte.

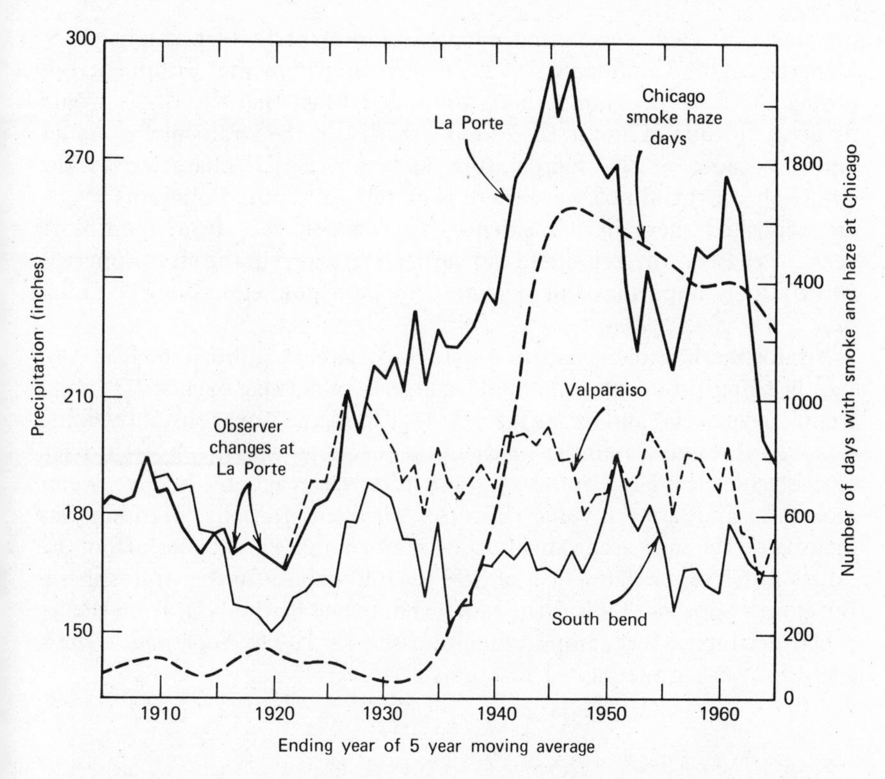

FIGURE 9.17 Precipitation values at selected Indiana stations and smoke-haze days at Chicago. (Note the way in which precipitation trends at La Porte follow the haze changes in Chicago. The results are plotted as 5 year moving averages.) (from HEW 1969 after Changnon 1968).

9.8 *The Supersonic Transport*

In 1971 Congress voted not to fund further construction of two prototype supersonic transport (SST) aircraft. Although the SST project appears to have been halted, there is speculation that it may be revived at some time in the future (see for example, 1972) and also the Anglo-French and Russian SST programs are still being pursued. It is worthwhile to review some of the arguments which are being raised against the SST, since they are quite relevant to the material in this chapter.

The original program involved a fleet of nearly 600 planes (including those of foreign manufacture) in operation by 1990, each making 5 flights per day, and each consuming 146,000 pounds of fuel per flight (U.S. Department of Commerce 1972). A key environmental argument opposing the SST program centers around the fact that the SSTs would fly at an altitude of about 16–20 km—that is, in the stratosphere. As we know, because of the temperature inversion which characterizes the stratosphere, the atmosphere there is extremely stable. Pollutants which are deposited there have residence times which vary from months to years. As has also been pointed out earlier, the ozone in the stratosphere is an extremely important shield against the damaging effects of UV radiation.

Stratospheric ozone exists in a state of dynamic equilibrium. It is constantly being produced as sunlight converts molecular oxygen (O_2) into atomic oxygen (O) and ozone (O_3). It is at the same time constantly being removed through a number of chemical reactions which convert it back to molecular oxygen. Some of these removal reactions involve water molecules and some involve oxides of nitrogen. Initially attention was focused on the increased rate of ozone removal that would result from the excess water vapor that a fleet of SSTs would deposit in the stratosphere. But it now appears that it is the removal of ozone by the NO_x from the jet exhausts which is more important (Johnston 1971; *The Supersonic Transport* 1972). A simple pair of reactions,

$$NO + O_3 \rightarrow NO_2 + O_2$$

$$NO_2 + O \rightarrow NO + O_2$$

has the net effect of

$$O + O_3 \rightarrow 2\,O_2$$

There are many more reactions involved but these demonstrate the catalytic ability of the NO_x pollutants to convert ozone back into molecular oxygen. Johnston calculates that a 2 year accumulation of NO_x from 500 SSTs could reduce the ozone shield by anywhere from 3 to 50%.

It is not yet possible to accurately predict the effects of such a reduction in ozone but a good deal of public attention has been directed to the possibility of an increase in human skin cancers as a result of the increased exposure to UV radiation.

There are other unresolved environmental dangers associated with a large fleet of SSTs. It is possible that the particulate and water vapor emissions could change the albedo which would alter the earth's climate. Still unsolved is the noise problem, both in terms of excessive noise around airports and in terms of sonic booms under the flight path.

We must also come to view such projects as the SST in terms of resources. While we are entering into a period of rapidly declining petroleum reserves, it is enlightening to calculate the extra amount of fuel required for a fleet of SSTs over an equivalent (same passenger carrying ability) fleet of 747s. Using the Department of Commerce figures just quoted, it would require about 1 million barrels of jet fuel per day to keep the SST fleet in operation. The fuel required, per passenger-mile, is three times as high for the SST as for a 747 (Rice 1972). That means an extra 670,000 barrels of jet fuel per day just to gain the extra speed of the SST. At the rate of 7 barrels of crude oil required for each barrel of jet fuel, this means an extra 4.7 million barrels of crude oil per day would be required. That is more than twice the amount of crude as would be delivered by the huge Alaska pipeline. The entire North Slope find of about 10 billion barrels could be used up in less than 6 years just to meet the *extra* demands of the SST. To waste these precious resources so that a very privileged few could save a couple of hours in the air would be the ultimate in immorality.

9.9 *Air Pollution Modeling*

One of the most important tools of an engineer is the technique known as modeling. Models are useful to predict the behavior of systems which are too complex to be amenable to exact analysis. In this case, the system to be modeled is an air shed over a city, and the problem is to relate air pollutant emissions to the resulting air quality. Such a model could be an aid to planning and decision making.

For example, we might ask what would be the effect of a new factory on air quality. The model might be useful in choosing a site for the factory so that the resulting change in air quality would be minimized. For another example, models can help to decide when agricultural burning should be allowed. Under given wind and inversion layer conditions, it is possible to estimate whether air quality standards will be exceeded if burning is allowed. In fact, models are used to predict episode-days of the type described in Section 9.5.

The most controversial application of modeling has arisen as a result of the section of the Clean Air Act, which requires the EPA to determine whether a state's implementation plan will bring air quality within Federal Primary Standards. To test the implementation plans the EPA has a choice of models, the simplest of which will be briefly described here.

In modeling, there is always a trade-off to be made between the exactness of the model and the difficulty in using it. Very simple models are usually appropriate for hand calculations and often give quite reasonable results. More complicated models invariably require the use of a digital computer. The following model is one of the simplest for relating emissions to air quality and is called the *box model*.

Consider a city whose rectangular boundary forms the base of a box-shaped air shed. The height of the box is equal to the mixing height above the city. We assume pollutants are being emitted from the city and also that the wind blows fresh air into one side of the box causing polluted air to be blown out the opposite side as shown in Figure 9.18. We then make the grossly simplifying assumption that pollutants instantly distribute themselves uniformly throughout the box. Let

V = box volume (volume)

C = pollution concentration in the box and leaving the box (weight/volume)

P = pollutant emission rate (weight/time)

Q = air flow rate (volume/time)

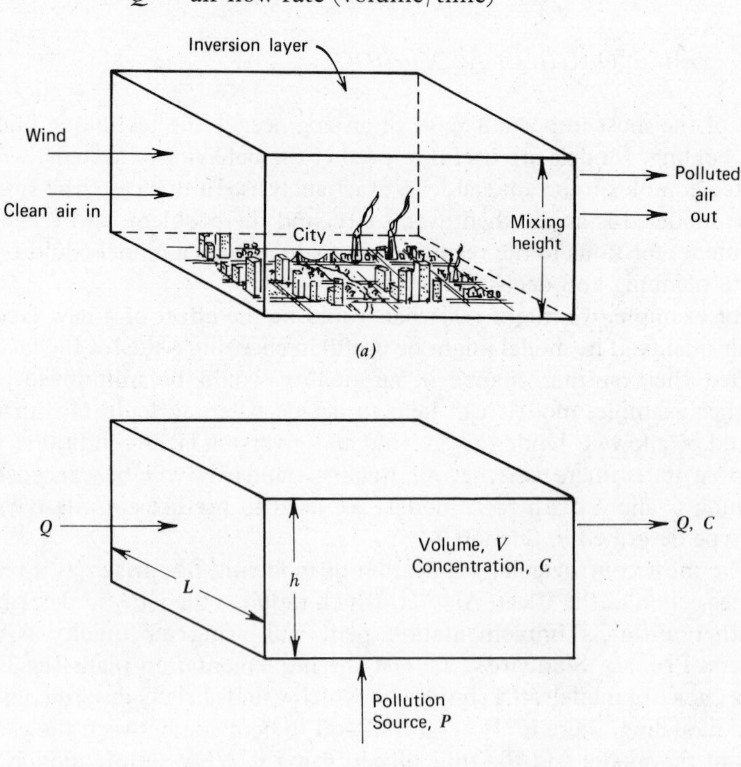

FIGURE 9.18 The box model for relating emissions to air quality.

Then we can write a simple mass balance equation:

$$\text{rate of change of pollution in box} = \text{rate of pollution entering the box} - \text{rate of pollution leaving box}$$

or,

$$V\frac{dC}{dt} = P - QC \tag{9-6}$$

If the initial concentration of pollution in the box is zero, then equation (9-6) has solution:

$$C = \frac{P}{Q}(1 - e^{-(Q/V)t}) \tag{9-7}$$

This is plotted in Figure 9.19, and we can see that the pollution concentration in the box climbs asymptotically to a final value of P/Q.

The quantity Q is a measure of the amount of ventilation that the box receives and is equal to the wind speed times the cross-sectional area of the box against which it blows:

$$Q = vLh$$

where

v = wind speed

L = base dimension of box

h = mixing height

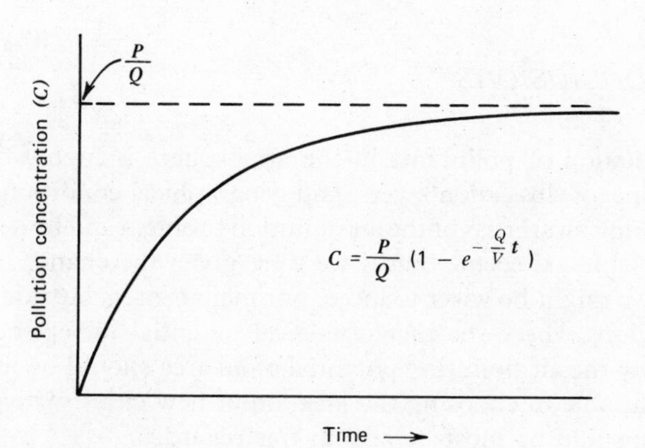

FIGURE 9.19 Increase in air pollution as predicted by the box model. The final value is proportional to emissions and inversely proportional to mixing height and wind speed.

The steady-state pollution concentration predicted by this model is therefore

$$C_{\text{final}} = \frac{P}{vLh} \tag{9-8}$$

This makes sense. The concentration increases if emissions increase, or if the inversion height or wind velocity decrease.

EXAMPLE 9.1 For a city having $L = 20$ miles, $h = 600$ feet, and $v = 5$ mph, and whose CO emissions equal 5000 tons per day, use the box model to calculate the steady-state CO concentration (assume none of the CO converts to CO_2).

$$v = 5 \text{ mph} = 8050 \text{ m/hr}$$

$$L = 20 \text{ miles} = 32{,}200 \text{ m}$$

$$h = 600 \text{ feet} = 183 \text{ m}$$

$$P = 5000 \text{ tons/day} = 190 \times 10^9 \text{ mg/hr}$$

$$C = \frac{P}{vLh} = \frac{190 \times 10^9}{8.05 \times 10^3 \times 3.22 \times 10^4 \times 183}$$

$$C = 4 \text{ mg/m}^3$$

The National Primary Standard for CO is 10 mg/m^3 averaged over an 8 hour period. In this example, if the wind speed were only 2 mph, the primary standard would be equalled.

9.10 *Conclusions*

The accumulation of pollutants in the atmosphere is highly dependent upon wind speeds, inversion layers, and geographical conditions. In view of the increasing awareness of the air pollution problem and its dependence on these variables, it seems that if we were given the chance to do it all over again, we might be wiser to locate our main centers outside of valleys and into regions where the meteorological potential for episode-days is less. Certainly the air pollution potential of an area should be included as a decision variable in choosing the location of new cities. The technique of modeling would be most helpful in this regard.

Our understanding of the earth and its systems is quite rudimentary in many ways. It is not possible to accurately predict the environmental effects of our activities. We may be bringing on an ice age or a melting of the ice caps, no one knows. We may, if man pursues the SST program, cause serious danger to all life on the planet—again, no one knows. The danger is that we will irrevocably and detrimentally alter our environment before we realize what we have done.

Bibliography

Bryson, R. A. (1971). All other factors being constant—theories of global climatic change. In T. Detwyler ed., *Man's impact on environment*. New York: McGraw-Hill.

Changnon, S. A., Jr. (1968). The La Porte weather anomaly—fact or fiction? *Bull. Am. Meteor. Soc.*, 49(1):4–11.

Holzworth, G. C. (1972). *Mixing heights, wind speeds, and potential for urban air pollution throughout the contiguous united states.* U.S. Environmental Protection Agency, AP-101, Research Triangle Park, N.C. Jan.

Johnston, H. (1971). Reduction of stratospheric ozone by nitrogen oxide catalysts from supersonic transport exhaust. *Science* 173(Aug.): 517.

Landsberg, H. (1962). City air—Better or worse. In *Air Over Cities Symposium,* U.S. Dept. Health Education & Welfare, Taft Sanitary Engineering Center, Cincinnati, Oh. TR A62-5.

Lovelock, J. E. (1971). Air pollution and climatic change. *Atmospheric Environment*. Vol. 5, June.

Lowry, W. P. (1967). The climate of cities. *Sci. Am.* Aug.

Manabe, S., and Strickler, R.F. (1964). Thermal equilibrium in the atmosphere with convective adjustment. *Jour. Atmos. Sci.* 21: 361–385.

Newell, R. E. (1971). The global circulation of atmospheric pollutants. *Sci. Am.* Jan.

Petterssen, S. (1969). *Introduction to meteorology*. 3rd ed. New York: McGraw-Hill.

Plass, G. N. (1959). Carbon dioxide and climate. *Sci. Am.* July.

Rice, R. (1972). System energy and future transportation. *Technol. Rev.* Jan.

Sauter, G. D. (1970). *Air Improvement Recommendations for the S.F. Bay Area.* Stanford-Ames Summer Workshop, Stanford, Ca.

Study of Critical Environmental Problems (SCEP) (1970). *Man's impact on the global environment*. Cambridge, Mass.: M.I.T. Press.

Study of Man's Impact on Climate (SMIC) (1971). *Inadvertent climate modification.* Cambridge, Mass.: M.I.T. Press.

The Supersonic Transport (1972). Hearings before the Subcommittee on Priorities and Economy in Government of the Joint Economic Committee, Washington, D.C. Dec. 27 and 28.

U.S. Department of Commerce (1972). *Environmental Aspects of the Supersonic Transport*. Report of the Panel on Supersonic Transport Environmental Research. Washington, D.C. May.

U.S. Department of Health, Education, and Welfare (1969). National Air Pollution Control Administration. *Air Quality Criteria for Particulate Matter*. AP-49. Washington, D.C. Jan.

Questions

1. Follow the procedure used in Section 9.2 to demonstrate that a super-adiabatic lapse rate creates an unstable atmosphere.
2. If the concentration of CO_2 in the atmosphere continues to increase at the rate of 0.2% per year, how many years would be required to double the concentration?
   ans. 350 years
3. Explain what is meant by the statement that NO_x acts as a catalyst in the conversion of ozone to molecular oxygen.
4. If the city in Example 9.1 is square, and starts with perfectly clear air, what would be the concentration of CO after one day under the conditions given in the example.
   ans. 3.6 mg/m^3
5. Using the box model, calculate the emissions of CO that would result in a steady-state value of CO equal to the primary standard of 10 mg/m^3. Assume a square city 35 miles on a side, with a wind speed equal to 5 mph, and a mixing height of 300 feet.
   ans. 11,000 tons/day
6. Using the box model, derive an expression for the pollutant concentration when the wind velocity is zero.
   ans. $C = (P/V)t$
7. Identify the following terms:

   (a) adiabatic lapse rate
   (b) stratosphere
   (c) fumigation
   (d) subsidence inversion
   (e) radiation inversion
   (f) greenhouse effect
   (g) heat island
   (h) Hadley cell

Chapter 10 Air Pollution Control Techniques

The quite different emission characteristics of mobile and stationary sources has led to the development of completely different control techniques for these two categories. The technology which is applicable to the control of motor vehicle emissions has been evolving rapidly in the last few years, as governmental regulations become more and more severe. If the requirements of the 1970 amendments to the Clean Air Act can be met, then we should be entering into a decade of increasing air quality in our cities. The gains are liable to be short lived, however, if the number of motor vehicles on the road continues to increase at its present rate.

Stationary sources present a different set of problems. Most of the control techniques in use today were developed more than 65 years ago to reduce particulate emissions. Significant technological advances still need to be made before we will be able to adequately control sulfur oxides and nitrogen oxides.

10.1 *Mobile Sources*

While ships, locomotives, and aircraft are also included in the mobile sources category, it is motor vehicles which are by far the most important in terms of total emissions. There were approximately 100 million gasoline-powered autos and light-duty trucks in use in the United States in 1973, and that number is growing by about 3.4% per year (EPA 1972b). If this rate were to continue for another 20 years, there would be 200 million light-duty vehicles in use. Recall that the three major pollutants emitted by motor vehicles are: carbon monoxide (CO), nitrogen oxides (NO_x), and hydrocarbons (HC). Motor vehicles accounted for 66 % of the 1970 emissions of CO; 48 % of the HC emissions; and 40 % of the NO_x.

In a vehicle which has no emission control equipment, essentially all the CO and NO_x are emitted from the tailpipe while the hydrocarbons are emitted partly from the exhaust, partly from crankcase blowby (gases which slip past the piston rings during the compression and power strokes of the engine cycle), and partly from evaporation, as shown in Figure 10.1.

The actual quantities of emissions from these various sources are highly dependent on the particular driving conditions encountered. For example, CO and hydrocarbon emissions *decrease* with increasing driving speed, while NO_x emissions remain relatively unaffected. While an engine is just idling, hydrocarbon and CO emissions are high but NO_x emissions are low. Table 10.1 qualitatively summarizes vehicle emissions under various conditions, and indicates the advantage to be gained by good urban traffic flow design which minimizes stop-and-go driving.

10.2 *Emission Standards*

The history of auto emission controls began in California in 1959 with the adoption of standards to control exhaust hydrocarbons and carbon

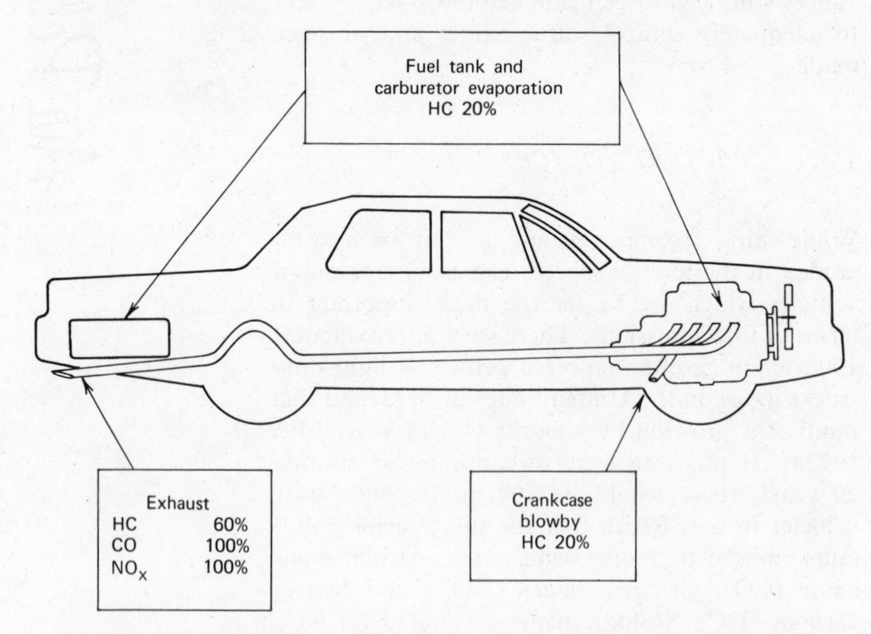

FIGURE 10.1 Approximate distribution of emissions by source for a vehicle with no emission control systems (HEW 1970a).

TABLE 10.1 Effect of Vehicle Mode on Emissions[a]

Condition		Exhaust				Blowby[b] flow[c]	Fuel system[d] flow,[c] HC	
			Concentration					
Vehicle	Engine	Flow	HC	CO	NO_x		Tank	Carburetor
Idle	Operating	Very low	High	High	Very low	Low	Average to Moderate	Moderate
Cruise								
Low speed		Low	Low	Low	Low	Moderate		Small
High speed		High	Very low	Very low	Moderate	High		Nil
Acceleration								
Moderate		High	Low	Low	High	Moderate		Nil
Heavy		Very high	Moderate	High	Moderate	Very high		Nil
Deceleration		Very low	Very high	High	Very low	Very low		Moderate
Soak								
Hot	Stopped	None	—	—	—	None	High	High
Diurnal		None	—	—	—	None	Moderate	Very low

[a]Source: HEW 1970a, after Brehob.
[b]Concentration of HC is high, CO low, and NO_x very low.
[c]Flows are at least one order of magnitude lower than the exhaust flow.
[d]For a vehicle not equipped with an evaporative emission control system.

monoxide, but these standards were not implemented until they became "technologically feasible" in 1966. In 1960, standards to control emissions from crankcase blowby were added but implementation was not required until 1963. Federal emission standards first became effective on 1968 models. Controls were also required on evaporative emissions from fuel tanks and carburetors, and by 1970 industry had reduced hydrocarbon emissions from new vehicles by almost three-fourths and CO emissions by about two-thirds. Unfortunately, as we shall see in the next section, these improvements in HC and CO emissions were partly made at the expense of increased NO_x emissions. Standards for NO_x were not required until 1971 in California, and not until 1973 for the rest of the country.

The Clean Air Act was written to require that by 1975 emissions of hydrocarbons and carbon monoxide from new vehicles must be 90% less than the emissions allowed in 1970. Similarly, by 1976, the NO_x emissions must be 90% less than the average of vehicles manufactured in 1971. The law allows a 1 year delay in these deadlines if it is adequately proven that the technology is not available to meet them. That delay has been granted by the Environmental Protection Agency.

One provision of the Clean Air Act allows states to use their own emission standards if they were set prior to March 30, 1966, and if they are more stringent than Federal standards. California is the only state that qualifies under this section. Table 10.2 lists the California and Federal emission standards for new, light-duty vehicles, prior to the granting of any delays. Some explanations are necessary to fully understand the table. The first standards were often written with emissions limited by pollutant *concentrations* in the exhaust (either as parts per million or as percent). Unless these concentrations are multiplied by the volume of exhaust gas to obtain total emissions, big engines would be allowed to pour more pollutants into the atmosphere than smaller ones. Later standards do not grant this advantage to large-displacement engines by expressing emissions as grams per mile. But, as Table 10.1 indicates, emissions are very dependent upon driving conditions, so it has been necessary to define standardized driving cycles. California at first used a seven-mode cycle as shown in Figure 10.2. This cycle is performed in sequence seven times, on a dynamometer.

The test which has superseded the seven-mode cycle in both California and Federal standards is the Constant Volume Sample or CVS procedure. This test simulates urban driving and is based on an elaborate study of Los Angeles traffic patterns. The test cycle is based on a typical drive of 7.5 miles at an average speed of 19.7 mph, taking 22.8 minutes and including 17 stops. The first standards require that this test be run once,

TABLE 10.2 New-Vehicle Standards, California and Federal (Under 6000 Pounds)[a,b]

Year	Standard	Cold-start test	Hydro-carbons	Carbon monoxide	Oxides of nitrogen
Prior to controls			850 ppm (11 g/mi)	3.4% (80 g/mi)	1000 ppm (4 g/mi)
1966–1967	State	7-mode	275 ppm	1.5%	no std.
1968–1969	State and Federal	7-mode			
		50–100 CID	410 ppm	2.3%	no std.
		101–140 CID	350 ppm	2.0%	no std.
		over 140 CID	275 ppm	1.5%	no std.
1970	State and Federal	7-mode	2.2 g/mi	23 g/mi	no std.
1971	State	7-mode	2.2 g/mi	23 g/mi	4 g/mi
	Federal	7-mode	2.2 g/mi	23 g/mi	—
1972	State	7-mode	1.5 g/mi	23 g/mi	3 g/mi
		or			
		CVS-1	3.2 g/mi	39 g/mi	3.2 g/mi[c]
	Federal	CVS-1	3.4 g/mi	39 g/mi	—
1973	State	CVS-1	3.2 g/mi	39 g/mi	3 g/mi
	Federal	CVS-1	3.4 g/mi	39 g/mi	3 g/mi
1974	State	CVS-1	3.2 g/mi	39 g/mi	2 g/mi
	Federal	CVS-1	3.4 g/mi	39 g/mi	3 g/mi
1975[d]	State	CVS-1	1 g/mi	24 g/mi	1.5 g/mi
	Federal	CVS-2	0.41 g/mi	3.4 g/mi	3 g/mi
1976[d]	State	CVS-1	1 g/mi	24 g/mi	1.5 g/mi
	Federal	CVS-2	0.41 g/mi	3.4 g/mi	0.4 g/mi

[a]Source: BAACPD (1972).

[b]Note that ppm is parts per million concentration; g/mi is grams per mile; 7-mode is a 137 second driving cycle test; CVS-1 is a Constant Volume Sample cold-start test; CVS-2 is a Constant Volume Sample cold-start test average with a Constant Volume Sample hot-start test, both with the Federal 22 minute driving cycle; the values in parentheses are approximately equivalent values; CID is cubic inch displacement.

[c]Hot seven-mode.

[d]One year delay on Federal standards has been granted. Interim standards for 1975 have been set as follows (in g/mi):

	U.S.	*Calif.*
HC	1.5	0.9
CO	15.0	9.0
NO_x	3.1	2.0

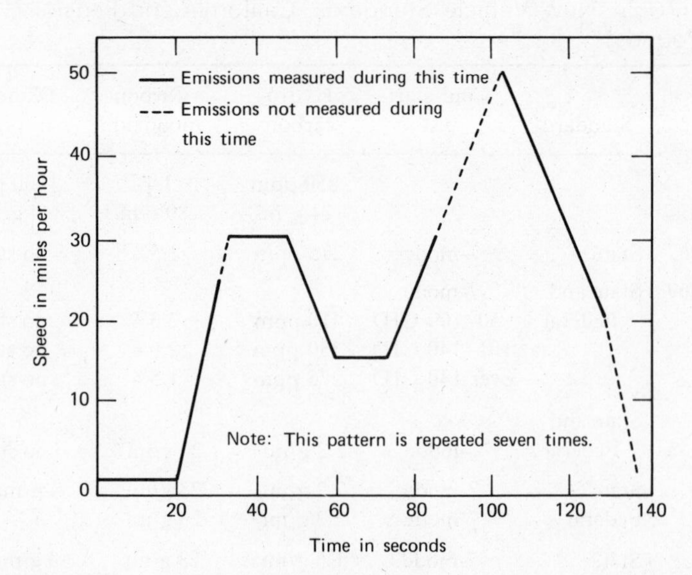

FIGURE 10.2 Driving cycle followed in seven-mode seven-cycle test. (General Accounting Office 1972).

from a cold start (CVS-1). The later standards are run twice, once with a cold start and once with a hot start, and a weighted average is taken (CVS-2). It is very important to state which test cycle is being used whenever standards or test results are being quoted. Unfortunately, this is not always done, making it impossible to compare different control schemes in many cases.

One of the difficulties encountered in trying to improve air quality by setting standards for new vehicles is that it takes a number of years before most cars on the road will be equipped with those controls. Figure 10.3 is a motor vehicle age distribution plot which indicates, for example, that it takes about five years after introduction of new controls for half the vehicles on the road to be so equipped.

Another difficulty has to do with the testing of new vehicles to show that they meet the emission standards. The Environmental Protection Agency has been criticized in the past for allowing a woefully inadequate certification procedure (Esposito 1970, General Accounting Office [GAO] 1972). Prior to 1972, testing was done by the manufacturers on a few hand-picked prototypes (selected by the manufacturers). Prototypes that did not meet the standards could still be certified if the average of all the prototypes were within the standards. Further, a deterioration factor was included in the averaging procedure which was supposed to account for the decrease in control efficiency with age and mileage. As actually used though the factor implied the control performance *increased* with age rather than decreasing. In 1971, 10.6 million cars were certified on the basis of tests of 242 prototypes (GAO 1972).

Procedures are changing, and the EPA hopes to begin assembly line testing after 1974. California has devised its own, simplified test which is now required of all new cars sold.

10.3 *Minor Modifications to the Internal Combustion Engine*

We can divide the discussion of the technological approaches to auto emission control into four categories: (a) minor modifications to internal combustion engines, (b) devices which are added onto the exhaust system, (c) changes in fuels, and (d) unconventional engines. In this section we will treat the minor modifications.

The HC, CO, and NO_x emissions are very dependent on the fuel-air mixture. The more air there is (leaner mixture) the more complete will be the combustion, and hence CO and HC emissions decrease. However, the improved combustion results in higher temperatures, so emissions of NO_x increase as the mixture becomes leaner. As long as standards did not

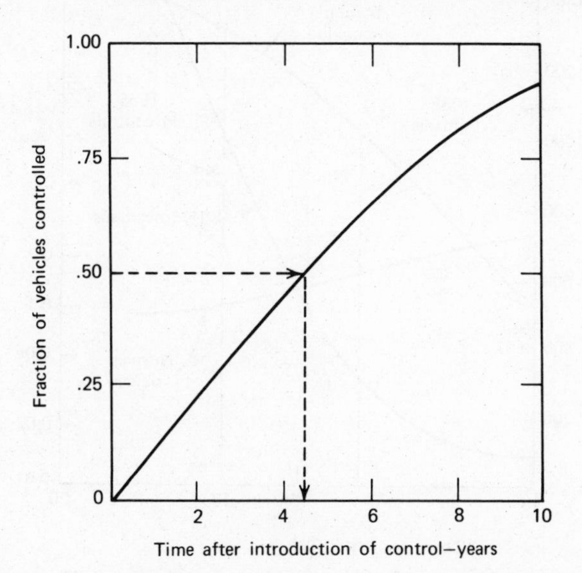

FIGURE 10.3 Motor vehicle population distribution, showing the fraction of vehicles with controls as a function of the time after introduction of the controls (Starkman 1971).

require control of NO_x, the control of HC and CO was rather simple, requiring only that the engine be modified to run smoothly on a leaner mixture. Figure 10.4 shows the variations in HC, CO, and NO_x emissions with mixture strength and the effects of the 1968 controls which reduced CO and HC but resulted in increased NO_x emissions. The zero reading on the fuel-air axis corresponds to the stoichiometric requirements for complete oxidation of the fuel (about 15 pounds of air for each pound of fuel).

Figure 10.5 indicates that for about a 10 year period, nitrogen oxide emissions in Los Angeles, from motor vehicles, will be greater than they would have been had there been no emission control program at all. Thereafter, total emissions are predicted to quickly decrease as new cars with good emission control systems replace the older more poorly controlled cars. However, it should be noted that by about 1985, if no further controls are required beyond the 1975–1976 standards, total emissions may begin to rise again due to the ever increasing number of vehicles which are predicted to be on the road. Hopefully, within the next 10 or 15 years

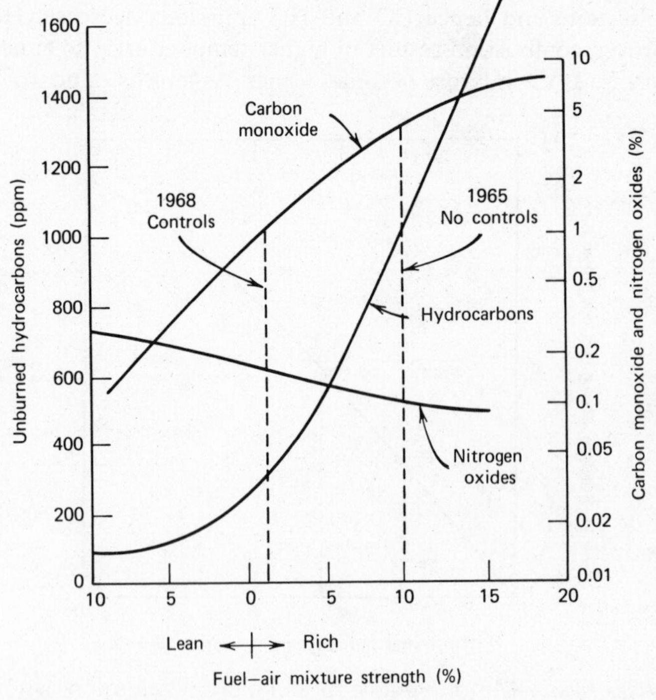

FIGURE 10.4 The effect of fuel-air mixture strength on auto emissions, showing the result of 1968 controls (Starkman 1971).

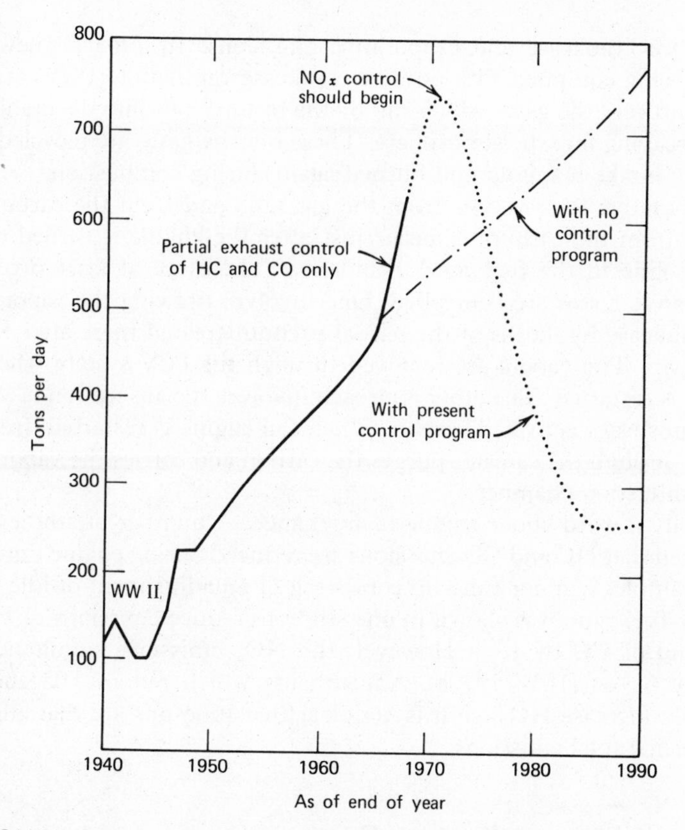

FIGURE 10.5 NO_x emissions from motor vehicles in Los Angeles, showing increase due to 1968 limited control program (CARB 1969).

enough mass transit systems can be built to cause an end to the growth in personal motor vehicle use.

There are other engine design parameters which can be varied to help control emissions. Retarding the spark reduces NO_x emissions and both NO_x and HC are reduced by decreasing the compression ratio. Unfortunately, retarding the spark and reducing the compression ratio results in some loss of power and fuel economy. On the positive side, lowering the compression ratio lowers the octane requirement of the fuel which will make it easier to phase the lead out of gasoline. Starting with 1973 vehicles, automobiles must be designed to run on 91 octane fuel—a requirement which is resulting in newer cars having lower compression ratios.

In addition to changing some of the internal combustion engine design parameters to help control exhaust emissions, it has also been necessary to reduce crankcase and evaporative emissions. In 1963 all new cars began being equipped with positive crankcase ventilation (PCV) systems which prevent the gases which slip by the piston rings into the crankcase, from escaping into the atmosphere. These blowby gases are recycled back into the intake manifold and burned again during combustion.

Evaporative losses occur from the gas tank and from the carburetor. Losses from the carburetor occur just after the engine is turned off, as the gasoline in the fuel bowl evaporates. There are at least two ways evaporative losses are controlled. One involves drawing the vapors into the crankcase by means of the partial vacuum created there after engine shutdown. The vapors are removed through the PCV system when the engine is restarted. The other approach involves the absorption of vapors in a canister of activated carbon. When the engine is restarted, fresh air drawn through the canister purges the carbon and carries the vapors into the combustion chamber.

Finally a word about engine maintenance. A number of studies have indicated that HC and CO emissions are reduced during engine tune-ups. For example, a minor tune-up consisting of an adjustment of idle speed and air-fuel ratio was shown in one study to reduce emissions of HC by 10% and of CO by 15%. However, the NO_x emissions simultaneously rose by 6.3% (HEW 1970a). Adjustments which reduce HC and CO typically increase NO_x, so it is not clear that tune-ups are that effective in reducing total emissions.

10.4 *Exhaust System Devices*

Besides minor engine modifications, there are devices which can be added to the exhaust system to help control emissions, and it appears that this is the way the auto manufacturers will try to meet the 1975–1976 standards.

One approach is to burn a rich mixture in the cylinders which gives favorable NO_x control and then add a *thermal reactor* onto the exhaust system to oxidize the HC and CO. The thermal reactor functions as a combustion chamber outside the engine, keeping the exhaust gases hot for a long enough time to more completely oxidize the CO and HC. Physically the device appears as an oversized exhaust manifold with an outside-air injection system. Another approach to exhaust-gas neutralization is the *catalytic converter* in which CO and HC are oxidized into harmless water vapor and carbon dioxide. A second catalyst can be used for reducing nitrogen oxides. The catalysts being considered are platinum or other transition metals. The device operates at a reasonably low temperature and can be removed from the engine compartment and placed in a more convenient location. One problem with catalytic converters is that lead from gasoline tends to coat the catalyst rendering it ineffective and this is one of the primary reasons for the attention being given to eliminating lead as a gasoline additive. The combination of a catalytic converter and a thermal reactor appears to be potentially an effective system for controlling vehicle emissions.

The third approach to be mentioned here is called *exhaust gas recirculation* in which a portion (up to about 25%) of the exhaust gases are returned to the intake manifold. The addition of this relatively inert gas reduces combustion temperatures without affecting the air-fuel ratio. Hence, NO_x emissions can be reduced without increasing HC and CO emissions. Such devices have been shown to reduce NO_x by as much as 80%.

Figure 10.6 shows a combination of some of these emission control techniques, in a system that might be considered somewhat typical of the present approach by American manufacturers for meeting the 1975 standards.

10.5 *Fuel Changes*

One approach to reducing motor vehicle emissions is to change the fuel which is burned. Some effort is being directed toward modifying present fuels and some toward substitution with new fuels.

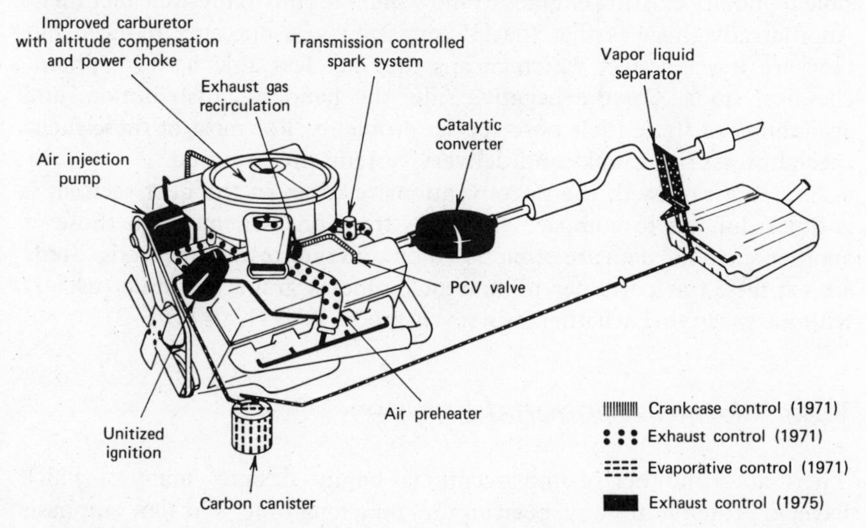

FIGURE 10.6 Typical approach to a 1975 emission control system assuming lead-free fuel (GAO 1972).

Present fuels may be modified to reduce their volatility and reactivity, but such adjustments must be made on an area-to-area basis because of differing climatic conditions and driving patterns. Reducing volatility, for example, can lead to difficult starting and warm-up in colder climates. One fuel modification which seems likely across the country is the gradual elimination of lead, for the following reasons: (a) the adverse health effects of lead, (b) its incompatibility with catalytic converters, (c) evidence which indicates that the accumulation of lead deposits in the cylinders leads to increased HC emissions. There is also evidence that with unleaded fuel such devices as exhaust systems and spark plugs will have longer lives, and engine oil change intervals could be extended, and that tune-ups would be needed less frequently. It will be necessary, though, to simultaneously lower the octane requirement of vehicles while finding a suitable technique for raising the octane of nonleaded fuels. Nonleaded fuels exist now but not in sufficient supply to meet the demand should a complete conversion be immediately required.

There is considerable interest being shown in more radical fuel changes. Changeovers to new fuels such as liquefied petroleum gas, LPG (primarily propane and butane); liquefied natural gas, LNG (mostly methane); compressed natural gas, CNG; or methanol (CH_3OH); would significantly reduce emissions. One advantage to using these fuels as the approach taken to meet more strict emission standards is that the basic internal combustion engine remains nearly unchanged. In fact, it is possible to modify existing engines to allow them to run on these cleaner fuels. Another advantage is that the HC emissions are not only reduced but they are less reactive, which means they are less able to form photochemical smog. On the negative side, the handling, distribution, and availability of these fuels pose special problems. For most of these fuels, special pressurized tanks and delivery systems are required.

As is the case with the unconventional engines in the next section, it is quite difficult to compare emissions from one scheme with those of another since the data are often based on different testing methods. Some are expressed as parts per million and some as grams per mile (usually without specifying whether it is seven-mode, CVS-1, or CVS-2).

10.6 *Unconventional Engines*

There are a number of unconventional engine designs, many of which have, in principle anyway, been around for a long time. The new emphasis on control of emissions has revived interest in these engines and the preliminary data from them has in some cases been rather spectacular.

The *gas turbine* has been in the development stage since the end of World War II. In its simplest form it consists of a compressor, combustion chamber, and a turbine as shown in Figure 10.7. Outside air is compressed and delivered under pressure to the combustion chamber where fuel is burned which causes the gas to expand through the turbine. Part of the turbine's output runs the compressor and part the vehicle. Emissions of CO and HC are quite low but there are still some problems with controlling NO_x. It appears these engines would be more costly and would probably be higher in fuel consumption that the internal combustion engine (ICE).

The *Wankel* is a rotary combustion chamber engine that is, at present, the most likely successor to the ICE. It has comparatively few moving parts, high power-to-weight and power-to-volume ratios, and is extremely smooth. The operation of the Wankel is illustrated in Figure 10.8. Emissions of NO_x and CO are relatively low but HC output is rather high unless something like an exhaust reactor is added, in which case the combination yields very good results.

The *steam engine,* which dates back to at least 1827, is an external combustion engine in which the water is heated to create steam which is allowed to expand through a turbine or piston system, thereby doing work. The steam is then condensed and the cycle begins again. Some engines operating in this cycle, called the Rankine cycle, use organic working fluids rather than steam. Extremely low emissions have been obtained.

Another old idea which is receiving a lot of attention now is the battery-operated *electric vehicle,* which dates back to the nineteenth century. The major advantage would be that the vehicle itself would be free of emis-

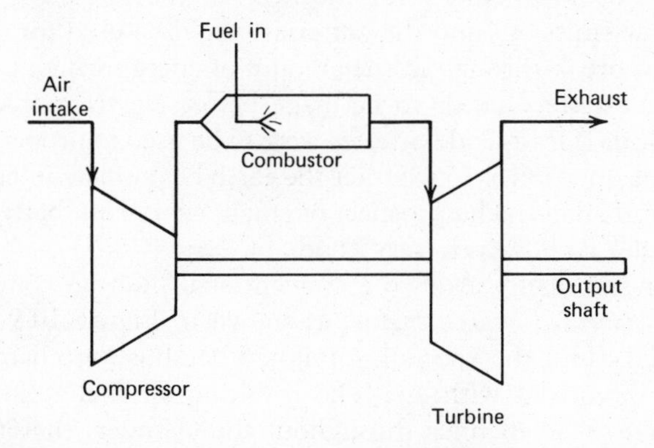

FIGURE 10.7 Schematic diagram of a simple gas-turbine engine (HEW 1970a).

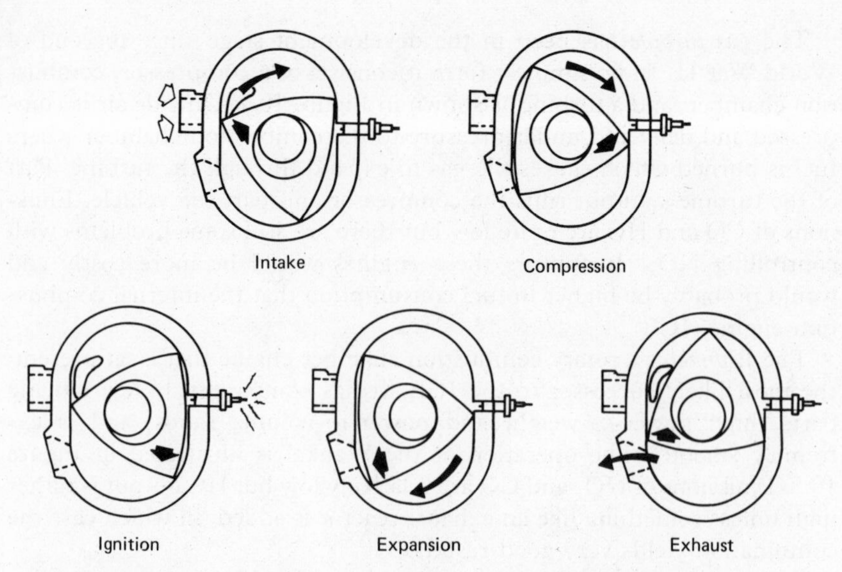

FIGURE 10.8 Sequence of Wankel rotary engine cycle events (HEW 1970a).

sions, although the energy to charge the batteries comes from a power plant somewhere which does cause pollution. The power plant's pollution is usually outside of the city anyway, so there may be some advantage. Also it is easier to control pollution from one large source than from thousands of little ones. The problem of course with battery vehicles is to obtain sufficient speed and range between recharges. A *hybrid* vehicle employing both a small heat engine and a battery has some advantages. The system could operate with the heat engine operating at constant speed for low emissions and the batteries would be used for acceleration. A significant break-through in the amount of energy which can be stored per pound of battery needs to be made before electric vehicles can become practical. If lead-acid batteries were to be used to propel all vehicles, there is some question as to whether the earth has sufficient lead resources to meet that demand. The prospect of small, short-run, battery-operated urban vehicles is, however, very good.

A less severe change from the conventional internal combustion engine is the *stratified-charge engine,* as shown in Figure 10.9. Fuel is injected directly into the specially designed combustion chambers rather than being premixed with air. The resulting air-fuel ratio varies in a carefully controlled manner throughout the cylinder, thereby reducing pollutant emissions. There is a good chance that this engine can meet the

1975–76 standards if it is coupled with the proper exhaust control devices. Also, in view of its good fuel economy and its similarity to conventional internal combustion engines, there is a very good chance, that the stratified-charge engine could become popular.

It remains to be seen whether any of these schemes will find wide acceptance. They all hold the potential to significantly reduce automobile emissions and hence cause an overall improvement in air quality. There is some concern that the Clean Air Act may not have allowed enough time for the more exotic schemes to be developed before the tougher standards come into effect. Very likely that is true, which will result in a couple of years of expensive, heavy, fuel-hungry vehicles using conventional internal combustion engines with lots of complicated add-on devices. However, the marketing advantage to be gained by the first manu-

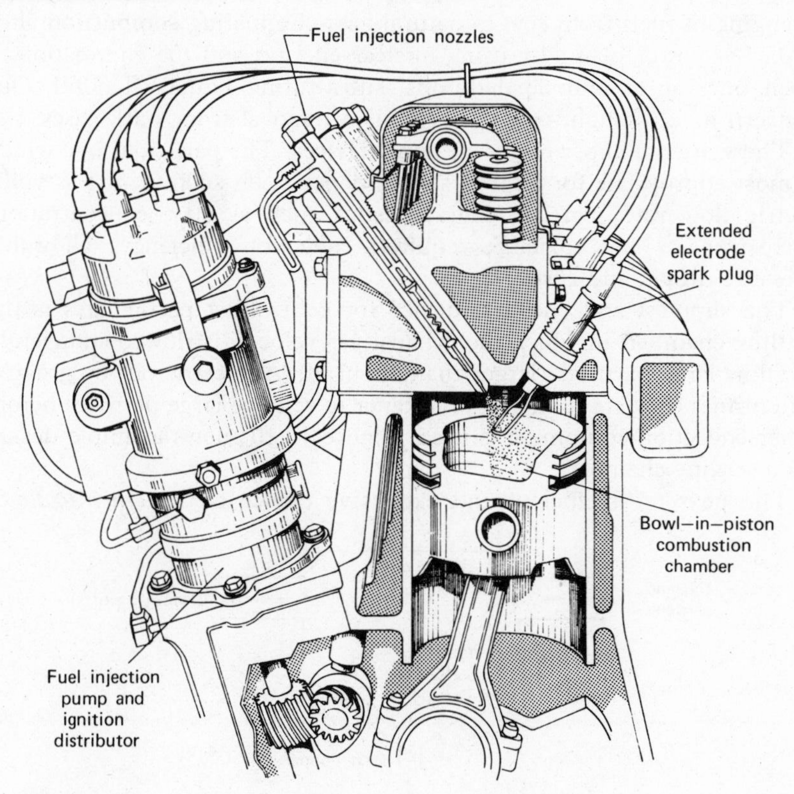

FIGURE 10.9 Stratified-charge internal combustion engine (GAO 1972).

facturer to perfect an efficient, low-pollution vehicle will surely spur its development. In the meantime, the buying public may learn to appreciate the advantage of small, low-powered, economical vehicles, or better still, may begin to demand adequate mass-transit.

10.7 *Control of Particulates from Stationary Sources*

About 97% of all particulate emissions are released by stationary sources. The two general approaches to reducing these emissions are: (a) changing the characteristics of the source so that fewer particulates are formed; or, (b) utilizing control equipment which removes particulates before they are released into the atmosphere. Examples of the first approach include the changing of fuels from coal to natural gas; eliminating combustion altogether at power plants by using nuclear energy; and the elimination of open burning at municipal dumps, substituting sanitary landfill. Our concern here, though, is with particulate removal from stack gases.

There are a number of gas cleaning devices. The particular one which is most appropriate for a given source depends on such factors as volumetric flow rate; particle characteristics such as size, concentration, corrosiveness, and toxicity; required collection efficiency; allowable pressure drops; and costs.

The simplest and cheapest device for controlling particulates is the settling chamber, in which the exhaust gas velocity is slowed sufficiently to allow large particles (larger than about 40 μ) to settle out by gravity. Oftentimes they are used as a precleaner to remove large particles before other collection equipment is used. Figure 10.10 shows a simple design for a settling chamber.

The next most efficient and expensive device is the *centrifugal*, or

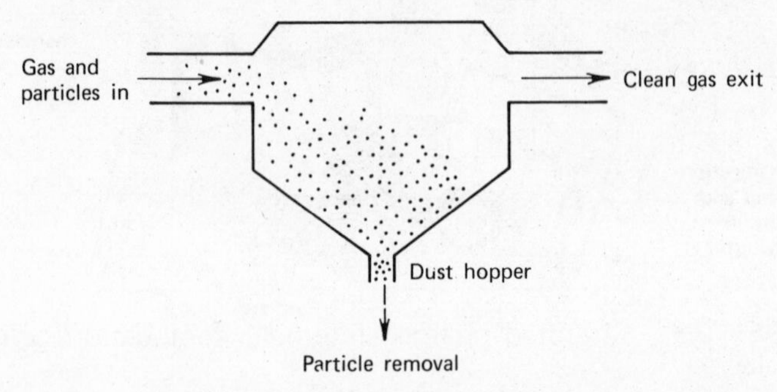

FIGURE 10.10 Settling chamber.

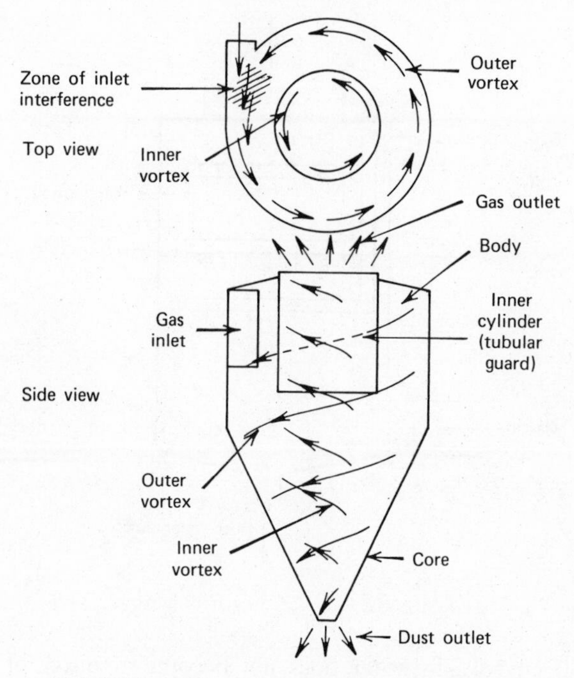

FIGURE 10.11 Dry centrifugal collector.

cyclone, separator. As shown in Figure 10.11, the gas is spun in the separator forcing the particles against the outer wall by centrifugal force. The particles then drop out of the bottom of the device while the cleaned gas exits at the top. Well designed, large diameter cyclones have collection efficiencies above 95% for particles larger than about 40 μ.* The efficiency drops to less than 50% for particles smaller than about 8 μ.

Scrubbers are collection devices in which the particles are washed out of the gas flow by a water spray. There are a number of different scrubber designs, one of the simplest being simply a tower in which the dirty gases rise from the bottom while water is sprayed down from a number of nozzles, as shown in Figure 10.12. Such towers are usually limited to the collection of particles larger than about 10 μ. Collection efficiency can be improved to about 96% for 2 to 3 μ particles by spinning the gas as in a cyclone separator, to increase the relative velocity between water droplets and the gas stream. The exiting gases are satured with water which results in a white steam plume. Special care must be exercised to ensure

*Percent collection efficiency = (weight of dust-in − weight of dust-out)/weight of dust-in × 100.

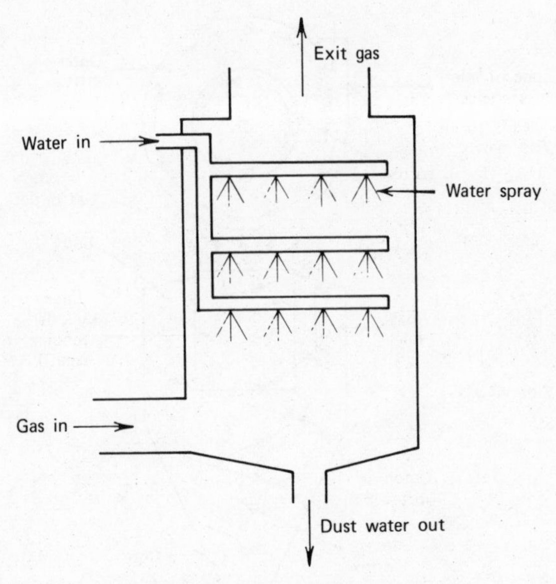

FIGURE 10.12 Spray tower.

that the collected waste water does not become a source of water pollution. This may require special settling tanks, chemical flocculation, or filtration units.

Electrostatic precipitators remove particles from the gas stream by causing them to pick up an electrical charge as they pass through a high-voltage direct-current corona. The charged particles then move under the influence of the electric field to a grounded collecting surface. They are removed from the collection electrode by gravity forces, by rapping, or by flushing with liquids. A common, tubular configuration of electrodes is shown in Figure 10.13. A high negative voltage is applied to the center electrode while the cylindrical, outer, collecting electrode is grounded. The principal advantage of precipitators is their high collection efficiency for small (even submicron) particles. They are the most common particulate control device used at large installations.

Fabric filter baghouses approach 100% efficiency for removal of particles as small as 1 μ and will even remove a substantial quantity of particles as small as 0.01 μ. As shown in Figure 10.14, the dirty gas is passed through a fabric filter bag which is supended upside down in the baghouse. A baghouse may contain several thousand of these bags distributed among several compartments. This allows individual compartments to be cleaned while the others remain in operation. Baghouses are quite expensive and the bags must be periodically cleaned and maintained.

Of primary importance in choosing the proper control device is the degree of emission control desired. Table 10.3 summarizes some average collection efficiencies of control equipment for various particulate sizes.

For control of large particles, inexpensive mechanical devices—settling chambers or cyclone separators—may be sufficient. However, to collect smaller particles, which are the most damaging to health, an electrostatic precipitator or baghouse may be called for.

Annualized costs of collection equipment are presented in Figure 10.15 as a function of the volume rate of gas to be processed. The annualized cost calculation is based on a 15 year depreciation of capital costs and annual capital charges (taxes, interest, insurance, etc.) equal to $6\frac{2}{3}\%$

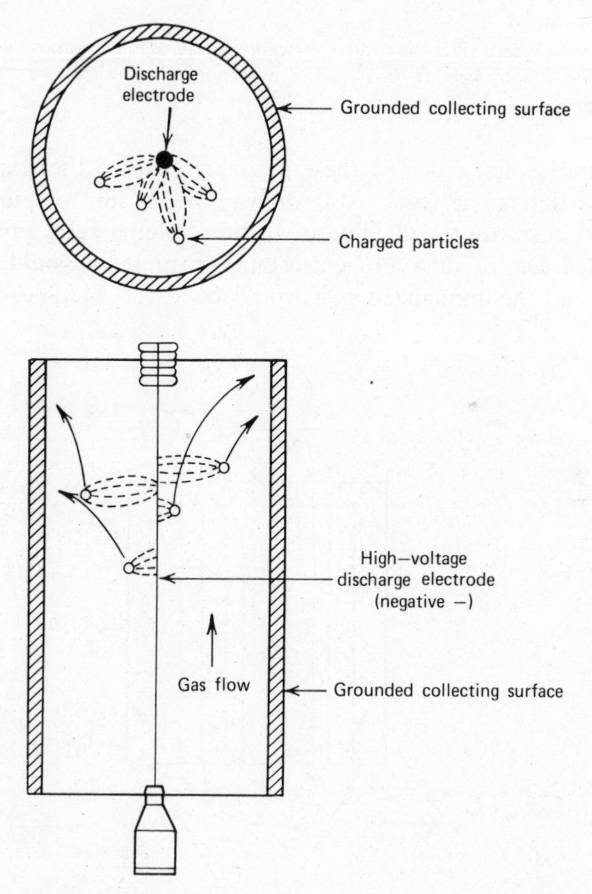

FIGURE 10.13 Schematic view of tubular electrostatic precipitator (HEW 1969a).

TABLE 10.3 Average Collection Efficiencies of Control Equipment for Various Particulate Sizes[a,b]

Type of collector	Overall	Particulate size range, μ 0–5	5–10	10–20	20–44	> 44
Baffled settling chamber	58.6	7.5	22	43	80	90
Long-cone cyclone	84.2	40	79	92	95	97
Spray tower	94.5	90	98	98	100	100
Electrostatic precipitator	99.0	97	99	99.5	100	100
Baghouse	99.7	99.5	100	100	100	100

[a]Source: EPA 1972a.
[b]Data based on standard silica dust with following particle size distribution: 0–5 μ, 20% by weight; 5–10 μ, 10% by weight; 10–20 μ, 15% by weight; 20–44 μ, 20% by weight; > 44 μ, 35% by weight.

of the initial capital cost. To these figures are added the annual operation and maintenance costs. Also shown are the sum of purchase and installation costs for a 200,000 cu. ft./min volume rate. For example, the installed cost of such an electrostatic precipitator would be around $300,000 and the annualized cost would be about $47,000.

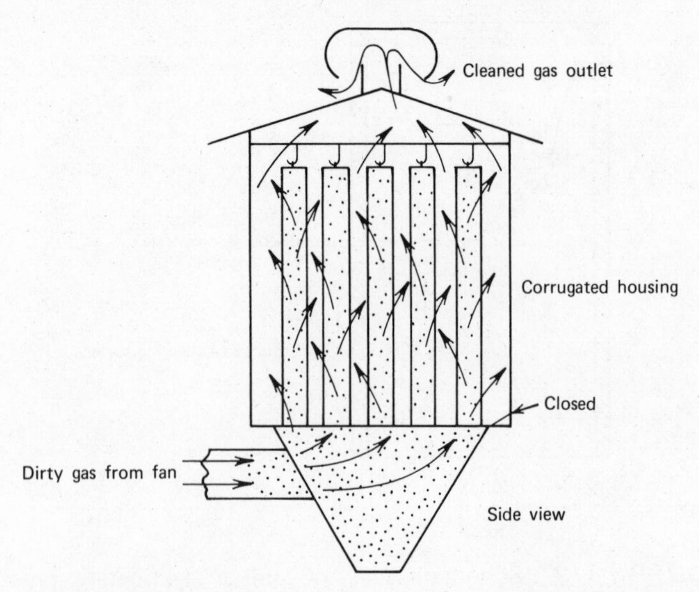

FIGURE 10.14 Closed pressure baghouse (HEW 1969a).

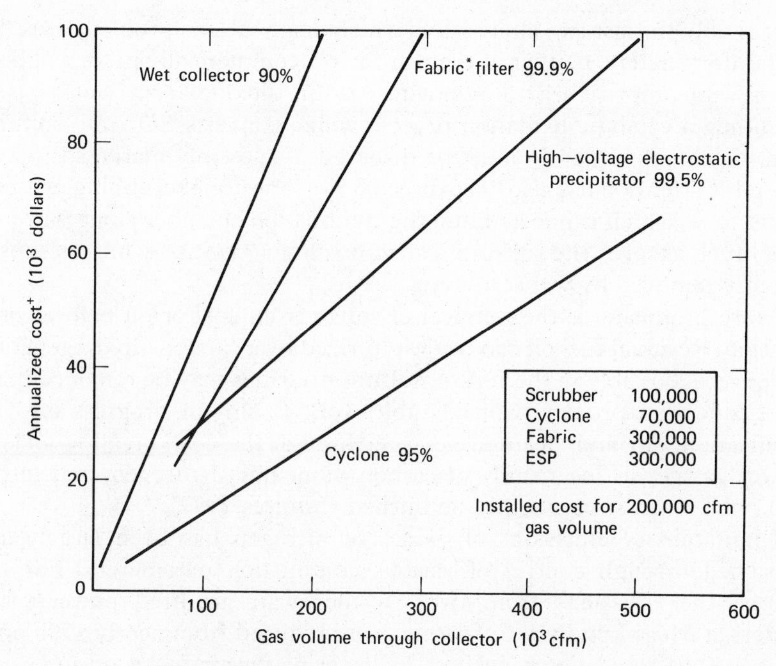

* High–temperature synthetic woven fibers, cleaned continuously and automatically.

\+ Annualized cost = 0.133 × total investment cost + annual operating and maintenance costs.

FIGURE 10.15 Annualized costs to purchase, install, and operate various gas cleaning devices (1968 prices) (HEW 1969a).

10.8 *Sulfur and Nitrogen Oxides Control*

Control of emissions of sulfur oxides is considerably more difficult than controlling particulates. As the following list of control alternatives indicates, many reduction techniques simply circumvent the problem.

- Switch to low-sulfur content fuels, especially during high-smog periods.
- Use other sources of energy such as nuclear or hydroelectric.
- Raise smokestacks so exhaust gases are diluted more before reaching ground level.
- Reduce fuel use by increasing combustion efficiency, lowering energy demands, limiting population.
- Remove sulfur from the stack gases after the fuel is burned.
- Remove sulfur from the fuel before it is burned.

It is only the last two alternatives which deal with the problem directly, and unfortunately, neither of them are as yet commercially proven. There are several approaches to removing sulfur dioxide from stack gases, including a catalytic oxidation process which converts SO_2 into sulfuric acid which subsequently must be disposed of (possibly marketed). Considerable attention has also been directed to a limestone scrubbing process. Limestone, which is injected into the combustion chamber along with the fossil fuel, absorbs the sulfur as calcium sulfate ($CaSO_4$) which is subsequently removed by wet scrubbing.

More promising is the removal of sulfur from coal or oil before combustion. Residual fuel oil can be desulfurized by a catalytic hydrogenation process or, possibly in the future, sulfur impurities may be removed in an oil gasification process. Considerable work is also in progress on coal gasification. Air and steam are driven through the coal, creating a clean power gas consisting mainly of carbon monoxide, hydrogen, and nitrogen, which can subsequently be burned (Squires 1972).

Limitation of emissions of oxides of nitrogen has been largely approached through control of various combustion parameters. For example, the emissions from two 750-megawatt gas-fired power plant boilers at Moss Landing, California, were reduced from nearly 1500 ppm to about 150 ppm, at nominal cost, by lowering the air-fuel ratio and using a two-stage combustion process that retains the gases for a longer time but at a lower temperature.

There are also schemes, with varied commercial success, for removing NO_x from stack gases. Some are public relations devices which merely convert the colored NO_2 gases into colorless NO, Of course, the NO may subsequently be converted back into NO_2 by the atmospheric photochemical reaction. Some schemes, though, convert NO_x gases into harmless elemental nitrogen, but there are no commercially proven processes.

10.9 *Air Pollution Costs*

The Environmental Protection Agency is required by the Clean Air Amendments of 1970 to annually submit estimates of the costs that will be incurred in carrying out the provisions of the Act. Estimates of the cumulative expenditures that would be required during the 10 year period 1971–1980, to meet existing standards, are presented in Table 10.4. As can be seen, the bulk of the expenditures must be made by the private sector, and most of that is for automobile controls. About half of the $61 billion which must be spent on automobiles is for the added cost of control equipment and the other half covers the increases in fuel con-

TABLE 10.4 Cumulative Air Pollution Control Expenditures Required to Meet Existing Standards, 1971–1980 (in Billions of 1971 Dollars)[a]

Sector	Capital investment	Operating costs	Cash flow
Public	1.1	6.8	7.9
Private			
Automobiles	31.5	29.5	61.0
Stationary	15.6	22.0	37.6
Total	48.2	58.3	106.5

[a]Source: CEQ 1972.

sumption and maintenance that will result from meeting the standards. Not included are expenditures that could be made for mass transit systems.

The Environmental Protection Agency has also estimated the costs of pollution damages in the United States. Table 10.5 shows the estimated damages for the year 1968, both by source of pollution and type of effect. The damages are in terms of additional costs of health care and impairment of human resources, reduction in residential property values, degradation of materials, and damage to vegetation and agricultural productivity. Of the total $16.1 billion, almost $8 billion is caused by stationary combustion sources, $1.47 billion by transportation, and $3.54 billion by industrial processes.

TABLE 10.5 National Costs of Pollution Damage, by Source and Effect, 1968 (Billions of Dollars)[a]

Effects	Stationary source fuel combustion	Transportation	Industrial processes	Solid waste	Miscellaneous	Total
Residential property	2.802	0.156	1.248	0.104	0.884	5.200
Materials	1.853	1.093	0.808	0.143	0.855	4.752
Health	3.281	0.197	1.458	0.119	1.005	6.060
Vegetation	0.047	0.028	0.020	0.004	0.021	0.120
Total	7.983	1.474	3.534	0.370	2.765	16.132

[a]Source: Barrett and Waddell 1973.

10.10 Conclusions

There are several approaches which can be taken toward the control of air pollution. The first is to merely move the source of pollution away from the city into an area where there are fewer people to be affected. Sadly, this approach is becoming not uncommon. The most extreme example is the huge power-generating complex being built in the Four Corners area in the Southwest, which has already become the single largest source of air pollution in the nation. We are sacrificing not only the quality of the air in what used to be one of the most pristine areas of the country, but also the land and water. The land is being strip mined for coal, and the water is being removed for cooling, which as was pointed out in Chapter 4, results in higher salt concentrations in the remaining river flow.

The second approach to controlling air pollution is by means of emission control devices. While temporary gains are possible, eventually the situation degenerates into an expensive race between technology and the rising gross national product. The prime example is, of course, the automobile. It will not be easy to meet the Federal emission standards, and the resulting vehicles will very likely suffer from a decrease in mileage and performance, and will certainly cost more. The gain in air quality will be significant but short-lived unless even more stringent controls are later imposed.

The alternative to the two previous approaches is to take a broader view of the problem and change the activities which cause the pollution. In the case of power plants this may mean reducing our energy demands and seeking alternate sources of energy such as the sun or the atom. In the case of transportation, it means eliminating our dependence on the automobile by substituting cleaner and more efficient mass transit systems. And, in general, it means quelling our insatiable appetites for the consumption of material goods.

Bibliography

Barrett, L. B., and Waddell, T. E. (1973). *Cost of air pollution damage: A status report*. Environmental Protection Agency. Research Triangle Park, N.C. AP-85. Feb.

Bay Area Air Pollution Control District (BAAPCD) (1972). *Air Currents*. Aug.

California Air Resources Board (CARB) (1969). *Control of Vehicle Emissions after 1974*. Tech. *Advisory Committee Report*, November 19.

Council on Environmental Quality (CEQ) (1972). *Environmental Quality, 3rd Annual Report*. Washington, D.C. Aug.

Esposito, J. C. (1970). Ralph Nader's Study Group on Air Pollution, *Vanishing air*. New York: Grossman.

General Accounting Office (GAO) (1972). *Cleaner engines for cleaner air: progress and problems in reducing air pollution from automobiles*. Comptroller General of the U.S. B-166506. May 15.

Purdom, P. W. ed. (1971). *Environmental health*. New York: Academic Press.

Satterfield, C. N. (1972). Nitrogen oxides: A subtle control task. *Technol. Rev.* Oct./Nov.

Squires, A. M. (1971). Capturing sulfur during combustion. *Technol. Rev.* Dec.

Squires, A. M. (1972). Clean power from dirty fuels. *Sci. Am.* Oct.

Starkman, E. S. ed. (1971). *Combustion generated air pollution*. New York: Plenum Press.

U.S. Environmental Protection Agency (EPA) (1972a). Office of Air Programs. *Compilation of Air Pollutant Emission Factors*. AP-42. Research Triangle Park, N.C. Feb.

U.S. EPA (1972b). Annual Report of the Administrator of the EPA. *The Economics of Clean Air*. Doc. No. 92-67. Washington, D.C. March 24.

U.S. Department of Health, Education, and Welfare (HEW) (1969a). *Control Techniques for Particulate Air Pollutants*. AP-51. Jan.

U.S. HEW (1969b). *Control Techniques for Sulfur Oxide Air Pollutants*. AP-52. Jan.

U.S. HEW (1970a). *Control Techniques for Carbon Monoxide, Nitrogen Oxide, and Hydrocarbon Emissions from Mobile Sources*. AP-66. March.

U.S. HEW (1970b). *Control Techniques for Nitrogen Oxide Emissions from Stationary Sources*. AP-67. March.

Questions

1. Why is lead being eliminated as a gasoline additive?
2. Explain the necessity for specifying the driving cycle used when quoting automobile emission figures.
3. What is wrong with saying electric vehicles would be pollution-free?
4. Can you tell from Figure 10.15 whether a 200,000 cfm electrostatic precipitator or fabric filter has higher operating and maintenance costs?

5. In California a motorist may be ticketed if the concentration of hydrocarbons or carbon monoxide in the exhaust exceed certain standards. Can you think of three good reasons why this test is inadequate?

6. If a vehicle is driven 1000 miles per month, and its emissions are equal to the 1974 Federal standards,

   (a) What would be the total weight of HC, CO, and NO_x emissions in 1 year?

   (b) What volume of air would be required to dilute the CO to 10 mg/m^3?

   ans. (a) 41 kg, 470 kg, 36 kg

   (b) 4.7×10^7 m^3

7. Explain the purpose and operation of the following:

   (a) Exhaust gas recirculation
   (b) Catalytic converter
   (c) Electrostatic precipitator
   (d) Thermal reactor
   (e) Baghouse
   (f) Scrubber
   (g) Carbon canister

Part IV

Energy and Raw Materials

Chapter 11

Energy Resources and Consumption

The availability of abundant, inexpensive energy in convenient form is essential to modern industrial societies. In fact, the per capita consumption of energy is probably the best single indicator of the level of material advancement in a country. It should be realized, however, that we are paying a heavy environmental price for that energy, and what is more, there is reason to question our ability to continue for a long period of time, at even present levels of consumption, let alone the much higher levels which are anticipated for the near future.

Interest in energy problems has been growing at a phenomenal rate. Towards the end of 1971 there appeared almost simultaneously a number of special issues in the semitechnical journals* describing what has come to be known as the "energy crisis." Since then, hardly a day goes by when we do not hear something about fuel shortages, brownouts and blackouts, radiation dangers, oil spills, strip mining, gas rationing, or the conflict between environmentalists and the oil and power companies.

We shall examine most of the factors contributing to the "crisis" in some detail in the next few chapters, but basically the main arguments center around the following three points: (a) the rapid growth in our consumption of energy; (b) the rapid depletion of the most desirable fuels; and, (c) the increasing environmental disruption associated with the acquisition and utilization of the sources of energy.

In this chapter attention will be focused on energy resources and the consumption of energy, while the following three chapters are concerned with the most important sector of energy use—the generation of electricity.

**Scientific American,* September, 1971; *Technology Review,* October/November, December, 1971, & January 1972; and *Bulletin of Atomic Scientists,* September, October, November, 1971.

11.1 Energy and Power

It is important in these next few chapters to bear in mind the difference between the two terms *energy* and *power*. Energy is the capacity to do work and power is the rate at which work is done. Remember also that *work* is the product of a force times the distance that the force is moved. For example, if a 10 pound block is raised 2 feet, then we have moved a 10 pound force through a distance of 2 feet so that 20 foot-pounds (ft-lb) of work have been done. The block has acquired 20 ft-lb of energy relative to its initial position. If the block is raised quickly, then more *power* is required than if it is raised slowly, even though the *energy* acquired by the block is the same in either case.

Energy can take many forms. The block, for example, acquires potential energy when it is raised and, if it is dropped, that potential energy is converted into motion, or kinetic energy. Mechanical forms, such as these, are only one way that energy can be stored; others include chemical, nuclear, electromagnetic, thermal, and electrical. Energy may be transformed from one form to another, but by the First Law of Thermodynamics, energy can neither be created nor destroyed. Thus in any transformation, it is theoretically possible to account for the disposition of all the energy that goes into the process. The Second Law of Thermodynamics can be phrased in many ways, but basically it states that no process can be 100% efficient in converting energy into useful work.

There are a number of different energy and power units which are commonly encountered in studying the energy problem. One of the most commonly used is the British Thermal Unit (Btu), which is equal to the amount of energy required to raise the temperature of 1 pound of water 1°F at 60°F. Since much of the discussion in the next few chapters deals with the generation of electricity, the electrical units will also be important. The electrical unit of power is the *watt,* which is 1 ampere of current flowing through a voltage drop of 1 volt. We will often use the *kilowatt* (kW) which is 1000 watts or the megawatt (MW), which is 1 million watts. The electrical energy unit which we will use is the kilowatt hour (kWh) which is simply 1 kilowatt of power operating for 1 hour.

Table 11.1 lists some conversion factors between various energy units. Included in the table are the approximate amounts of energy contained in 1 barrel of oil (42 gallons), 1000 cubic feet of natural gas, and 1 metric ton (1000 kg) of bituminous or anthracite coal. Sometimes energy is measured in coal equivalents. For example, the 1970 per capita consumption of all forms of energy in the United States was equivalent to 11.13 metric tons of coal, which equals about 90,000 kWh. At the other extreme, Burundi had a per capita energy consumption of 0.009 metric tons of coal, or 73 kWh.

TABLE 11.1 Some Common Energy Units

Energy	Btu	kWh
1 kilowatt hour (kWh)	3,412	1
1 British thermal unit (Btu)	1	2.94×10^{-4}
1 kilocalorie (kcal) = 1 food calorie (Cal)	3.97	1.16×10^{-3}
1 horsepower-hour	2,545	0.746
1 foot-pound (ft-lb)	1.28×10^{-3}	3.8×10^{-7}
1 barrel petroleum (bbl) (42 U.S. gallons)	5.8×10^{6}	1,700
1000 cubic feet of natural gas	1.03×10^{6}	300
1 metric ton (2200 lbs) bituminous or anthracite coal	27.3×10^{6}	8,100

To try to get a feel for these units, one Btu is enough energy to raise a 2500 pound automobile about 1/3 foot, while 1 kWh is sufficient to toss it over the Empire State Building. In terms of power, a typical lightbulb is rated at about 100 W which is equivalent to the power generated by a man walking up stairs at about one step per second, or is about equal to the power required to leisurely ride a bicycle at about 10 mph (Rice 1972).

11.2 *Energy Consumption*

The extreme differences in the living standards around the world can be illustrated by the wide range in annual per capita energy consumption. In 1970, the per capita consumption in the United States was 8.4 times that of the rest of the world; was 20 times that of China; was 34 times that of the average developing country; was 59 times that of India; and about 1000 times that of Burundi, Rwanda, Upper Volta, Yemen, Nepal, and Portuguese Timor (United Nations 1972). Table 11.2 lists some energy consumption data for various regions of the world. While the United States in 1970 had only 5.7% of the world's population, it was consuming 33.4% of the world's energy expenditure.

While we in the United States are already consuming far more than our share of these precious resources, the gap between us and the rest of the world is predicted to widen extremely rapidly as shown in Figure 11.1. It should be noted that these are per capita consumptions. That is, not only is consumption going to increase because of our population growth, but it is predicted to increase much more rapidly because of the increased demand by each person.

This prediction of future energy consumption in the United States, and all subsequent predictions were made by the U.S. Department of Interior (1972) and are based on the following assumptions. The annual

TABLE 11.2 1970 Energy Consumption by Region[a]

Region	Total energy consumption, 10^6 metric ton coal equivalent	Per capita energy consumption, kg coal equivalent	Percent of world's population	Percent of world's energy consumption
World	6,843	1,897	100	100
North America	2,472	10,923	6.3	36.2
Centrally planned economies—Europe	1,534	4,404	9.7	22.4
Western Europe	1,351	3,791	9.9	19.8
Far East	526	473	30.7	7.7
Centrally planned economies—Asia	440	543	22.5	6.4
Central America–Caribbean	144	1,170	3.4	2.1
South America	112	705	4.4	1.6
Africa	109	311	9.7	1.6
Middle East	78	738	2.9	1.1
Oceania	78	4,034	0.5	1.1

[a]Source: United Nations 1972.

growth rate of the Gross National Product (GNP) is assumed to be 4.3% until 1980, and 4.0% thereafter. Population is assumed to increase to 280 million in the year 2000. Industrial production is assumed to be increasing at an annual rate of 5% up to 1980, and 4.4% thereafter. Also included in the forecast are the limitations expected in fuel availability; increasing prices of fuels; several major technological innovations (coal gasification, sulfur oxides control, breeder reactors); a shift in life styles towards a more service-oriented economy; and finally, the professional judgments of the Department's energy specialists.

Let us examine the energy consumption in the United States both by source and by use. The five sources of energy and the percent contribution to our total consumption in 1971 were: petroleum (44.2%), natural gas (33.0%), coal (18.2%), hydropower (4.0%), and nuclear power (0.6%). Figure 11.2 shows the projected increases in these sources up to the year 2000. As can be seen, there will be very little increase in hydropower as most of the good sites have already been used. The greatest increases are projected for petroleum and nuclear power. Nuclear power is predicted to increase by more than 100-fold from 1971 to 2000 (405 trillion Btu to 49,230 trillion Btu). The total consumption of energy is predicted to double in the U.S. in the next 19 years which corresponds to an average annual rate of increase of 3.7%.

Figure 11.3 shows the following uses for energy in the United States in 1971: The generation of electricity (25.4%); transportation (24.6%); industrial (29.3%); and, household and commercial (20.7%). Of the 15.5×10^{15} Btu which went into the generation of electricity, only 5.5×10^{15} Btu (or 35.5%) is actually available as electrical energy. The other 64.5% is lost as heat and is the cause of the thermal pollution associated with fossil-fueled and nuclear power plants. The growth rate of

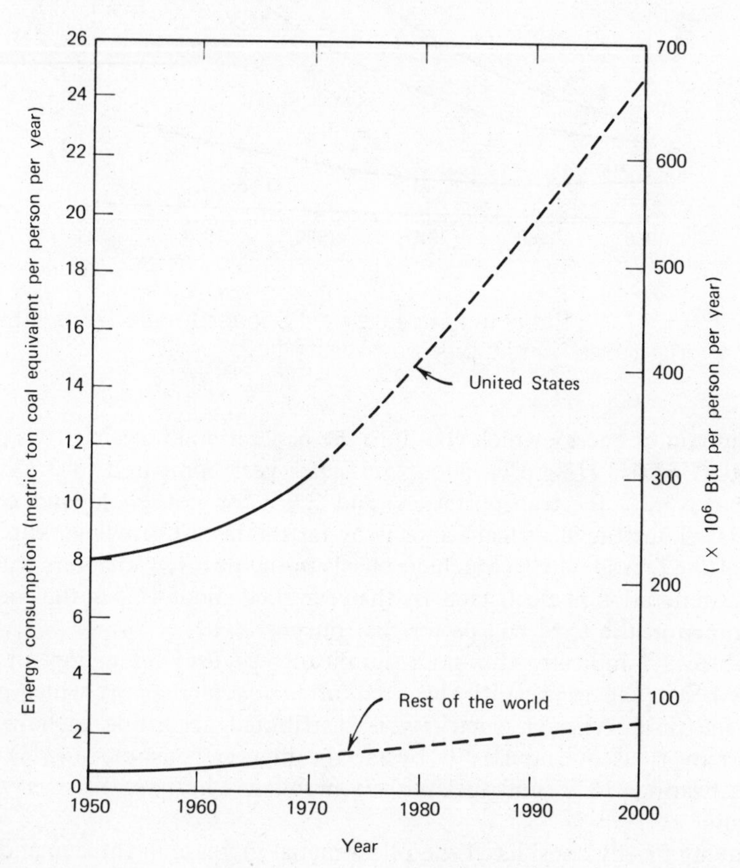

FIGURE 11.1 Comparing U.S. per capita energy consumption to that of the rest of the world (U.S. projection based on Dupree and West 1972; rest of world based on Malenbaum 1973).

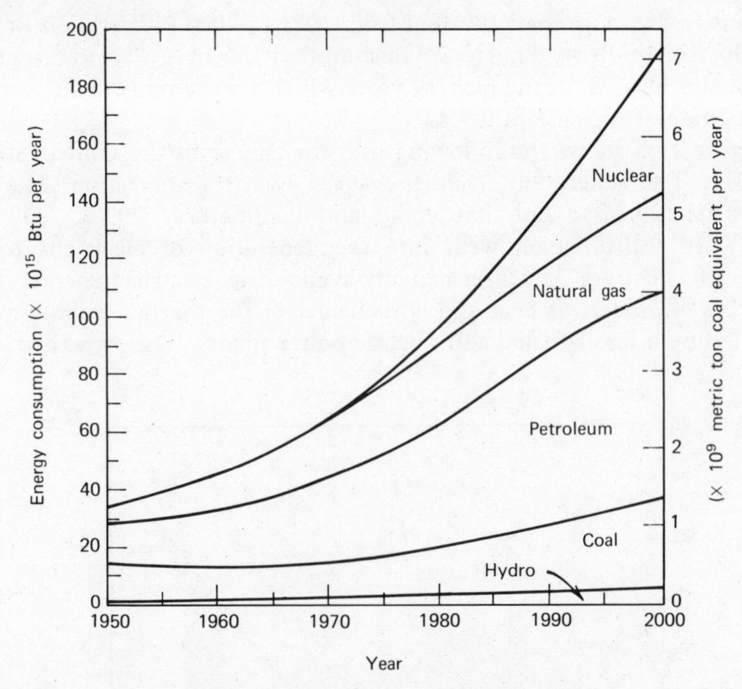

FIGURE 11.2 Projected increase in U.S. consumption of energy by source (based on Dupree and West 1972).

the amount of energy which goes into the generation of electricity is predicted (USDI 1972) to be about 5.6% per year compared to 3.3% for industry, 3.2% for transportation, and 2.9% for household and commercial. Thus the electrical sector is by far the fastest growing, which is part of the reason why so much emphasis in the next few chapters will be on the generation of electricity. By the year 2000 about 42% of the energy consumed in the U.S. will be for that purpose.

Table 11.3 indicates the most significant *end* uses for energy in the United States in 1968. With this accounting scheme, energy which goes into the production of electricity is distributed according to how the electricity is used. Industry is by far the biggest consumer (42 %) but space heating (18 %) and fuel for automobiles (14 %) are also very important.

Having briefly considered the phenomenal increase in the demand for energy expected in the United States in the coming years, we must now compare these projections with the estimated supplies of energy resources. The resulting picture is not at all encouraging.

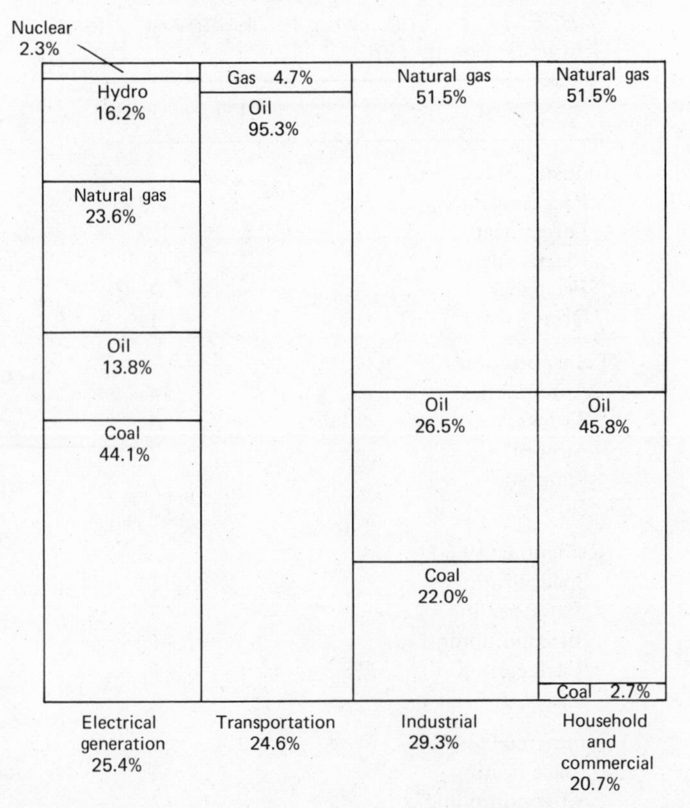

FIGURE 11.3 Energy consumption in United States in 1971 by use category and by source (Dupree and West 1972).

11.3 *Petroleum Resources*

Let us first consider the situation in the United States, where petroleum provides about 44% of our total energy. How much petroleum is left in this country; how much is left in the world; and who has got what is left? Before we can answer these questions we must make the distinction between the terms *resources* and *reserves*. The terms are often used rather loosely and there are slightly varying definitions in the literature. Suffice it to say that *reserves* are quantities that can be reasonably assumed to exist and which are producible with existing technology and under present economic conditions. Reserves can be further broken down into proven and inferred reserves. The term *resources* includes not only reserves, but also deposits not yet discovered as well as those which have

TABLE 11.3 End Uses for Energy in the United States in 1968[a]

Use	Percent of total
Industry (42 %)	
Process steam	17
Direct heat	12
Electric drive	8
Feedstock	4
Other	1
Transportation[b] (25 %)	
Automobiles	14
Trucks	5
Aircraft	2
Railroads	1
Other	3
Residential (19 %)	
Space heating	11
Water heating	3
Air conditioning	1
Refrigeration	1
Other	3
Commercial (14 %)	
Space heating	7
Air conditioning	2
Water heating	1
Refrigeration	1
Other	3

[a]Source: Stanford Research Institute (SRI) (1972).
[b]Transportation breakdown from Hirst (1972).

been identified but are not recoverable under present technological and economic conditions.

As Table 11.4 indicates, the United States has proven onshore reserves of about 41 billion barrels of petroleum (crude oil and natural gas liquids). Included in this figure are the approximately 10 billion barrels of crude oil reserves on Alaska's North Slope. Indicated reserves plus undiscovered resources producible with current economics and technology amount to 277 billion barrels onshore and 197 billion barrels offshore (in depths less than about 200 meters).

TABLE 11.4 Estimated Petroleum Resources for U.S., Billions of Barrels[a,b]

	Onshore	Offshore	Total
Proved reserves as of January 1, 1971	41	6	47
Indicated reserves and undiscovered resources producible with current economics and technology	277	197	474
Para and submarginal resources	1,155	1,257	2,412
Total resource base	1,473	1,460	2,933

[a]Source: USDI 1972.
[b]Crude oil and natural gas liquids.

At 1971 consumption rates of about 5.5 billion barrels per year, the total proved reserves could be depleted in about $8\frac{1}{2}$ years. It is interesting to consider the huge Alaskan find which is unofficially estimated at from 10 to 20 billion barrels making it 2 to 4 times as large as the East Texas field, which was the largest in the U.S. previously. At the 7.7 billion barrel consumption rate estimated for 1980 (which is approximately when the pipeline could go into full operation) the total find could be depleted in less than 3 years. However, the pipeline is only being designed to carry 2 million barrels per day, or 0.73 billion barrels per year. At that rate the Alaskan oil may last for perhaps 25 years but it would supply only a small percentage of our total demand each year.

How then are we planning to meet the accelerating demand for petroleum? Figure 11.4 presents the estimates of the Department of the Interior (Dupree and West 1972). In 1971, the nation imported one-fourth of its total consumption of 5.5 billion barrels. By 1985 it is estimated that only 37% of the demand will be supplied by domestic sources within the conterminous United States with an additional 8% coming from Alaska. Barring some significant advances in our ability to exploit shale oil, tar sands, and the liquefaction of coal, the remaining 55% must come from imports. By 2000, imports could account for 70% of our demand. At the 1970 average price of $3.18 per barrel,* these imports could increase the nation's balance of payment deficits by $4.5 billion in 1971, $15.5 billion in 1985, and $29 billion in the year 2000.

Thus the nation is just beginning a period of rather extensive dependence on foreign sources for its petroleum. As shown in Figure 11.5, most

*The price of imported crude oil jumped to over $9 per barrel during the 1973 Arab oil embargo.

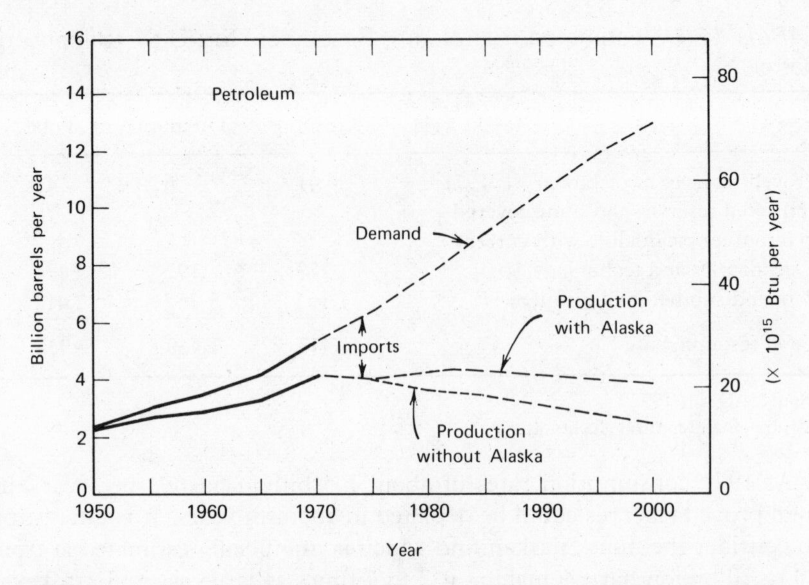

FIGURE 11.4 U.S. domestic production and demand of petroleum (including natural gas liquids). Production does not include unconventional sources such as shale oil, coal liquefaction, and tar sands whose development may become feasible after 1980. (Past data from National Commission on Materials Policy 1972; projections from Dupree and West 1972.)

of our imports in 1970 came from relatively secure sources in the Western Hemisphere—Venezuela and Canada. Very little came from the Middle East. Figure 11.6 indicates that by 1980 imports from the Middle East will increase ten-fold to about 6.5 million barrels per day (2.4 billion barrels per year), making it the major source of imports. Note too, that by this estimate (Shell Oil Company 1972) while North America will increase its consumption by 11 million barrels per day during the 10 year period, all of Africa (57% larger population) will add only 2.2 million barrels per day.

The majority of the world's crude oil reserves are located in the Middle East in Saudi Arabia, Kuwait, Iran, and Libya. Table 11.5 indicates total world reserves of about 632 billion barrels and world production of 18.5 billion barrels per year (USDI 1973a). If world production continued at this rate, these reserves would be consumed in 34 years; if consumption grows at the expected annual rate of 4% (Malenbaum 1973) these reserves could be consumed in 21 years.* It should be realized, however, that future discoveries and changes in technology and economics can bring more oil into the reserves category thereby extending their lifetime.

Hubbert (1969 1971) has a very effective way to illustrate the extremely short lifetime of fossil fuel *resources*. There are only a certain number of barrels of oil which can ever be produced and once they are consumed they will never be replaced. If a plot of production versus time is drawn, it must have certain characteristics. After starting at zero, production will most likely rise exponentially as the most easily found reserves are exploited. Then, as the production costs increase due to the difficulty of finding and extracting the resource, the production rate will slow down and eventually return to zero as the resource is depleted. Many curves would fit the above loose description, but the one proposed by Hubbert is the simple bell shaped production curve of Figure 11.7.

Whatever curve is chosen, the area under the curve must correspond to the total amount of oil that will ultimately be produced. In the figure, the curves correspond to two different estimates of total crude oil. Although these estimates differ by a factor of over 1.5, they reach their maximums within 10 years of each other, and both start to decline by 2000. Also indicated is the time period during which 80% of the resource will be consumed. By these estimates 80% of the total amount of crude oil that will ever be produced, resources which took millions of years to be formed, will be consumed within 64 years—less than a single lifetime!

There are several potential developments which could extend the petroleum reserves significantly. Research is in progress to derive synthetic petroleum from coal which is much more abundantly available throughout the world, and most especially so in the United States. The second alternative is to develop oil shale resources. Although no oil shale is presently being produced in the U.S., future technology and economics may make it possible to exploit these rather large resources. The most promising deposits occur as part of the Green River formation in areas of Colorado, Utah, and Wyoming. There are in this formation some 80 billion barrels of oil in the most accessible high grade deposits, with the total supply being estimated at 600 billion barrels (USDI 1972).

The third potential future source of crude oil is deposits of asphalt-

*The number of years (T) to deplete reserves (R) when present consumption is P, and consumption is predicted to increase at the average annual rate (r) is given by:

$$T = \frac{1}{r} \ln \left(\frac{rR}{P} + 1 \right)$$

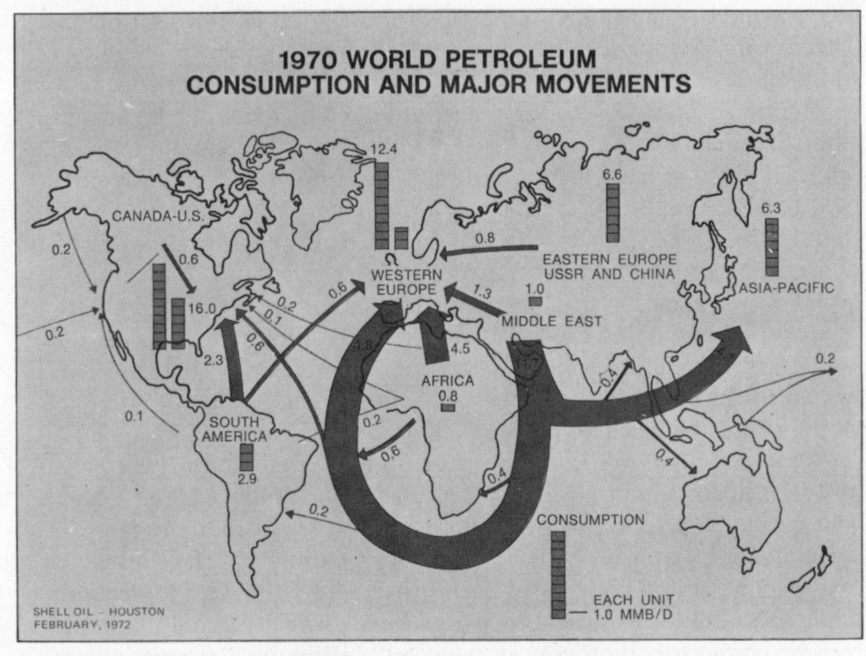

FIGURE 11.5 1970 World petroleum consumption and major movements (Shell Oil Company 1972).

TABLE 11.5 World Production and Proved Reserves of Crude Oil, 1972 in Billions of Barrels[a]

Country	Production	Proved reserves
Saudi Arabia	2.1	145.3
Communist countries (except Yugoslavia)	3.2	98.2
Kuwait	1.2	66.0
Iran	1.8	55.5
United States	3.5	38.1
Libya	0.8	25.0
Venezuela	1.2	13.9
Other Free World	4.7	189.9
World total	18.5	631.9

[a]Source: USDI 1973a.

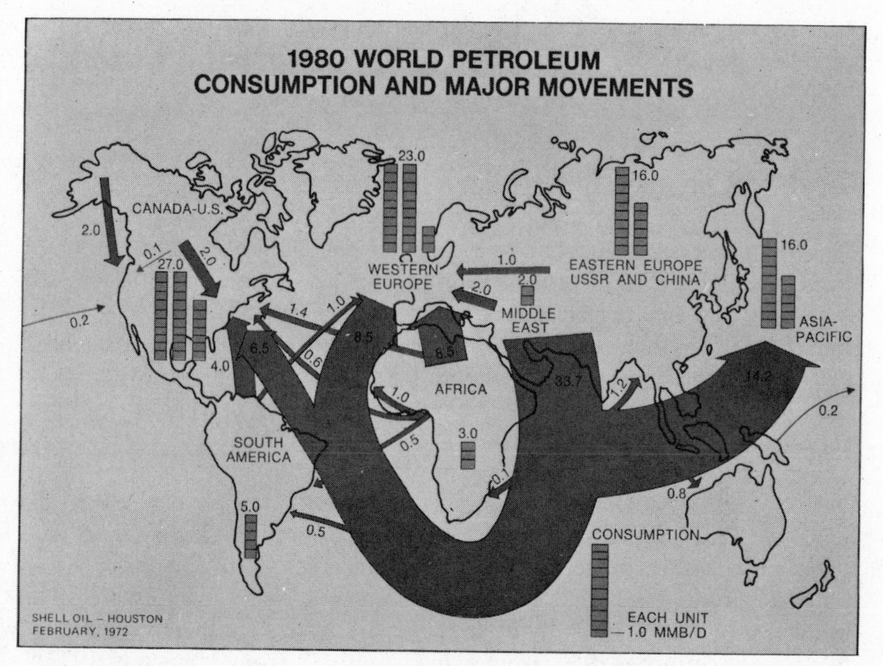

FIGURE 11.6 1980 World petroleum consumption and major movement (Shell Oil Company 1972).

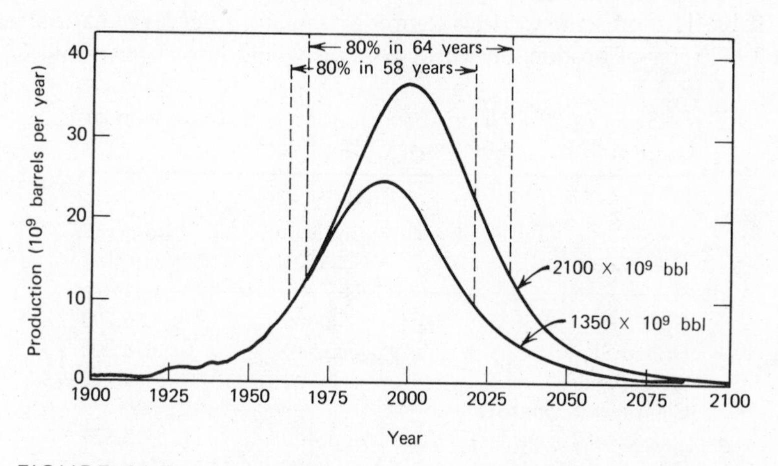

FIGURE 11.7 Complete cycle of world crude oil production for two estimates of the total amount which will ultimately be produced. From Hubbert, M. K., "The Energy Resources of the Earth."

bearing rocks known as tar sands. Again, production is not yet commercially feasible but there are considerable potential resources, especially in Canada.

11.4 *Natural Gas Resources*

Natural gas provides about one-third of the nation's energy supply. From the point of view of air pollution, it is the cleanest of all the fossil fuels, which is one reason why it has also experienced the fastest growth in demand. Another factor is that its price has been held at an artificially low level by the Federal Power Commission, which has not only increased demand but has probably also limited the incentive to explore for new supplies.

From 1950 to 1972, the average annual rate of growth was about 6 %, but it appears that that rate is going to drop to about 2 % per year from now on. This sudden decrease is not being caused by a change in demand, but rather reflects the dropping reserves in this country. Importing natural gas is difficult. Until recently, it was transportable only in its gaseous form, and only by pipeline. Imports were therefore limited to adjacent countries (95% from Canada and 5% from Mexico in 1972). It is now possible to liquefy the gas (liquefied natural gas, LNG) for transportation by specially designed tankers. Extensive trade in LNG is expected to develop shortly.

Table 11.6 presents worldwide production and reserves of natural gas. At 1972 rates of production world reserves would last 41 years and U.S.

TABLE 11.6 World Production and Reserves of Natural Gas in Billions of Cubic Feet[a]

Country	Marketed production 1972	Reserves 1971
U.S.S.R.	8,000	636,000
United States	22,910	278,000
Netherlands	2,000	88,210
Communist countries (except Yugoslavia)	1,400	15,827
Other Free World	5,800	680,935
World total	42,930	1,755,240

[a]Source: USDI 1973a.

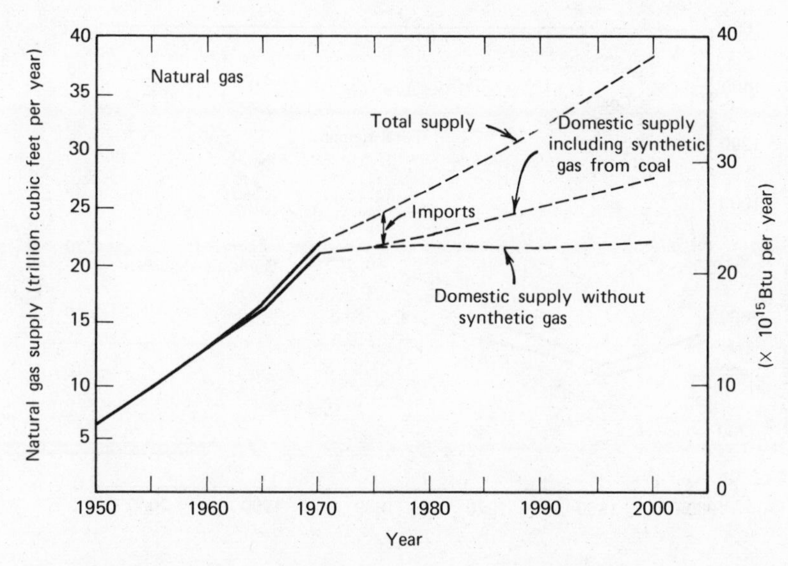

FIGURE 11.8 U.S. domestic supply and total supply of natural gas. (Dupree and West 1972).

reserves would last 12.2 years. This ratio of reserves to production has dropped in the U.S. from above 30 years just after World War II, to about 22 years in 1958, to 12.2 years in 1972.

Figure 11.8 shows the expected future growth in total supply of natural gas to the United States and also the expected domestic production. Included in the figure is the potential effect of present research efforts in the area of coal gasification. If successful, future imports could be decreased by about one-third.

11.5 *Coal Resources*

The only fossil fuel which seems assured to remain a major source of energy much beyond the year 2000 is coal. The United States is in the fortunate position of having about one-third of the world's recoverable reserves—enough to last about 1600 years at present rates of consumption. About 83% of the rest of the world's reserves are located in Communist countries. Coal presently supplies about 18% of the nation's total energy consumption. Even though U.S. consumption is expected to increase $2\frac{1}{2}$ times by 2000, the percentage of total energy is expected to

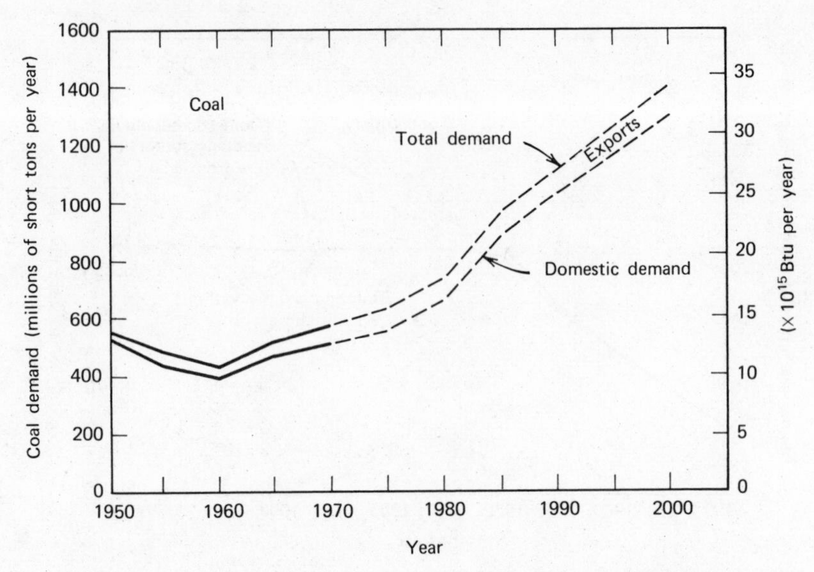

FIGURE 11.9 U.S. domestic demand and total demand for coal including bituminous, lignite and anthracite coal (Dupree and West 1972).

remain pretty much the same. As shown in Figure 11.9, the U.S. is a net exporter of coal.

Coal resources are typically categorized by *rank* and by *grade*. The rank is based on the energy content and the percentage of fixed carbon. Anthracite and bituminous coals have the highest heat content followed by subbituminous coal and then lignite. The energy content of lignite is typically only about half that of bituminous coal or anthracite. The grade of coal is based on the content of ash, sulfur, and other deleterious constituents. Table 11.7 presents some data on the distribution of the various ranks according to sulfur content. Bituminous (including subbituminous) coal is by far the most important in this country, comprising about 98% of our production and 72% of our reserves.

While coal resources are the nation's primary store of fossil-fuel energy, their exploitation is severely limited by the adverse environmental effects associated with their mining and combustion. It does appear that processes will be developed in the near future for reducing atmospheric emissions of sulfur oxides, either by gasifying the coal prior to combustion or by removing the sulfur from the flue gas after combustion. Certainly, in view of the predicted increase in the nation's use of coal, concentrated efforts should be devoted to solving the sulfur problem as soon as possible.

TABLE 11.7 U.S. Coal Reserves by Rank and Location

Rank	Percent of U.S. reserves[a]	States where major reserves are located[b]	Distribution by sulfur content, %[c]		
			Low sulfur 0–1%	Medium sulfur 1–3%	High sulfur > 3%
Anthracite	1	Pennsylvania	97.1	2.9	—
Bituminous coal	46[d]	Illinois, West Virginia, Missouri, Colorado, Pennsylvania	29.8	26.8	43.4
Subbituminous coal	26[d]	Montana, Wyoming, Alaska, New Mexico	99.6	0.4	—
Lignite	27	North Dakota, Montana	90.7	9.3	—

[a]From USDI 1973a (% by weight not energy content).
[b]From U.S. Congress 1969.
[c]From USDI 1973b.
[d]Assuming 35% of the bituminous coal listed in (*a*) is subbituminous.

As Table 11.7 indicates, the majority of the most desirable low sulfur coal is located in the states of North Dakota, Montana, Wyoming, Colorado, and New Mexico—far from the large urban centers which are going to want to use it. Either the coal must be shipped many miles or it must be utilized near the mine site. Shipping the coal by train would be quite expensive. For example, to fuel a typical 1000 MW power plant serving a city of about 1 million would require a 200-car freight train every single day. Since almost two-thirds of the domestic consumption of coal is for power plants anyway, power companies are choosing to build their facilities near the mines and then ship the power to the cities by means of long transmission lines. As a consequence, some of the few remaining areas where the air is still clean are beginning to get the pollution that belongs to the city.

Whether or not adequate control over stack emissions can ever be attained, the environmental problems associated with mining would still need to be solved. Coal is produced either from underground mines or by strip (surface) mining. Strip mining is a process wherein access to the coal is obtained by simply ripping away the ground cover above the coal (overburden) and then extracting the coal, which lies in seams typically 18 inches to about 7 feet thick at depths of from about 15–60 feet. Huge earth-moving equipment is used such as a gigantic 220 cubic yard drag-line weighing 27 million pounds which can remove 325 tons of overburden in one pass.

More than 40% of present production of coal is obtained from surface mines. That percentage is increasing not only because it is far more economical but also because much of the desirable low-sulfur coal lies in beds just under the surface (largely in those Western states mentioned in Table 11.7).

Of the 3.2 million acres in the U.S. which had been disturbed by surface mining by 1967, 1.3 million were related to coal production (the other mining was for sand and gravel, stone, gold, phosphate, iron, and clay). These 1.3 million acres were, roughly, evenly divided between the two types of strip mining: "area" stripping on relatively flat land; and, "contour" stripping in rolling or mountainous terrain. Area stripping leaves the terrain looking like a washboard and contour stripping leaves raw exposed gouges along the contours of the mountain. By 1980 the area

The rapid expansion of strip mining has been made possible by the development of giant earth moving equipment such as this shovel with its 115 cubic yard bucket. (Photograph by E. Dotter.)

strip mined is expected to have increased to about 4300 square miles.

The overturning of the land not only results in destruction of the natural beauty of an area, but the whole ecology is upset. The fertile cover soil is lost and extensive reclamation would be required to return it to even minimal usefulness. Good farmland lost this way is essentially permanently lost. Unfortunately there are no economic gains to be made by reclaiming an area so typically stripped areas have been abandoned. In 1965 about two-thirds of the stripped area still required reclamation (U.S. Congress 1970).

Reclamation, if applied, could reduce these effects and the cost would be nominal relative to the price of the coal removed. Strict new Federal laws need to be written and enforced to ensure adequate protection of the environment. Some interesting approaches, easing several environmental problems at once, have been proposed. One by Eliassen (1971), suggests that the trains which are shipping coal into an urban area be filled with the urban area's solid waste on the return trip to the mine, to be disposed of there. Dried sewage sludge could be used as a soil conditioner to help in the reseeding efforts as is now being tried in Chicago. Some of their sewage sludge is being used to help reclaim some 7000 acres of abandoned unreclaimed, stripped land, about 200 miles southwest of Chicago.

11.6 *Non Fossil-Fuel Energy Resources*

There are several non fossil-fuel energy resources which are going to become increasingly important in the future. While they will ease the demand on our precious fossil fuel reserves, it must be pointed out that their application lies only in the generation of electrical power.

The growing concern for our rapidly depleting supplies of petroleum and natural gas (but not coal) and the need to reduce air pollution, are factors which are contributing to the sudden proliferation of uranium-fueled nuclear fission reactors. In Chapter 13, the operation of these reactors will be explained and their environmental implications will be examined, but for now we will simply ask about uranium reserves.

Estimates of the quantities of uranium reserves are highly dependent upon the price that can be paid. The going price is just under $8 per pound of uranium oxide, U_3O_8, and according to the 1970 report to the Joint Economic Committee of Congress, "the economic feasibility of present nuclear reactors depends upon a price of $10 per pound or less." Present reserves in the U.S. of under $10 U_3O_8 are estimated at 330,000 short tons (USDI 1973a). Federal Power Commission estimates (1971) of cumulative uranium (U_3O_8) requirements indicate that these reserves could be used up by about 1980–1985. Cumulative requirements up to the year 2000 will be about 1.8 million tons if present design burner reactors continue to be used.

It turns out that present reactors are highly inefficient in terms of their ability to extract energy from naturally occurring uranium. This is because they use only the isotope ^{235}U (92 protons + 143 neutrons) which comprises only about 0.711% of naturally occurring uranium. Work is presently underway to design reactors which will be able to use the much more abundantly occurring isotope, ^{238}U, which comprises 99.283% of natural uranium. Called *fast-breeder reactors,* they would extend the lifetime of our reserves of uranium by a factor of at least 140, making it possible to supply power for hundreds of years. There are, however, some unique dangers associated with these reactors which will require very careful design of safeguards (see Chapter 14).

As will be discussed in Chapter 14, there are two sources of energy which could supply our electrical needs indefinitely—fusion power and solar power. Fusion power, which is the source of the sun's energy, is only in the research stage today, with predictions of the date of its commercial feasibility usually centering around the year 2000. Solar power, it appears, could be technologically and economically feasible somewhat sooner, with some estimates indicating that it could take over a substantial portion of the country's electricity demand by 2000.

11.7 *Energy Conservation*

Clearly there is something a little absurd about an energy crisis occurring in a country in which each person consumes $8\frac{1}{2}$ times as much energy as the average for the rest of the world. Must we really double our already gluttonous energy consumption in the next 18 years? There are a number of relatively straightforward, and relatively painless, conservation measures which could slow our growth considerably. What is really needed is a reversal of present growth patterns, but that is not likely to ever occur voluntarily.

In the area of transportation, which accounts for about 25% of the nation's energy consumption, the overall fuel efficiency can easily be improved. Certain trends need to be halted or reversed. While it is much more efficient to ship freight by train than by truck and, in turn, much more efficient by truck than by airplane, present growth patterns are favoring the inefficient modes. Total fuel consumption by aircraft is increasing faster than that of trucks, which is in turn increasing faster than that of trains. This information is presented in Figures 11.10 and 11.11.

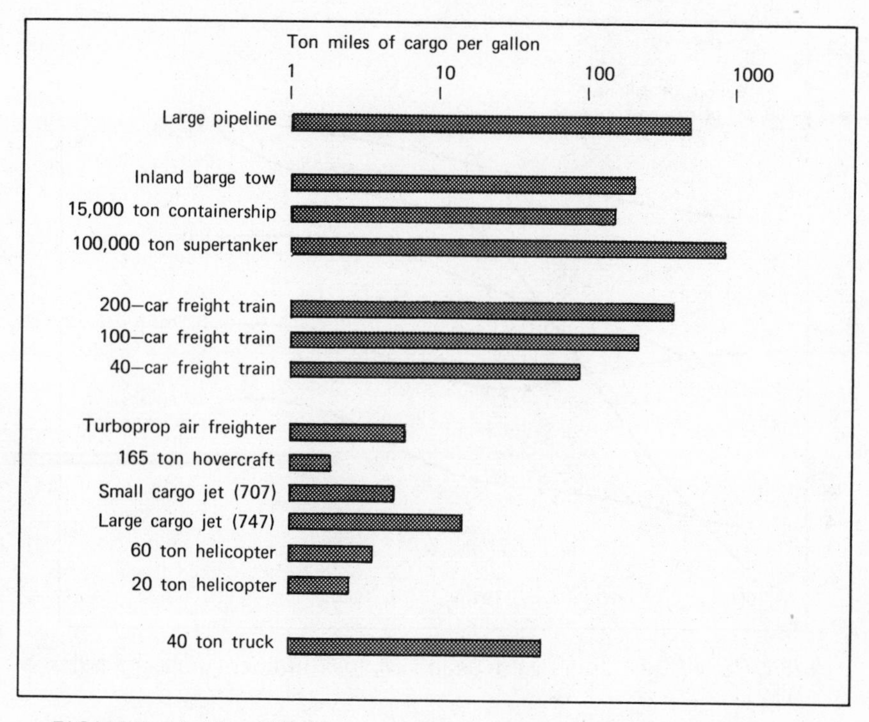

FIGURE 11.10 Efficiency of fuel usage to transport cargo by various carriers (Rice 1972).

The efficiency in transporting passengers has similarly seen a decline. The automobile is one of the least efficient modes of transport and yet it is predominant, accounting for 55% of all transportation energy in 1970. About 54% of all car trips are less than 5 miles long—easy bicycling distance (OEP, 1972). These short trips account for 15% of all automobile fuel consumption, or about 320 million barrels of oil per year (compared to our total oil consumption of about 5,500 million).

The fuel economy of automobiles is closely related to vehicle weight and driving speed, both of which could be decreased without undue hardship to the general public. The typical 5000 pound automobile gets about 9 miles to the gallon while a small 2000 pound car typically gets about 25 miles per gallon.* It takes about 25% less energy to propel a vehicle at 50 mph than at 70 mph. Lowering speed limits and taxing vehicles based on fuel consumption would be simple enough steps to take.

*From data published in the Federal Register, May 2, 1973, based on the driving cycle used in the Federal emissions test.

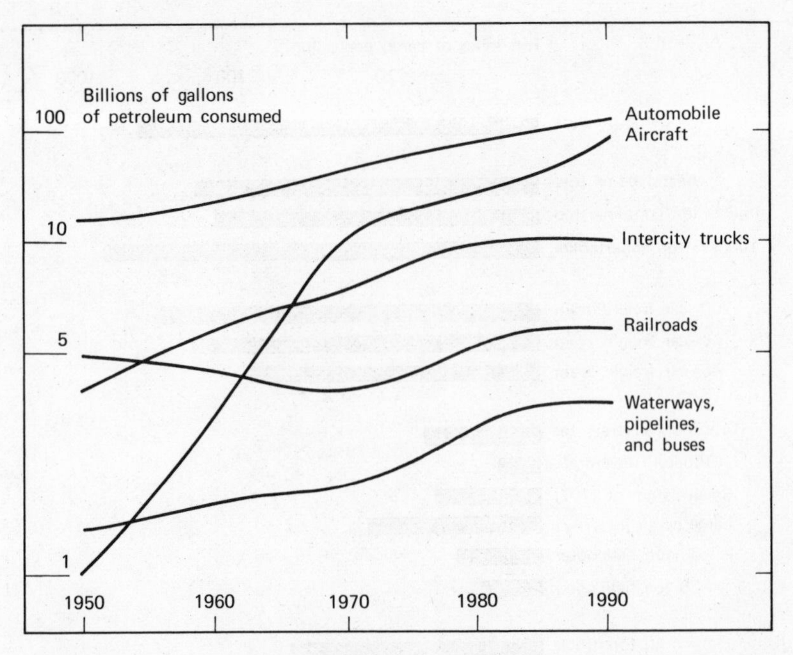

FIGURE 11.11 Increased usage of less efficient carriers (Rice 1972).

More importantly, our dependence on the automobile itself must be eliminated—not only to reduce fuel consumption but to improve urban air quality and congestion. The present approach to reducing emissions by adding control equipment to vehicles may help air quality but it simultaneously results in increased fuel consumption.

Figure 11.12 shows the efficiency with which various transportation systems can transport people, as measured in passenger-miles per gallon (pmg). For example, a vehicle carrying one passenger, getting 12 miles per gallon, would have a "net propulsion efficiency" of 12 pmg. The same vehicle with two people would have an efficiency of 24 pmg. Unless an automobile is loaded with passengers, it is much more efficient to utilize mass transit systems.

To get people out of their automobiles and into public transportation systems will require that public systems be inexpensive, convenient, and fast. To discourage automobile usage, various taxes and inconveniences can be imposed. For example, gasoline and parking taxes and tolls can be raised and the increased revenue used to subsidize mass transit systems.

The OEP (1972) has estimated that conservation measures such as these

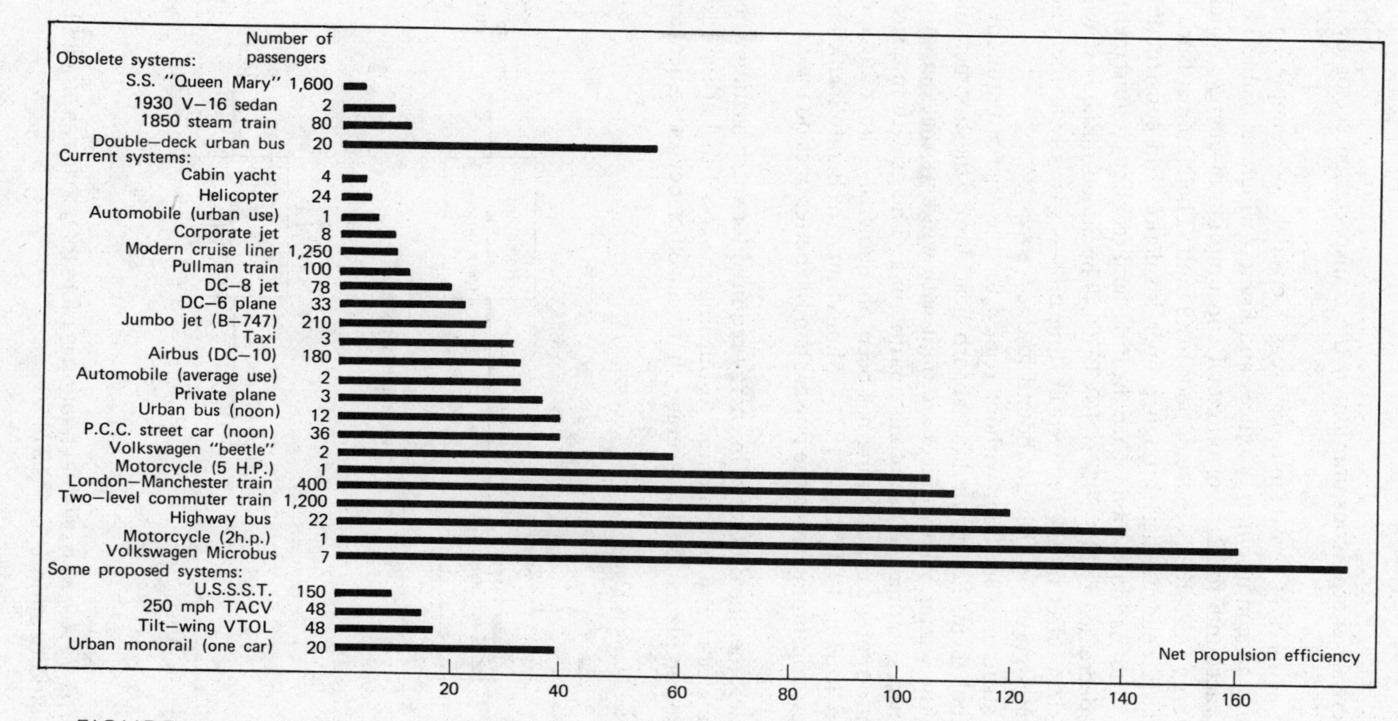

FIGURE 11.12 Passenger transportation efficiencies measured in passenger-miles per gallon of fuel (Rice 1972).

could decrease energy consumption in the transportation sector by as much as 25%.

Similar gains can be made in other areas. Consider, for example, space heating and air conditioning of offices and homes, which, as Table 11.3 indicates, accounts for 21% of our total consumption of energy. Much of this energy is wasted because of poor insulation. The National Bureau of Standards believes that through proper insulation and construction practices, the nation's total space heating and cooling requirements could be decreased by 40–50% (OEP 1972). The added initial cost will be more than offset by the reduced heating bills—especially as fuel and electric rates go up, as they are bound to do.

Many structures are built without regard for the energy required to heat or cool them. Large windows absorb heat during the summer and lose it in the winter. Many offices are built with windows that cannot be opened, necessitating the use of air conditioning. The growth in energy consumption for air conditioning has been phenomenal—about 20% per year from 1967 to 1970 (Cook 1971). Many units are highly inefficient, the poorest consuming twice the power than the better ones do for equivalent cooling.

Consider, too, the difference in energy required for space heating when it is done electrically as compared to burning the fossil fuel at the residence. Power plants, on the average, have an efficiency of somewhere

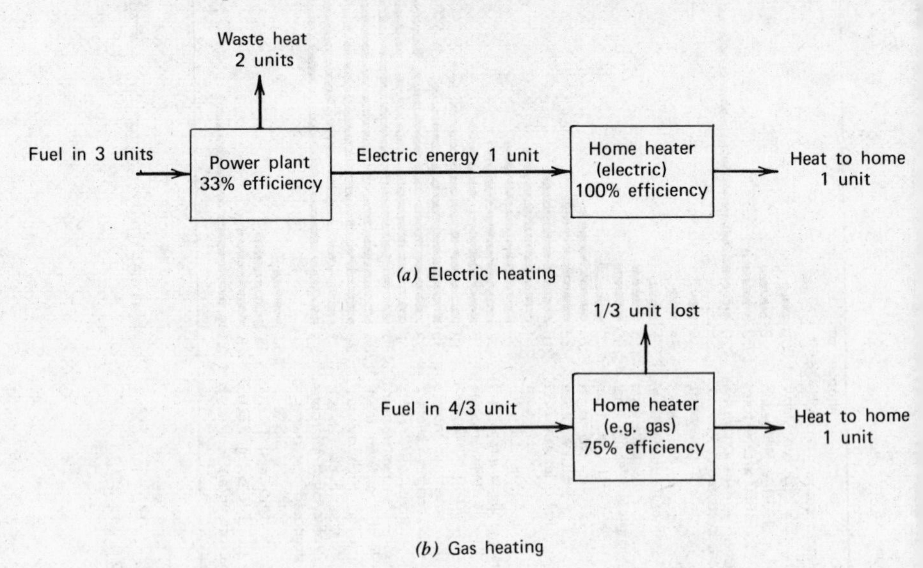

FIGURE 11.13 Fossil-fuel heating is more than twice as efficient as electric heating (Cook 1971).

around 33%—that is, only 1/3 of the energy given off by the fuel source ends up as electric energy. As shown in Figure 11.13 that electric energy is then converted to heat in the home at nearly 100% efficiency, giving an overall efficiency of about 33%. If the same space heating is accomplished by burning the fuel in the home, the loss in efficiency at the power plant is avoided and an overall efficiency of about 75% can be achieved. So space heating with fossil fuels can be more than twice as efficient as with electricity. Also, the lower temperature at which these fuels are burned in the home reduces the NO_x emissions per unit of fuel.

Unfortunately, because installation is cheaper, new homes are increasingly being constructed with electric heating. From 1960 to 1968, the number of electrically heated homes increased from 0.7 million to 3.4 million, corresponding to an annual growth rate of about 20%.

Industry also can take many measures to decrease their energy consumption. At present they are insufficiently motivated, partly because of the strange utility rate structure which, in effect, encourages the use of power by decreasing unit costs as consumption increases. Industry pays, on the average, less than half as much per unit of electric energy as does a residential user, as shown in Table 11.8.

The OEP (1972) has prepared estimates of the total amount of energy which could be saved if measures of the type discussed above would be instituted. Their suggested programs are based on minimizing the consumption of energy while providing the same or similar service to the consumer. As shown in Figure 11.14, the growth which is anticipated by 1990 without these measures, could be decreased by about 53%. Unfortunately, while these gains are truly significant, growth would still be continuing. The consumption rate that would have occurred in 1985 would, even with the conservation program, be reached in about 1993, so all that would be gained is about 8 years' time.

TABLE 11.8 Average Cost to Consumer for Electrical Energy in 1971[a]

Sector	Average cost in cents per kilowatt hour
Industry	1.10
Commercial	2.20
Residential	2.32

[a]Federal Power Commission, 1972.

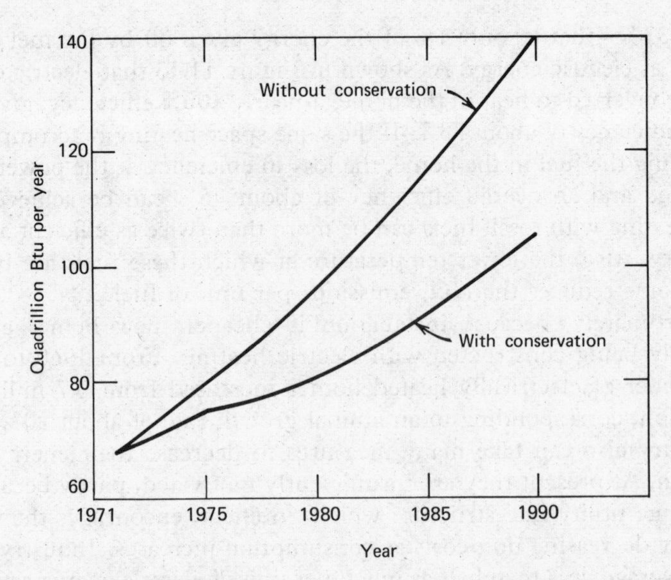

FIGURE 11.14 Projections of total U.S. energy consumption with no conservation measures and with the long-term measures suggested by the Office of Emergency Preparedness (1972).

11.8 *Conclusions*

In spite of the fact that the per capita energy consumption in the U.S. already far exceeds that of any other country in the world, estimates indicate that we may double that rate within the next 18 years. Our way of life seems to be dependent not only on a high level of consumption, but on an ever-increasing level as well. In the past we have been largely self-sufficient in terms of energy resources, but we have just passed the maximum in our rate of domestic production of oil and natural gas, and we are entering an era of growing dependence on foreign sources.

The implications of a dependence on Middle Eastern oil are difficult to predict with confidence, but they appear to be enormous. The billions of dollars which will be required to purchase the oil will greatly worsen our balance of payments problems, and just what the oil-producing nations will do with the money is subject to speculation. For example, if they reinvest it in the U.S. they will control a large segment of our economy. By the early 1980s they would be able to purchase, outright, a corporation the size of General Motors *every year!*

By controlling such a critical resource, the oil-producing countries will have great influence in the course of international politics. Our alignment with Israel may be shifted, and relations between competing countries in the "Free World" will also be strained as we bargain against each other for the remaining reserves; and there will always be the danger that one will act militarily to maintain access to the resources.

The alternatives include increasing the exploitation of our own domestic resources such as the oil in Alaska and the coal in the West. Alaskan oil will not last long at present and future high rates of consumption and it might be better to save it as a reserve for the future. Coal will very likely become the main source of the complex hydrocarbons which are essential not only for energy purposes but also for the many petrochemical and lubricating functions which are now satisfied by petroleum. However, the resulting environmental problems could be overwhelming.

The other alternative is to decrease consumption. Politically acceptable programs for conserving energy can reduce growth significantly, but to cause consumption to actually decrease would require a revolutionary change in our way of life.

If it were just a matter of trading a polluted environment for some additional luxuries, some might consider it a fair bargain. But it is immoral for us to waste these precious resources at the expense of the less-developed peoples of the world, and indeed at the expense of our own future generations.

Bibliography

American Society of Civil Engineers (1972). Chicago reclaiming strip mines with sewage sludge. *Civil Eng.—ASCE.* Sept.

Cook, E. (1971). The flow of energy in an industrial society. *Sci. Am.* Sept.

Dupree, W. G. Jr., and West, J. A. (1972). *United States energy through the year 2000.* U.S. Department of the Interior. Washington, D.C., Dec.

Eliassen, R. (1971). Power generation and the environment. *Bull. Atom. Sci.* Sept.

Federal Power Commission (1971). *The 1970 National Power Survey*, Washington, D.C., Dec.

Federal Power Commission (1972). *Statistics of Privately Owned Electric Utilities in the United States, 1971.* Washington, D.C. Oct.

Hirst, E. (1972). *Energy Consumption for Transportation in the U.S.* ORNL-NSF-EP-15, March.

Hubbert, M. K. (1969). In National Academy of Sciences, *Resources and man.* San Francisco: W. H. Freeman.

Hubbert, M. K. (1971). The energy resources of the earth. *Sci. Am.* Sept.

Malenbaum, W. (1973). *Materials Requirements in the United States and Abroad in the Year 2000*, National Commission on Materials Policy, Washington, D.C. March.

National Commission on Materials Policy (1972). *Towards a National Materials Policy*, Interim Report, Washington, D.C. April.

National Commission on Materials Policy (1973). *Toward a National Materials Policy, World Perspective*, 2d Interim Report, Washington, D.C. Jan.

Population Reference Bureau (1970). *1970 World Population Data Sheet*. Washington, D.C.

Rice, R. (1972). System energy and future transportation. *Technol. Rev*. Jan.

Shell Oil Company (1972). *The National Energy Position*. Paper No. 2-2064-72, July.

Stanford Research Institute (SRI) (1972). *Patterns of Energy Consumption in The United States*. Office of Science and Technology, Executive Office of the President, Washington, D.C. Jan.

Summers, C. M. (1971). The conversion of energy. *Sci. Am*. Sept.

United Nations (1972). *World Energy Supplies, 1961–1970*. Statistical Papers Series J No. 15, New York.

U.S. Congress (1969). *Environmental Effects of Producing Electric Power*, Hearings, Joint Committee on Atomic Energy, Part I, October/November.

U.S. Congress (1970). *The Economy, Energy, and the Environment*, Background Study for Joint Economic Committee of Congress, Washington, D.C. Sept.

U.S. Department of the Interior (USDI) (1967). *Surface Mining and our Environment*. Washington, D.C.

USDI (1972). *First Annual Report of the Secretary of the Interior Under the Mining and Minerals Policy Act of 1970*. Washington, D.C. March.

USDI (1972). *United States Energy, A Summary Review*. Washington, D.C. Jan.

USDI (1973a). Bureau of Mines. *Commodity Data Summaries*. Washington, D.C. Jan.

USDI (1973b). Geological Survey. *United States Mineral Resources*. Professional Paper 820. Washington, D.C.

Questions

1. List as many ways as you can to reduce energy consumption.

2. Using Table 11.1, calculate the amount of coal, oil, or natural gas required to fuel a 1000 MW (1 million kW) power plant for 1 day assuming a power plant efficiency of: (a) 100%; (b) 40%.

   ans. (a) 2960 MT, 15,000 bbls, 80 million ft^3

   (b) 7400 MT, 37,500 bbls, 200 million ft^3

3. Suppose the Alaska pipeline, carrying 2 million barrels of oil per day, is used only to fuel power plants. How many 1000 MW power plants operating at 40 % efficiency could be kept continuously operating? Assuming each power plant supplies the electrical needs of 1 million people, how many people could be serviced?

   ans. 53; 53 million

4. Derive the following expression for the number of years (T) to deplete a quantity of reserves (R), when the present rate of consumption is P, and consumption grows exponentially at the rate (r).

$$T = \frac{1}{r} \ln\left(\frac{rR}{P} + 1\right)$$

5. Use the equation given in Question 4 to estimate how long it would take for the U.S. alone to consume the world's petroleum reserves. Assume present consumption of 6 billion barrels per year and an annual rate of increase of 3 %.

   ans. 47 years

6. At $3.18 per barrel of petroleum, $0.17 per thousand cubic feet of natural gas, $6.90 per metric ton of coal (1970 prices), and 0.5¢ per kWh, calculate the unit price of each of these energy sources in dollars per million British thermal units.

   ans. $0.58; $0.165; $0.25; $1.47

7. Explain the difference between power and energy. Which of the following units measure energy and which power: watt-second, inch-ounce, megawatt, British thermal units per second.

8. Suppose you consume 3000 food calories per day. What average amount of power, measured in watts, does this correspond to?

   ans. 145 W

Chapter 12

Electrical Energy and Power

In 1950, 14.6% of all energy consumed in the United States went into the production of electricity. By the year 2000 that figure is predicted to increase to 42%. The *per capita* consumption of electrical energy is predicted to increase from 2161 kilowatt hours (kWh) in 1950 to 32,210 kWh in the year 2000—a 16 fold increase in just 50 years! The growth rate in the electrical sector can best be described as spectacular.

In view of its importance, not only in terms of gross energy consumption but also in terms of environmental effects, we shall devote three chapters to the discussion of electrical energy. This chapter is concerned with the growth and environmental effects of electricity production and consumption, from a reasonably general point of view. A good portion of the chapter is devoted to thermal pollution.

As we shall see, most of today's electrical power is generated by fossil-fueled steam-electric power plants, but a major portion of future growth will be supplied by nuclear power. The unique set of environmental questions which are arising from this shift to nuclear power will be discussed in Chapter 13. Chapter 14 will then present some alternative sources of energy for future electrical production.

12.1 *Electric Generating Capacity*

Recall that the important electrical units of power are kilowatts (kW, 1000 watts) and megawatts (MW, 1 million watts). A typical large steam-electric unit in 1955 produced about 300 MW but the average of all units in operation was only 35 MW. By 1968 the largest individual unit was 1000 MW and the average of all units in operation was 66 MW (Federal Power Commission 1971). Several generating units may be combined at a single steam plant, such as the 4 unit, 3200 MW Monroe plant of the Detroit Edison Company which was the largest in the nation in 1973. It will be convenient, and reasonably accurate, to consider the typical large steam power plant of today to have a capacity of 1000 MWe (1×10^9 W, or 1 million kW, or 1 kMW—the "e" signifies this is *e*lectrical power out of the plant as opposed to thermal power into the plant). A single such power plant, operating at full capacity, 24 hours a day, would be able to supply the 1973 electrical needs of 1 million people, so it is a convenient unit.

Figure 12.1 shows the electric-power generating capacity in the United States from 1950, projected out to 2000.* The installed capacity in 1971 was 368 kMWe and growing at an annual rate of about 7%. Only about 2.4% of the installed capacity was nuclear (8.7 kMWe); 15.2% was hydroelectric power (55.9 kMWe); and the bulk of all generation was from fossil-fuel sources (coal, natural gas, and oil), comprising 82.4% of the total (302.8 kMWe).

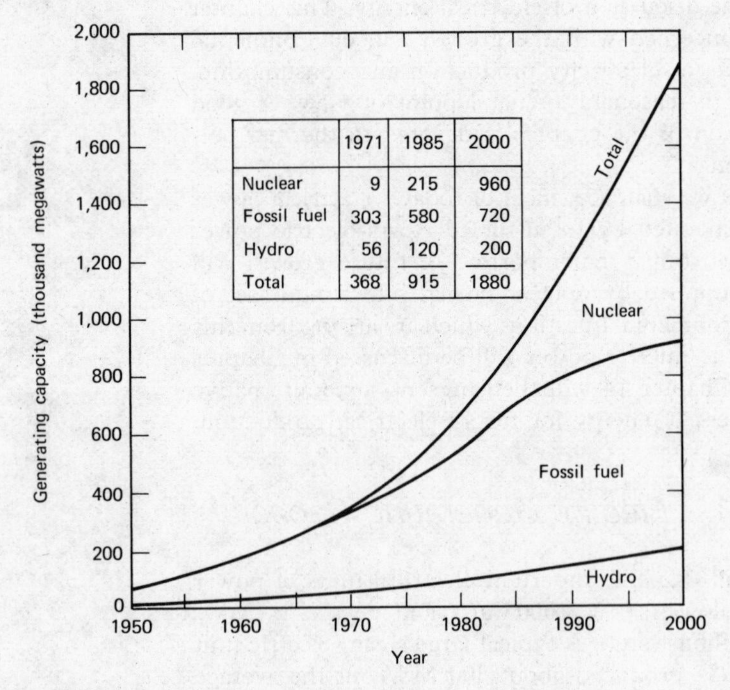

FIGURE 12.1 Projected electrical generating capacity by energy source (Dupree and West 1972).

*The projections are USDI estimates based on the growth parameters outlined in the last chapter.

Except for a few years during the depression, the annual growth rate has held at about 7% ever since the inception of the electric power industry in the 1880's resulting in a doubling of capacity every 10 years. As Figure 12.1 indicates, while nuclear power plants supplied almost none of the capacity in the early 1970s (2.4% in 1971), by the year 2000 they may be supplying just over 50%. To do so will require a more than 100-fold increase in installed capacity in just 29 years.

There is alarm in some quarters that we may be rushing these nuclear plants into production before their safety can be assured. If in 10 years or so, we discover that they are unsafe, the country will already be so dependent upon them for power that it would cause severe disruption to society to have to abandon them, hence the urgency to establish their credibility immediately.

While Figure 12.1 shows the amount of electric power which the nation *could* generate, it must be realized that this equipment is not operating at full capacity. The demand for energy is not constant but shows hourly, daily, and seasonal fluctuations, and it is necessary to have an installed capacity which is at least as great as the largest peak in demand. Figure 12.2 shows the weekly load curve for a typical large utility which points out the higher demand during the day than the night, and the higher weekday demand than weekend demand. The difference between the installed capacity and the peak demand is called the "reserve" or "margin." While utilities like to keep this margin at about 20 %, combinations of circumstances such as hot weather (which increases the air-conditioning load) and equipment outages can cause it to shrink to zero resulting in "brownouts" (decreased voltage on the line) or "blackouts" (no voltage at all). The summer margin in 1972 was 20.5% while the December margin was 36.1% (Edison Electric Institute 1973).

Also shown in Figure 12.2 is the "base load," which is always present, an "intermediate" load which is present for at least 12 hours on weekdays, and the "peak" load. The usual practice is to use the new, large fossil-fueled or nuclear steam-electric units to meet the base load, since they operate best at high and continuous loads. The older fossil-fueled plants often serve the intermediate load and hydroelectric plants, because their output is the most easily varied, are often used for peaking.

Since it is the anticipated peak load which determines the generating capacity which must be available, it would be worthwhile to smooth out the load curve to lower the peaks and raise the valleys. The same total amount of energy could be generated with less equipment. One suggested way to accomplish the smoothing would be to charge higher rates during the peak hours of the day so that users would be encouraged to shift consumption into the late night if possible.

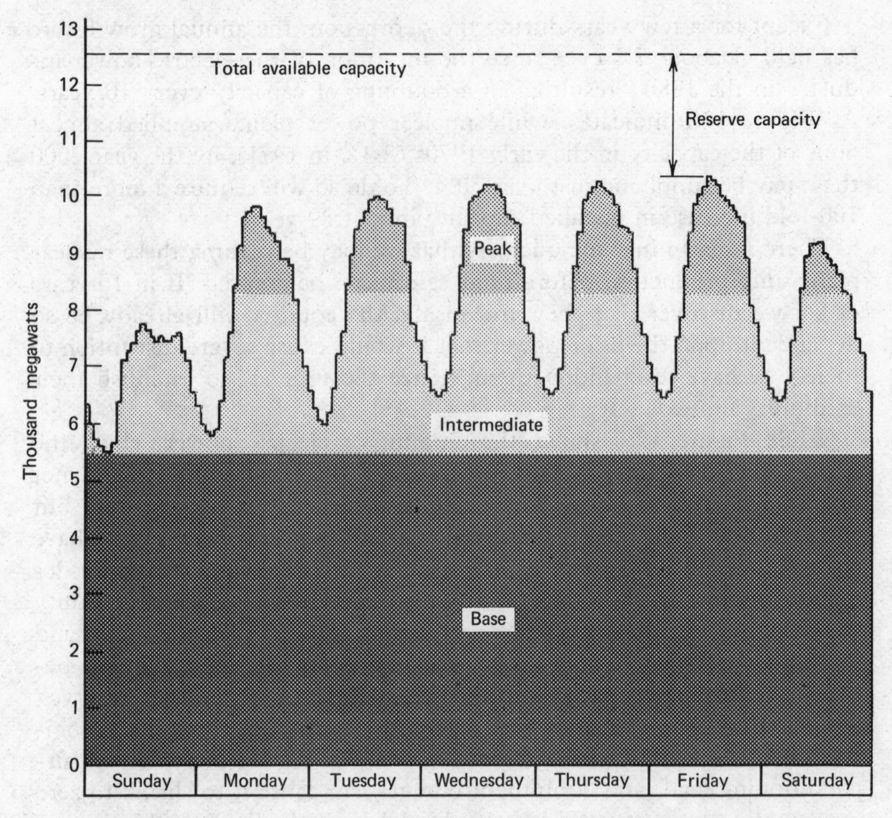

FIGURE 12.2 Example weekly load curve, indicating reserve capacity and base load. (Federal Power Commission 1971).

Another approach is to use some of the idle capacity at night to pump water uphill into a reservoir. Then during the peak load, the water is drained from the reservoir through turbines to generate extra power to meet the daytime demand. *Pumped storage* schemes of this type require about 3 units of energy for every 2 units that they deliver. Thus, while the economics of pumped storage make sense, the additional pollution and resource consumption required per unit of energy delivered detracts from the desirability of such projects.

12.2 *Electric Energy Consumption*

While Figure 12.1 showed the growth in the nation's generating capacity (power), Figure 12.3 shows the increase in electric energy consumption. Consumption is growing at about 6.1% per year. Also plotted is per capita consumption, which is growing nearly as fast. The 1971 per capita consumption of electric energy of 7800 kWh is predicted to increase by 415% to 32,210 kWh by the year 2000. This increase can easily be termed fantastic and just as easily can be called crazy.

Figure 12.4 shows the predicted energy inputs to the electrical sector. The contribution by natural gas has already passed its maximum and the contribution by oil will peak in about 1985. Both of these fossil fuels are progressively becoming less important for power generation. The consumption of coal is going to increase significantly and will be the dominant source of energy until about 1988. This rapid growth in coal-fueled power plants which is often overlooked when the need for nuclear power plants is discussed, demonstrates the urgent need for further research into processes which will make them environmentally acceptable.

The categories of electric power use, and their corresponding annual consumptions are listed in Table 12.1. As can be seen, while the total consumption is predicted to increase greatly, the percent of consumption in each sector is predicted to remain fairly constant. Nearly half of the nation's consumption is by industry and one-quarter is by residential users.

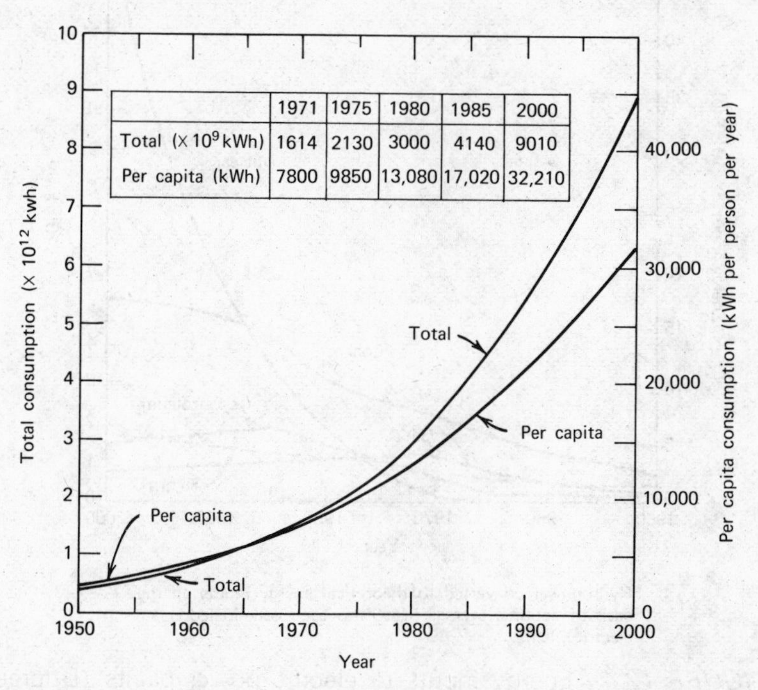

	1971	1975	1980	1985	2000
Total ($\times 10^9$ kWh)	1614	2130	3000	4140	9010
Per capita (kWh)	7800	9850	13,080	17,020	32,210

FIGURE 12.3 Electric energy consumption (Dupree and West 1972).

TABLE 12.1 Annual Consumption of Electric Energy by Class of Use, Contiguous U.S.[a]

Category	1965		1980		1990	
	10^9 kWh	Percent	10^9 kWh	Percent	10^9 kWh	Percent
Industrial[b]	538	47	1,384	43	2,536	42
Residential[c]	270	23	791	25	1,467	25
Commercial	190	16	577	18	1,138	19
Miscellaneous and losses	160	14	450	14	837	14
Total	1,158	100	3,202	100	5,978	100

[a]Source: Based on Federal Power Commission 1971.
[b]Includes industrial in-plant generation.
[c]Includes residential use on farms.

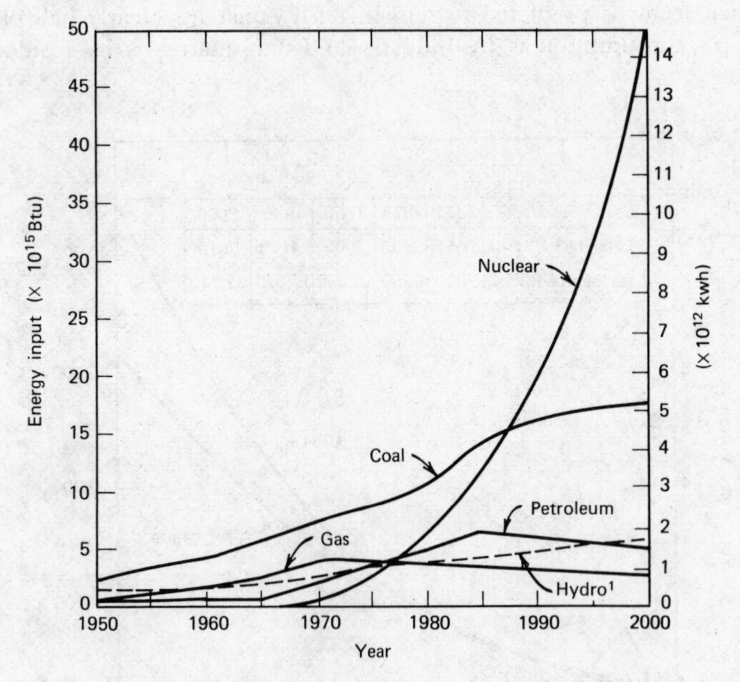

[1] Hydropower converted to theoretical energy inputs on the basis of national average heat rates for fossil–fueled steam–electric plants.

FIGURE 12.4 Energy inputs to electric power plants (Dupree and West 1972).

12.3 Residential Consumption

The growth in the average household consumption of electric energy can be attributed to several important factors. A significant factor in the increase is the growing number of all-electric homes. The number of such dwellings is projected to increase from 4.2 million in 1970 to 24 million in 1990, in spite of the gross inefficiency associated with electric home heating, as was pointed out in the last chapter (Federal Power Commission 1971). The average electric energy consumption in an all-electric home is nearly three times as great as the average residential customer. In 1970, the average residential customer consumed 7000 kWh, which, at 2.32¢/kWh, would cost $162, while the average all-electric home consumed 20,000 kWh, which, at the same rate, would cost $464.*

A second factor contributing to residential electrical energy consumption is the increasing number of major household appliances. Typically as one product begins to reach market saturation, something new is introduced which then continues the overall pattern of growth. By 1960, near saturation had already been reached for such appliances as vacuum cleaners and clothes washers, and every home had a black-and-white television. The present major growth items include air conditioners, clothes dryers, dish washers, and waste disposals.

Besides increasing the number of appliances in each household, a further growth factor is the increasing energy consumption of each appliance. For example, a color television requires 40% more power than a black-and-white television; a frostless refrigerator consumes about 67% more energy than a frosty one; and to go from a normal refrigerator-freezer to a frostless one increases energy consumption by 61%. Table 12.2 lists some typical home appliances with the power requirement for each; the typical total amount of energy consumed by each in one year (determined by multiplying the power consumption by an estimate of the number of hours the appliance is used each year); and the energy cost per year based on the average residential rate of 2.32¢/kWh. As can be seen, the gadgets such as electric toothbrushes and carving knives consume a negligible amount of energy because they are used so little. Also note that a hair dryer with a power requirement of 381 W uses less energy per year than an electric clock which is rated at 2 W. The difference, of course, being the grossly different number of hours each appliance is used per year.

The list in Table 12.2 indicates that energy consumption is highest

*Actually, because of the decrease in unit cost with increasing consumption, the figure should be slightly lower.

TABLE 12.2 Approximate Power and Energy Requirements for Selected Household Electric Appliances Under Normal Use, 1969[a]

Appliance	Average wattage	Estimated kilowatt hours consumed annually	Annual[b] cost, dollars
Air conditioner (window)	1,566	1,389	32.00
Bed covering	177	147	3.40
Broiler	1,436	100	2.30
Carving knife	92	8	0.19
Clock	2	17	0.40
Clothes dryer	4,856	993	23.10
Coffee maker	894	106	2.46
Dishwasher	1,201	363	8.42
Fan (window)	200	170	3.94
Food blender	386	15	0.35
Food freezer (15 cu.ft.)	341	1,195	27.70
Food freezer (frostless 15 cu. ft.)	440	1,761	40.90
Food waste disposer	445	30	0.70
Frying pan	1,196	186	4.32
Hair dryer	381	14	0.32
Heat pump	11,848	16,003	371.00
Heater (radiant)	1,322	176	4.08
Hot plate	1,257	90	2.09
Iron (hand)	1,088	144	3.34
Lights (5 bulbs, 5 hrs per day)	500	910	21.10
Radio	71	86	2.00
Radio-phonograph	109	109	2.53
Range	12,207	1,175	27.30
Refrigerator (12 cu. ft.)	241	728	16.90
Refrigerator (frostless, 12 cu. ft.)	321	1,217	28.20
Refrigerator-freezer (14 cu. ft.)	326	1,137	26.40
Refrigerator-freezer (frostless, 14 cu. ft.)	615	1,829	42.50
Sewing machine	75	11	0.25
Shaver	14	18	0.42
Television (black and white)	237	362	8.40
Television (color)	332	502	11.65
Toaster	1,146	39	0.91
Tooth brush	7	5	0.11
Vacuum cleaner	630	46	1.07

[a]Source: Modified from Edison Electric Institute data given in *1970 National Power Survey*, 1971.

[b]Calculated at 2.32 ¢/kWh.

TABLE 12.2 Continued

Appliance	Average wattage	Estimated kilowatt hours consumed annually	Annual[b] cost, dollars
Waffle iron	1,116	22	0.51
Washing Machine (automatic)	512	103	2.39
Washing machine (nonautomatic)	286	76	1.76
Water heater (standard)	2,475	4,219	99.50
Water heater (quick recovery)	4,474	4,811	111.60
Water pump	460	231	5.36

for those applications where electricity is used for heating and cooling. In fact, of the total residential electric energy consumption in 1968, 18% was for refrigeration, 16% for water heating, 12% for space heating, and 11% for air conditioning. The biggest nonheating uses were lighting, 11%; and television, 9.5%, as shown in Table 12.3.

In fact, of the total amount of energy consumed in residences, both from direct combustion of fossil fuels and from home electric energy consumption, 85.9% is used simply for space heating, water heating, and air conditioning.* With very minor advances in technological development, these three end uses for energy could easily be met by individual home solar energy devices. Dividing the total residential energy consumption of 2689 billion kWh by the number of residences, 60.4 million, suggests the average total energy consumption per residence is equivalent to about 45,000 kWh per year, or 120 kWh per day. As we shall see in Chapter 14, this much solar energy daily falls on a square 15 ft × 15 ft on an average, reasonably sunny day in the United States.

12.4 *The Generation of Electricity*

Let us shift our attention now to a discussion of how electric energy is generated so that we can understand the environmental problems posed by the various generation systems.

Common to any of the presently used systems is the turbine-generator pair. For our purposes, it is not particularly useful to go into any of the

*This is energy actually consumed in homes and does not include the energy lost at the power plant which generates the electricity.

TABLE 12.3 Energy Consumption in Residences in 1968 from Fossil Fuel and Purchased Electricity[a,b]

Use	Primary energy[c] (fossil fuel) billion kilowatt hours	Purchased electric energy,[d] billion kilowatt hours	Total energy, billion kilowatt hours	Percent of total
Space heating	1820	48	1868	69.5
Water heating	330	65	395	14.7
Cooking	110	28	138	5.1
Refrigeration	1	73	74	2.8
Air conditioning	1	45	46	1.7
Lighting	—	44	44	1.6
Television	—	38	38	1.4
Clothes Drying	20	15	35	1.3
Other	—	51	51	1.9
Total	2282	407	2689	100

[a]Source: Office of Science and Technology 1972.
[b]60.4 million residences, total.
[c]Converted from Btu at 3412 Btu/kWh.
[d]Does not include power plant losses.

technological details of their design, but a basic understanding is helpful. Turbines are devices which convert the energy of a moving fluid—usually water or steam—into the rotational energy of a shaft. At a dam, water flows through the turbine blades rotating the shaft, while at a "steam plant" (no matter what type of fuel is being used to heat the water) high-pressure steam is allowed to expand through the turbine blades causing the rotation. Figure 12.5 shows schematically the function of a turbine.

The second component, the generator, converts the rotational energy

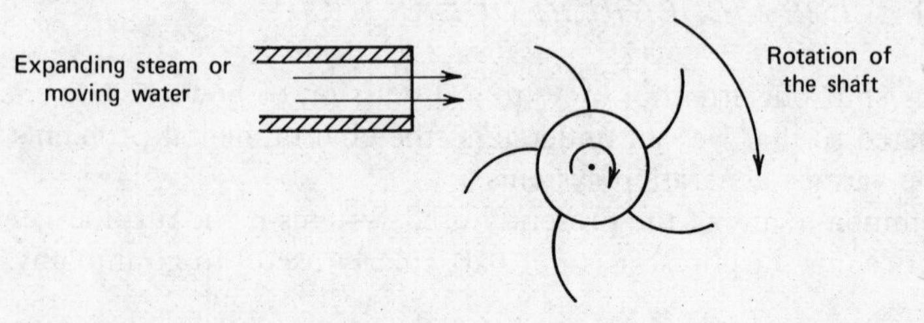

FIGURE 12.5 In a turbine the energy in the working fluid is converted to rotational energy of a shaft.

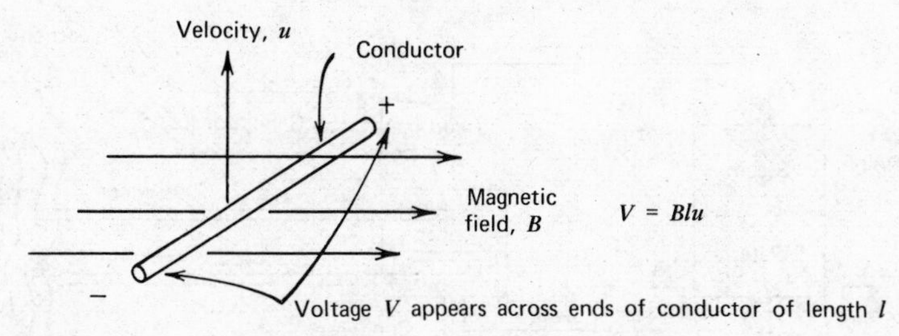

FIGURE 12.6 Principle of a generator: Moving a conductor through a magnetic field produces a voltage.

of the turbine shaft into electrical energy, in accordance with the principles discovered by Michael Faraday in the early nineteenth century. Any time an electrical conductor, such as a piece of copper wire, is moved through a magnetic field, a voltage will be developed across the ends of the wire, the voltage being proportional to the length of the wire, the strength of the magnetic field, and the speed that the wire is moved through the field (Figure 12.6).

Turbines and generators are coupled together; the turbine receiving its energy from the steam (or water), the generator receiving its energy from the turbine. The output from the generator is an electrical voltage. For transmission efficiency, this voltage is raised to perhaps 345 or 500 kilovolts (kV) in a transformer and then put out onto the transmission lines. Figure 12.7 shows a sketch of the system.

While Figure 12.7 shows enough equipment to generate electricity at a hydroelectric plant, a steam electric plant needs additional equipment to handle the steam, as shown in Figure 12.8. In the boiler, heat, either from fossil-fuel combustion or nuclear fission,* converts the water into high-pressure, high-temperature steam (perhaps 3500 pounds per square

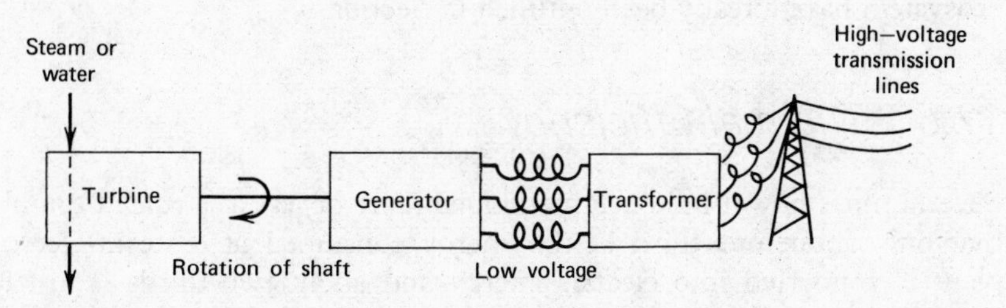

FIGURE 12.7 Turbine, generator, transformer, and transmission line for generating and distributing electrical power.

*Actually, nuclear reactors can be somewhat more complicated, but the principles are the same (see Chapter 13).

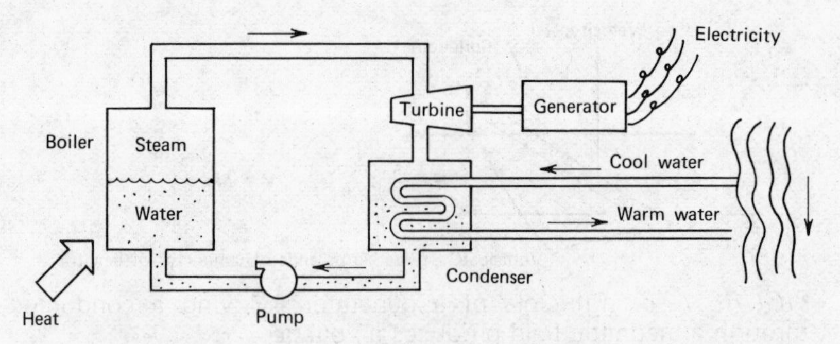

FIGURE 12.8 Main components of a steam power plant.

inch (psi) and 1000° F). The steam is fed to the turbine where it is allowed to expand through the blades, rotating the turbine shaft. The steam leaving the turbine is condensed back into liquid water by passing it over a network of cold pipes in the condenser. The condensate is then returned to the boiler, closing the cycle. Reusing the same water over and over again in the steam cycle not only conserves water but also makes it easier to maintain its purity, which is essential to avoid corrosion and the buildup of mineral deposits in the system.

The temperature of the condenser cooling pipes is kept low by passing large volumes of cold water through them. This cooling water, which is drawn from a river, lake, or ocean, gets heated in the condenser and is returned to its origin perhaps some 20° F higher in temperature. It is not unusual for nearly two-thirds of the energy which is liberated by the fuel in the boiler to end up in this cooling water. When the warmed-up cooling water is returned to the river (or wherever), the result is called thermal enrichment, thermal addition, or thermal pollution, depending on your point of view. The effects of raising the temperature of an aquatic ecosystem have already been outlined in Section 5.7.

12.5 *Thermal Efficiency*

Several times now it has been mentioned more or less as a rule of thumb, that only about one-third of the energy consumed at a steam-electric plant is converted into electric energy and about two-thirds is lost as heat. This is, of course, an example of the Second Law of Thermodynamics which tells us we cannot expect to convert all of the potential energy in the fuel into useful work, but why is the efficiency so low?

The steam-electric plant we have just described is an example of a *heat engine,* a subject studied at some length in thermodynamics. Figure

12.9 shows a theoretical heat engine operating between two heat reservoirs, one at temperature T_h and one at T_c. An amount of energy Q_h is transferred from the hot reservoir to the heat engine. The engine does work W and rejects an amount of waste heat Q_c to the cold reservoir. The analogous quantities in the steam plant are also shown in the figure. Heat Q_h is supplied to the boiler by the fuel, work W is done by the turbine, and waste heat Q_c is rejected to the environment from the condenser.

The most efficient engine which could possibly operate between the two heat reservoirs is called a *Carnot* engine after the French engineer Sadi Carnot. For a Carnot engine, the following ratio holds (and in fact can be used as a definition of absolute temperature):

$$\frac{Q_h}{Q_c} = \frac{T_h}{T_c} \tag{12-1}$$

In equation (12-1), T_h and T_c are absolute temperatures, that is they are measured using the Rankine scale or the Kelvin scale. Conversions from Centigrade to Kelvin, and Fahrenheit to Rankine are:

$$°K = °C + 273.15$$

$$°R = °F + 459.67$$

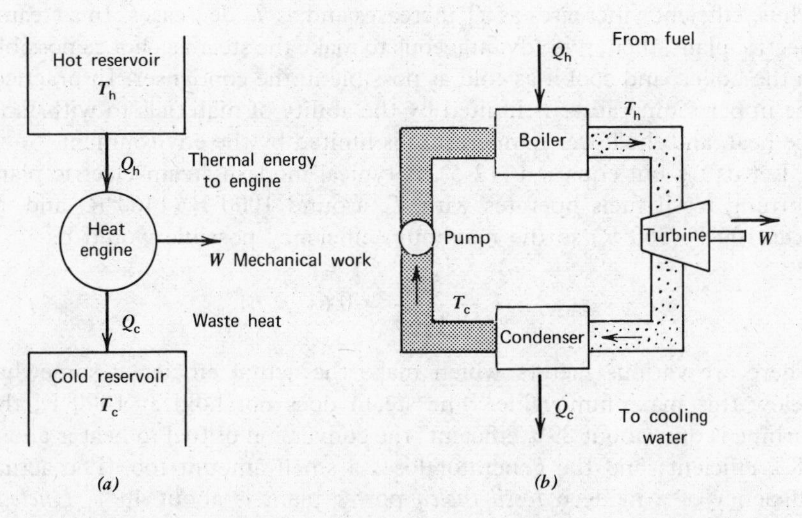

FIGURE 12.9 A heat engine converts a portion of the thermal energy received from a hot reservoir into mechanical work with the remainder becoming waste heat rejected to a cold reservoir: (*a*) Theoretical heat engine; (*b*) corresponding steam-electric plant.

The efficiency of any heat engine is simply the ratio of the amount of work produced, W, to the amount of energy put into the system, Q_h:

$$\text{efficiency} = \varepsilon = \frac{W}{Q_h} \tag{12-2}$$

The First Law of Thermodynamics states that energy is always conserved, so we can write

$$Q_h = W + Q_c \tag{12-3}$$

Combining equations (12-3) and (12-2) gives the following efficiency relationship, good for any heat engine:

$$\varepsilon = \frac{Q_h - Q_c}{Q_h} = 1 - \frac{Q_c}{Q_h} \tag{12-4}$$

By substituting equation (12-1) into (12-4), we obtain the following expression for the maximum efficiency possible for any heat engine operating between the two temperatures T_h and T_c.

$$\varepsilon_{max} = 1 - \frac{T_c}{T_h} \tag{12-5}$$

Thus, efficiency increases as T_h increases and as T_c decreases. In a steam-electric plant, then, it is advantageous to make the steam as hot as possible in the boiler, and cool it as cold as possible in the condenser. In practice, the upper temperature is limited by the ability of materials to withstand the heat, and the lower temperature is limited by the environment.

Let us try out equation (12-5). A typical modern steam-electric plant burning fossil fuels operates with T_h around 1000°F (1460°R) and T_c near 100°F (560°R) so the maximum efficiency possible would be

$$\varepsilon_{max} = 1 - \frac{560}{1460} = 0.61 = 61\%$$

There are various factors which make the actual efficiency somewhat below this maximum value. The steam does not hold at 1000°F; the turbine is only about 89% efficient; the conversion of fuel to heat is about 88% efficient; and the generator loses a small amount too. The actual efficiency of a modern *fossil-fueled* power plant is about 40 %. *Nuclear* power plants do not operate at as high a steam temperature and so their efficiency is somewhat lower, being more like about 33%. The total mix of new and old fossil-fueled plants and nuclear plants, actually in operation at the present time, has an efficiency of close to 33 %.

Thus for every 3 units of energy that is expended at a steam-electric power plant, 2 end up as waste heat and 1 is converted to electric energy. From a conservation viewpoint, any time the consumption of 1 unit of electric energy can be eliminated, 3 units of total energy will be saved.

12.6 *Cooling Water Requirements*

As was pointed out in Chapter 4, the amount of cooling water required to carry away the waste heat from power plants already exceeds the amount withdrawn for any other purpose. The need for cooling water is one of the primary considerations in siting new power plants.

Let us compare the amount of cooling water required by a modern 1000 MWe fossil-fueled plant operating at 40% efficiency with that required by a 33% efficient, 1000 MWe nuclear plant. For the fossil-fueled plant, 40% of the input energy is converted to electric energy, 10–15% is lost to the air by the stack gases and boiler, and the remaining 45–50% is removed by the cooling water. For the nuclear plant, 33% is converted to electric energy, about 3–5% is lost to the air, and the remaining 62–64% ends up in the cooling water (Krenkel, Bradley, Hastings, Jaske, Mihursky 1972).

Figure 12.10 illustrates the power flow through the two plants and points out that while the nuclear plant is only 7% less efficient, its cooling water must remove about 60% more waste heat, which is why thermal pollution is most often associated with nuclear plants.

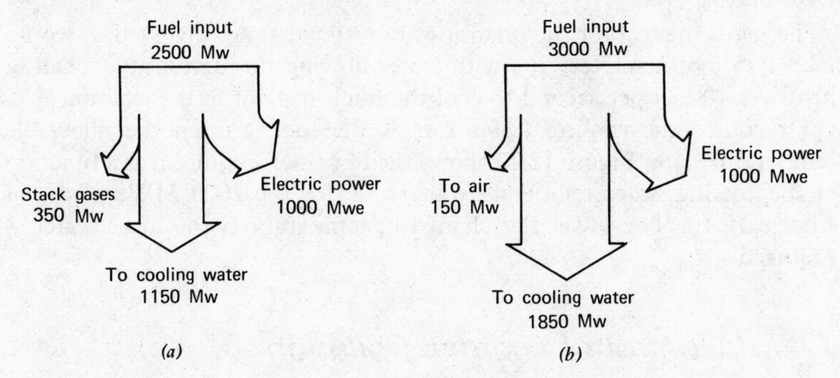

FIGURE 12.10 Comparison of fossil and nuclear power plants for heat lost to cooling water when both generate 1000 MW of electric power: (*a*) Fossil at 40% efficiency; (*b*) nuclear at 33% efficiency.

EXAMPLE For the 1000 MWe nuclear plant of Figure 12.10 operating at full capacity, calculate the minimum rate at which cooling water must pass through the condenser to keep the cooling water temperature rise from exceeding 15°F.

Solution: The cooling water must carry away 1850 MW, or 1.85×10^6 kW of waste heat, which is equivalent to

$$1.85 \times 10^6 \text{ kW} \times 3413 \text{ Btu/kWh} = 6.3 \times 10^9 \text{ Btu/hr}$$

By definition, 1 Btu can raise the temperature of 1 pound of water by 1°F, so to raise 1 pound by 15°F would require 15 Btu. Therefore the amount of cooling water required is

$$6.3 \times 10^9 \text{ Btu/hr} \times \frac{1}{15 \text{ Btu/lb}} = 420 \times 10^6 \text{ lb/hr}$$

Converting this answer to the more convenient unit of cubic feet per second (cfs), gives

$$420 \times 10^6 \text{ lb/hr} \times \frac{1}{62.4 \text{ lb/ft}^3} \times \frac{1}{3600 \text{ sec/hr}} = 1850 \text{ cfs}$$

It is interesting to note that, for a 15°F rise in cooling water temperature, the numbers conveniently work out about 1 cfs of water required for every megawatt of power to be removed. Thus, for example, the fossil-fueled plant of Figure 12.10a would require about 1150 cfs of cooling water, whose temperature would rise 15°F, to remove the 1150 MW of waste power.

To get some feel for the quantities of water that are involved, it would take a 15 foot diameter pipe with water flowing through it at a speed of just over 10 feet per second to cool the nuclear plant in the example.

Of course the cooling water rate is dependent upon the allowable temperature rise. Figure 12.11 shows the flow rates required, as a function of the cooling water temperature increase, for the 1000 MWe plants of Figure 12.10. The lower the change in temperature, the more water is required.

12.7 *Control of Thermal Pollution*

So far we have implied that the cooling water is extracted from some suitable source, say a river, passed through the condenser where its temperature is increased by anywhere from 10 to 30°F, and returned to the source body of water. Eventually the warmed sink gives up this

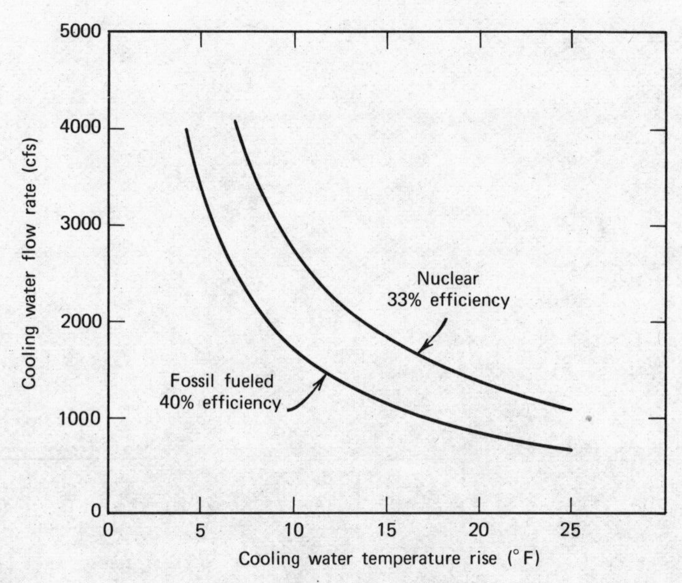

FIGURE 12.11 Cooling water requirements as a function of temperature change for 1000MWe steam-electric plants of Figure 12.10.

extra heat to the atmosphere. Such a system, which is referred to as "once through" cooling, may cause unacceptable environmental changes.

It is possible to control thermal pollution by passing the heated cooling water through a *cooling pond* or *cooling tower* and then returning it to the condenser for reuse as shown in Figure 12.12:

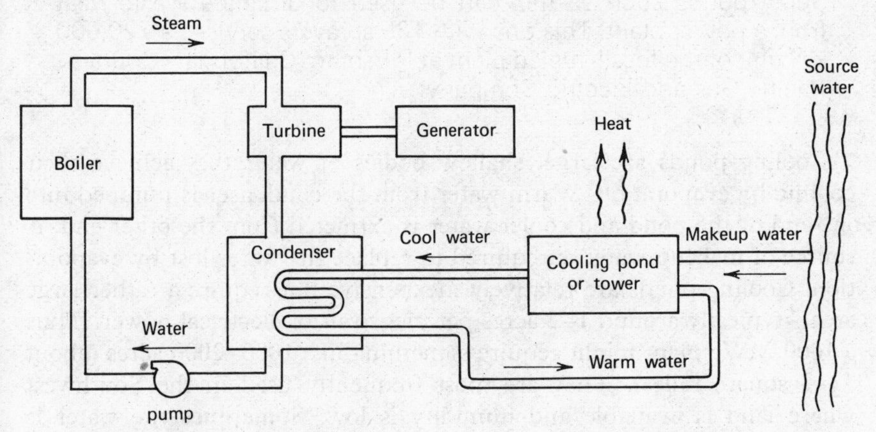

FIGURE 12.12 Thermal pollution control with a cooling pond or cooling tower.

Spray ponds such as this can be used to dissipate waste heat from a power plant. This one with 135 sprayers services a 720,000 kW unit of the fossil-fueled plant at Pittsburg, California. (Courtesy Pacific Gas and Electric Company.)

Cooling ponds are large, shallow bodies of water that achieve their cooling by evaporation. Warm water from the condenser is pumped into one end of the pond and cooler water is extracted from the other end. A source of makeup water is required to replace the water lost by evaporation. Cooling ponds are relatively inexpensive but require a rather large area—typically around 1–2 acres per megawatt of electrical power. Thus a 1000 MWe plant might require something like 1000–2000 acres (about $1\frac{1}{2}$–3 square miles). They are most frequently used in the Southwest where land is available and humidity is low. Sometimes the water is mechanically sprayed into the air to encourage the evaporative cooling, in which case it is called a "spray pond."

Cooling towers can be classified as *wet* or *dry*. Wet cooling towers achieve cooling by evaporation, and so, like cooling ponds, require a source of makeup water. Dry cooling towers, on the other hand, cool by conduction and convection, in much the same way that an automobile radiator cools. The cooling system is closed and there is no loss of water. For economic reasons, dry cooling towers are very seldom used for power plants. There is one servicing a small 20 MW unit in Wyoming and there are a few slightly larger ones outside of the country.

Wet cooling towers can be classified as *natural* draft or *mechanical* draft. In a mechanical draft tower, air is forced through a spray of the warm water, by large motor driven fans. The towers are smaller and less expensive to build than natural draft ones, but their operating costs are high. Until recently cooling towers were of the mechanical draft type, but many new plants are using huge, hyperbolic, natural draft towers similar to the one shown in Figure 12.13. The warm water from the condenser is sprayed over baffles, which encourages evaporation, and is then allowed

These wet, natural-draft cooling towers at the Rancho Seco nuclear generating plant are 425 feet high and 325 feet across at their base. (Photograph by W. A. Masters.)

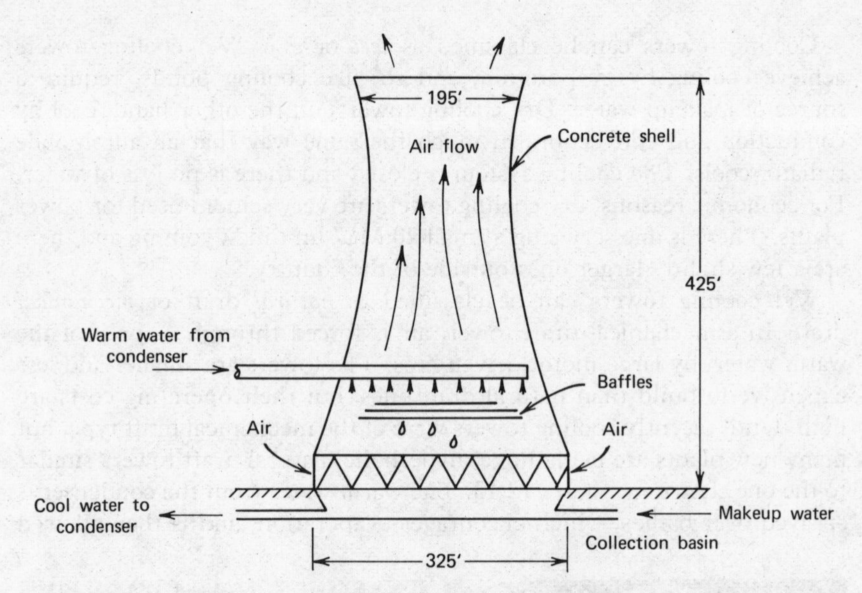

FIGURE 12.13 Wet, natural-draft, cooling tower. Dimensions are those at the Rancho Seco tower outside of Sacramento, California.

to drip down to the collection basin. Outside air is naturally drawn in at the base of the tower to replace the less dense, warm, moist air which is rising out of the tower.

An outside source of makeup is required to compensate for the water lost to evaporation (about 3%), as well as water lost during "blowdown." Blowdown is the continuous or periodic flushing of the cooling system to remove solids and chemicals which accumulate in the circulating cooling water. This flushing can become a water pollution problem unless special treatment is provided. These towers also create huge plumes which can contribute to local fogging and icing problems.

Table 12.4 summarizes some cost estimates for the various cooling systems just described. The estimates are calculated for a 40% efficient fossil-fueled plant and a 33% efficient nuclear plant, assuming a 20°F rise in cooling water temperature within the condenser. Thus for example, a typical natural draft cooling tower for a 1000 MWe nuclear plant would have a capital cost of about $11.5 million and, if the power plant operated at full capacity all year around, would have a total annual cost about $4.08 million. While these costs seem very high, the additional cost to the consumer, above once-through cooling, would be only about 1.49 %.

TABLE 12.4 Cooling System Costs for 40% Efficient Fossil-Fueled Steam Electric Plant and 33% Efficient Nuclear Plant, Assuming 20°F Cooling Water Temperature Change[a]

Cooling system	Equipment capital cost, $/KW		Operation and maintenance costs, $/KW-Yr.		Total cost, $/KW-Yr.		Additional[b] cost to consumer, %	
Cooling system	Fossil	Nuclear	Fossil	Nuclear	Fossil	Nuclear	Fossil	Nuclear
Once through	5.00	5.24	0.6	1.00	1.05	1.47	0.00	0.00
Cooling pond	6.50	7.50	0.76	1.24	1.34	1.92	0.16	0.26
Spray pond	7.60	8.10	0.90	1.50	1.58	2.23	0.30	0.43
Natural-draft wet tower	7.50	11.50	1.20	2.00	2.92	4.08	1.17	1.49
Mechanical-draft wet tower	7.20	9.40	1.88	2.66	3.23	4.20	1.25	1.56
Mechanical-draft dry tower	13.00	15.00	1.88	2.66	4.45	5.41	1.94	2.26

[a]Source: EPA 1969.

[b]Above cost of once through system.

TABLE 12.5 Emissions for 40% Efficient, Fossil-Fueled Steam Electric Power Plants, in Pounds per Megawatt Hour[a]

Fuel	NO_x	SO_2	Particulates[b]
Coal	6.8	13S[c]	57[d]
Oil	5.9	8.9S[c]	0.57
Natural gas	3.2	0.003	0.12

[a]Source: Derived from Duprey, 1968.
[b]Assuming no emission control.
[c]S is % sulfur, e.g. 2% sulfur coal would emit 26 lb/MWh SO_2.
[d]Assuming 10% ash in coal.

12.8 *Environmental Effects (Fossil Fuel)*

Let us summarize the environmental effects of fossil-fueled steam electric power plants. There are a number of environmentally damaging operations which occur even before the fuel is burned: strip mining, acid

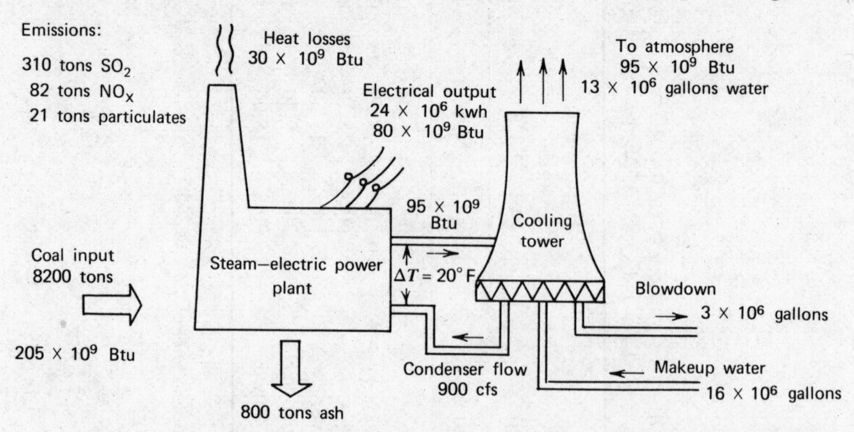

FIGURE 12.14 Daily input/output for a 40% efficient, 1000 MWe coal-fired steam-electric power plant with cooling tower, operating at full capacity. *Assumptions:* 25,000,000 Btu/ton coal, 10% ash, 2% sulfur; precipitator 97% efficiency; 40% efficient power plant; emissions based on 1968 NAPCA air pollutant emission factors.

mine drainage, oil spills, refinery emissions, etc. But specifically, at the power plant site, we have identified two major problem areas—air and thermal pollution. The amount of air pollution depends on the type of fuel being burned and the degree of emission control. Table 12.5 clearly indicates the desirability of natural gas as a fuel as compared to oil or coal.

Figure 12.14 summarizes the daily, on-site, air and thermal emissions from a 1000 MWe coal-fired plant operating at 40% efficiency. The plant's waste heat is given up to the atmosphere, mostly in the cooling tower. About 16 million gallons per day of makeup water is required to balance the 13 million lost out of the cooling tower and the 3 million required for blowdown.

12.9 *Conclusions*

Not only is total energy consumption in the United States growing rapidly, but the percentage of that energy which is being used to fuel electric generating plants is also increasing. The result is a doubling of electric energy consumption every 10 years.

To keep up with this growth in demand is going to become increasingly difficult. The most desirable fossil fuels, natural gas and oil, are becoming more difficult to acquire, and the other presently available options, coal and uranium, pose serious environmental problems.

There are some bright prospects for the future, however, if we can successfully develop any of the promising options which are being researched today. If a nonpolluting, abundant source of energy can be harnessed for the generation of electricity, it would become highly desirable to shift as much of our energy demand as possible into the electrical sector, to relieve the pressure on fossil fuels. However, the shift is already occurring, and because of the low conversion efficiency from thermal energy to electric energy, much of the fuel's energy content is being wasted.

Not only are the fuels not being utilized as efficiently as they might, but the wasted energy raises the heat load which the environment must dissipate. Fortunately, relatively inexpensive cooling devices are available for passing the heat directly into the atmosphere, which eliminates the aquatic thermal pollution. At some future date, if consumption continues its present growth rates, climatic changes may result from the increased atmospheric heat burden but that is not yet a problem.

Bibliography

Duprey, R. L. (1968). *Compilation of Air Pollutant Emission Factors*, USHEW, Public Health Service Pub. 999-AP-42, Durham, N.C.

Dupree, W. G., Jr., and West, J. A. (1972). *United States energy through the year 2000*. U.S. Department of the Interior. Washington, D.C. Dec.

Edison Electric Institute (1973). Electric industry set new highs in generation, capability to meet customers' demands in 1972. *Edison Electric Institute Bull.* Jan./Feb.

Federal Power Commission (1971). *The 1970 National Power Survey*. Washington, D.C. Dec.

Federal Power Commission (1972a). *Electric Power Statistics*. Washington, D.C. May.

Federal Power Commission (1972b). *Statistics of Privately Owned Electric Utilities in the United States 1971*. Washington, D.C. Oct.

Harleman, D. R. F. (1971). Heat—the ultimate waste. *Technol. Rev.* Dec.

Hossli, W. (1969). Steam turbines. *Sci. Am.* April.

Keilman, L. and Brown, L. (1971). Rancho Seco unit number 1, Design Considerations for a 900 MW Nuclear Unit at an Arid Site. ASME Winter Annual Meeting. Washington, D.C. Nov.

Krenkel, P. A., Bradley, L. B., Hastings, V. S., Jaske, R. T., and Mihursky, J. A. (1972). *The Water Use and Management Aspects of Steam Electric Power Generation.* Prepared for the National Water Commission, National Tech. Info. Service, Springfield, Va., Accession No. PB 210355, May.

Office of Science and Technology (1972). *Patterns of Energy Consumption in the United States*. Executive Office of the President, Washington, D.C. Jan.

U.S. Environmental Protection Agency (EPA) (1969). *A Survey of Alternate Methods for Cooling Condenser Discharge Water, Large-Scale Heat Rejection Equipment*, Dynatech R/D Company, Project No. 16130 DHS, Cambridge, Mass., July.

Questions

1. Table 12.2 indicates an average black-and-white television requires 237 W of power, and yearly consumes 362 kWh of energy. How many hours per day is this average television turned on? How many of those hours are worth watching?

   ans. 4.2 hours

2. Look at some of your old home electric bills and see if you can determine your average monthly electric energy consumption. If you cannot find the rate that your utility is charging, assume it is 2.3¢/kWh. How does your consumption compare to the national average of about 600 kWh per month? Using Table 12.2 can you account for most of your consumption?

3. Compare the 35¢ purchase price of a 750 hour, 100 W light bulb with the cost of the energy to keep it lit, over its lifetime.
ans. $1.72

4. What does it cost for the electricity to prepare your breakfast if you fry some eggs and bacon in an electric frying pan for 15 minutes, make some coffee in an electric pot (15 minutes), burn some toast (5 minutes), and top it off with a 10 minute waffle cooked in an electric waffle iron? (Use Table 12.2 and assume 2.3¢/kWh).
ans. 1.85¢

5. Electric energy can be utilized in a water heater at an efficiency of about 92%. Including power plant losses, estimate the overall efficiency from fuel to water heat. How does it compare to utilizing natural gas directly, at an efficiency of 64%?
ans. 31% overall

6. Explain the statement, "Saving 1 Btu of electric energy conserves 3 Btu of total energy."

7. The *heat rate* is defined to be the amount of heat input in Btu, required to produce a net output of 1 kWh of electric energy. The national record annual heat rate for a turbine-generator is 8534 Btu/kWh. What efficiency does this correspond to? What heat rate would be equivalent to a 33% efficient power plant?
ans. 40%, 10,239 Btu/kWh

8. If 50% of residential electric energy consumption could be eliminated, approximately what percent of total U.S. energy consumption would be saved? What if 50% of industrial electric energy could be eliminated?
ans. 3%, 5.9%

9. What is the Carnot efficiency for a power plant with steam temperature of 900°F and condenser temperature of 100°F? Which would

improve efficiency more: Adding 50° F to the steam temperature or subtracting 50° from the condenser temperature?

ans. 58.8%, subtract 50° F

10. For the 1000 MWe fossil-fueled plant in Figure 12.10, calculate the rate of flow of cooling water if the water temperature increases by 20° F in the condenser.

ans. 870 cfs

11. Calculate the daily air pollution emissions for the power plant in Figure 12.14 if natural gas is the fuel.

ans. 38.4 tons NO_x ; 0.036 tons SO_2 ; 1.44 tons particulates

Chapter 13

Radiation and Nuclear Power

The electric power industry is in the beginning stages of a radical changeover from the use of fossil fuel energy sources to a reliance on the energy liberated by nuclear fission. As was pointed out in the last chapter, the installed capacity of nuclear power plants is predicted to increase more than 100-fold between 1971 and 2000, and its share of the total capacity would correspondingly increase from 2.4 to over 50%.

Nuclear power plants offer several important advantages over conventional fossil-fueled plants—they produce no air pollution of the type we have previously discussed, there are no oil spills, strip mines, or acid water pollution, and they do not contribute to the depletion of the world's fossil fuel reserves. However, the blessings are mixed, in that along with the generation of power, nuclear fission also produces dangerous radioactive waste products which must be kept out of the environment for many hundreds of years.

13.1 *Atomic Structure*

As a simple but adequate model, we may consider the atom to consist of a dense nucleus, which contains a number of neutrons and positively charged protons, surrounded by a cloud of negatively charged, orbiting electrons. For an electrically neutral atom, the number of electrons equals the number of protons.

Nearly all of the mass of the atom is contained in the nucleus, which has an estimated density of around 100 million tons per cubic centimeter! Matter must therefore consist mostly of empty space. In fact, if we picture the nucleus as being the size of a tennis ball, the electrons would be revolving around it at an average distance of about 1000 feet.

The chemical element to which an atom belongs is determined by the number of protons in the nucleus—called the *atomic number*. The *mass number* of an atom is the sum of the number of protons and neutrons. Atoms which have the same number of protons, and hence are the same chemical element, but differing numbers of neutrons, are called *isotopes* of the particular element. The element of most concern in this chapter is uranium which is characterized by having 92 protons in the nucleus. Uranium has several isotopes, one of the most important is used below to explain the notation:

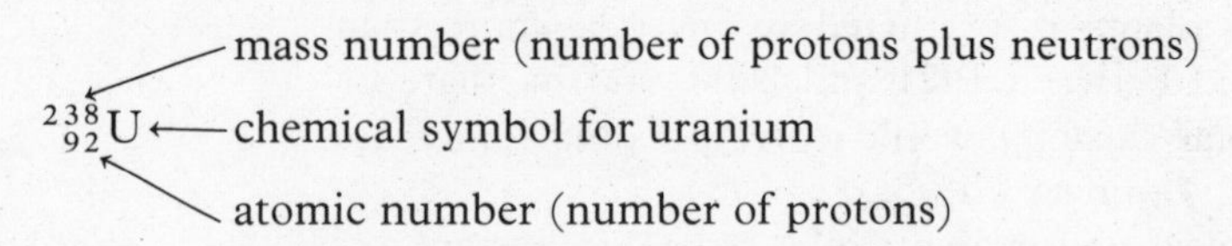

This notation is redundant in that the chemical symbol U implies 92 protons in the nucleus, so usually the notation for this isotope is abbreviated to ^{238}U. The number of neutrons in ^{238}U is 238 − 92 = 146. Another important isotope of uranium is ^{235}U which has 143 neutrons. Naturally occurring uranium consists of 99.283% ^{238}U, 0.711% ^{235}U, and 0.006% ^{234}U. Present fission reactors use the scarce isotope ^{235}U while future "breeder" reactors are being developed to use the more abundant ^{238}U.

13.2 Radioactivity

Some atomic nuclei are unstable, that is, *radioactive*, and during the spontaneous changes which take place, various forms of *radiation* are emitted. While all elements having more than 83 protons (bismuth) are naturally radioactive, it is possible to artifically produce unstable isotopes, or *radionuclides*, of virtually every element in the periodic table.

During radioactive decay, various particles may be emitted from the nucleus, as well as short wavelength electromagnetic radiation, called *gamma* (γ) rays.

Alpha (α) particles consist of two protons and two neutrons which are held together by strong nuclear forces (essentially a helium nucleus). When an unstable nucleus emits an alpha particle, its atomic number decreases by 2 units and its mass number decreases by 4 units. Gamma radiation may also accompany the decay. The following example shows the decay of plutonium into uranium:

$$^{239}_{94}Pu \rightarrow {}^{235}_{92}U + {}^{4}_{2}\alpha + \gamma$$

As an alpha particle passes through an absorbing material, its energy is gradually dissipated through interaction with the constituent atoms.

Its positive charge attracts electrons in its path, raising their energy levels or completely removing them from their nuclei (ionization). Alpha particles do not penetrate matter very deeply and can be stopped with an ordinary sheet of paper. The skin is sufficient protection for sources which are external to the body but taken internally, such as by inhalation or ingestion, they can be extremely dangerous.

Beta (β) particles are electrons that are emitted from an unstable nucleus as a result of the transformation of a neutron into a proton. As a result, the mass number remains unchanged while the atomic number increases by one. A gamma ray may or may not accompany the transformation. The following example shows the decay of strontium-90 into yttrium:

$$^{90}_{38}\mathrm{Sr} \rightarrow {}^{90}_{39}\mathrm{Y} + \beta$$

As negatively charged beta particles pass orbital electrons in matter, the coulomb repulsive force between the two can raise the energy level of the electrons or even kick them free of their corresponding atoms. While such ionizations occur less frequently than with alpha particles, beta penetration is much deeper. Alpha particles travel less than 100 μ* in tissue, while beta particles may travel several centimeters. They can be stopped by aluminum approximately one centimeter thick.

Gamma rays have no charge or mass, being simply a form of electromagnetic radiation. As such they can be characterized by a wavelength (or frequency) in the same way that other, more familiar forms of electromagnetic energy can, such as light rays or radio waves. Gamma rays and x rays, which differ only in their origin, have very short wavelengths in the range 10^{-11}–10^{-7} cm. Having such short wavelengths means they are capable of causing ionizations and hence are biologically damaging. These rays are very difficult to contain and may require several inches of lead shielding.

All of these forms of radiation are dangerous to living protoplasm. The electron excitations and ionizations which are caused by the radiation, cause molecules to become unstable, resulting in the breakage of chemical bonds and other molecular damage. The chain of chemical reactions that follows results in the formation of new molecules that did not exist before the irradiation. Then, on a much slower time scale, the organism responds to the damage which has been done, and it may, for example, be a matter of years before the final effects become visible, as in the induction of cancers.

One of the important parameters which characterizes a given radioactive isotope is its *half-life,* which is the length of time required for 50%

*Recall, 1 $\mu = 1 \times 10^{-6}$ m.

of the atoms to decay. For example, if we start with 100 million atoms of an isotope which has a half-life of 1 year, we would expect 50 million atoms to remain after 1 year, 25 million after 2 years, and so on. Note that this is a statistical parameter and as such is only valid for large numbers of atoms. The half-lives of some of the isotopes of interest to us are: cesium-137, 30 years; strontium-90, 28.8 years; iodine-131, 8.05 days; and plutonium-239, 24,360 years.

There are a number of commonly used radiation units, which, unfortunately can easily be confused. The *curie* (Ci) is the basic unit of decay rate; 1 ci corresponds to the disintegration of 3.7×10^{10} atoms per second which is approximately the decay rate of 1 g of radium. While the curie is a measure of the amount of radiation being emitted by a source, it tells us nothing about the radiation dose which is actually absorbed by an object. The *roentgen* (R) was designed to measure the effects of x rays or gamma rays *only*, and is defined in terms of the number of ionizations produced in a given amount of air.

Of more interest is the amount of energy actually absorbed by some material—be it bone, fat, muscle, or whatever. The *rad* (*r*adiation *a*bsorbed *d*ose) corresponds to an absorption of 100 ergs of energy per gram of any substance and has the further advantage that it may be used for any form of radiation, alpha, beta, gamma, or x. For water and soft tissue the rad and roentgen are approximately equal.

Another unit, the *rem* (*r*oentgen *e*quivalent *m*an) has been introduced to take into account the different biological effects that various forms of radiation have on organisms. Thus, for example, if a 10 rad dose of beta particles produces the same biological effect as a 1 rad dose of alpha particles, both doses would have the same value when expressed in rems. This unit is rather loosely defined making it difficult to convert from rads to rems, but for our purposes we may consider rads and rems to be numerically approximately equal for x rays, gamma rays, and beta rays. The rem will be the unit we use most.

13.3 *Effects of Radiation Exposure*

When discussing the effects of exposure to radiation there are many factors to be considered. For high radiation doses received by the whole body, the effects can be observed almost immediately and hence are relatively easy to characterize. Chronic exposure to low-level radiation, on the other hand, produces much more subtle effects which may not be observable for many years or even until future generations.

Acute exposures of the whole body to high levels of radiation, such as might occur following an atom bomb blast or some sort of nuclear accident, results in a potentially lethal illness called radiation sickness. The probability of death depends on the level of exposure as shown in Figure 13.1. For exposures above about 650 rads, death is almost certain, usually within a few hours or days. The probability of death drops to 50% for exposures of around 400 rads. This is usually expressed as the $LD_{50(30)}$ value which means the *l*ethal *d*ose which would be expected to kill 50% of the exposed organisms, within 30 days after exposure. For sublethal doses of around 100–250 rads, nausea, vomiting, and diarrhea would be expected within hours, followed by a latency period of several days during which no specific symptoms occur. A period of general malaise and fatigue with vomiting and diarrhea would then be followed by a gradual disappearance of symptoms. No deaths would be expected for doses of less than about 100 rads, but there may be delayed effects including a higher incidence of cancer, leukemia, sterility, and cataracts, and a reduction in lifespan could result.

Low-level exposures can cause *somatic* and/or *genetic* damage. Somatic damage may be observed in the organism which is exposed, while genetic damage, by increasing the mutation rate in chromosomes and genes,

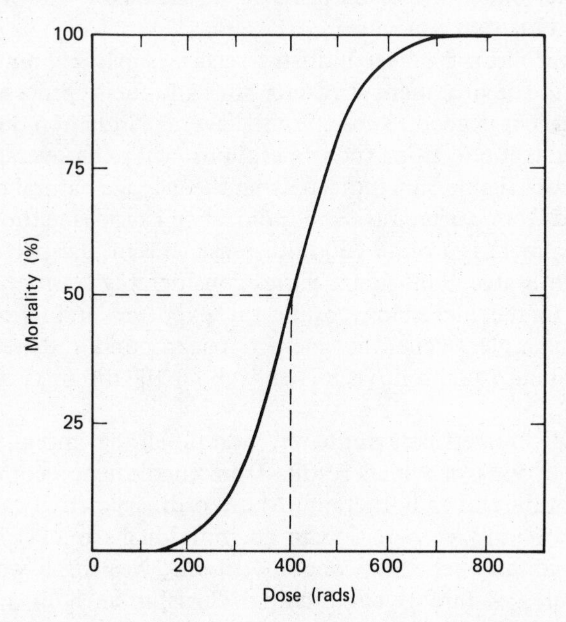

FIGURE 13.1 Human mortality curve, showing a 400 rad dose would be expected to result in 50% mortality (Lindop 1971).

affects future generations. Another complicating variable is the exposure rate. For example, does a single 10 rad dose cause the same amount of damage as 10 doses of 1 rad each spread over time? Genetic damage seems to be relatively independent of dose rate since, for all practical purposes, mutations are irreversible. However, for somatic effects, it seems a sequence of several smaller doses allows time for some repair to take place and so is less damaging than a single, larger dose.

13.4 *Chronic Exposure and Nuclear Power*

The biosphere is continuously exposed to a low level of ionizing radiation, the effects of which are largely unknown. Most of this radiation results from natural causes but there is a component that is added by man's activities.

There is somewhat of a controversy over whether there exists some level of radiation exposure below which there is no damage. If such a threshold exists it might be true that the man-made additions are not harmful. However, there is considerable evidence to indicate that there is no threshold and that any exposure is damaging, so the prudent thing to do is to act under the assumption of no threshold and minimize all unnecessary radiation exposures.

The nuclear electric power industry releases small amounts of radioactivity into the environment at reactor sites and fuel reprocessing plants. For perspective it is good to compare the average radiation dose received by the general public from such operations, with the average received from all sources. Table 13.1 indicates that the average natural background dose received from cosmic rays and radioactive isotopes naturally present in the environment is around 130 millirems* (mrem) per person per year in the United States. This figure varies considerably from place to place due largely to the increasing cosmic-ray exposure at higher altitudes. Thus, for example, while the average background radiation level in Colorado is about 250 mrem/yr, it is only about 100 mrem/yr in Louisiana (Klement 1972).

Man-made sources contribute an additional 81 mrem/yr, on the average, about 92% of which results from medical procedures such as diagnostic x rays and radiotherapy. Many ordinary activities contribute measurable amounts of radiation to an individual's intake, including a single cross-country jet flight, about 4 mrem; wearing a watch with a luminous dial, 1–4 mrem/yr; watching television an hour a day, about

*The prefix "milli" means one-thousandth, thus 1000 mrem = 1 rem.

TABLE 13.1 Whole-Body Annual Radiation Doses in the United States, 1970 and 2000 in Millirems per Person per Year[a]

Radiation source	1970	2000
Natural	130	130
Medical procedures	74	88
Radioactive fallout	4	5
Miscellaneous (luminous watch dials, television, air travel, etc.)	2.7	1.1
Occupational exposure	0.8	0.9
Nuclear electric power	0.05	0.5
Other AEC activities	0.015	0.012
Total[b]	211	225

[a]Source: Klement, Miller, Minx, Shleien 1972.
[b]Totals may not agree due to rounding.

5 mrem/yr. Compared to the total average exposure of 211 mrem/yr, the 0.05 mrem/yr contribution of the nuclear power industry is insignificant.

It must be realized that, while the average dose to the public from the nuclear power industry is only about 0.05 mrem/yr (0.5 mrem in 2000) some people will receive higher exposures. However, even at the "fence post" (property line) of a reactor facility, exposures must be less than 5 mrem per year.

There has been considerable controversy over the setting of national standards for allowable radiation doses. Present guidelines are set at 170 mrem/yr as the allowable average exposure to the public, over and above medical and natural background levels.

Critics argue that the standard is set far too high. Gofman and Tamplin (1971), for example, of the Lawrence Radiation Laboratories, calculate that an additional 32,000 cancer deaths per year might be expected if everyone in the United States were exposed to the allowable limit of 170 mrem/yr. They contend that the standard should be reduced by at least a factor of 10.

Proponents of nuclear power counter by saying that the exposure rate used by Gofman and Tamplin in their calculation is unrealistic and that nuclear power will come nowhere near laying such an additional radiation burden on the public. That is not the point however. The point is not that actual emissions are excessive but that the standard itself is set too high and ought to be lowered to ensure that the industry does not become negligent in the future.

It should be pointed out that there are also standards for maximum permissible concentrations (MPCs) of radioactive isotopes in air and water. However these standards are of questionable value due to the difficulty in predicting the effect of biological concentration in the food chain. Organisms can concentrate certain radionuclides in much the same way that was mentioned earlier for DDT. Of particular importance are iodine-131 which concentrates in the thyroid gland; strontium-90 which concentrates in the bone; and cesium-137, which concentrates in the muscles. These three are commonly released in reactor effluents, and may be concentrated by a factor of thousands in the aquatic food chain or in the important air-to-soil-to-cow's milk-to-man food chain.

13.5 *Nuclear Fission*

With that introduction to radioactivity, we can now proceed to a discussion of nuclear power itself. Nuclear reactors obtain their energy from the heat that is produced when uranium atoms are split, in the process called nuclear *fission.* In the fission reaction, a neutron is captured by uranium-235, momentarily converting it into ^{236}U. Uranium-236 almost instantaneously breaks into two *fission fragments,* releasing two or three neutrons, some gamma rays, and a small amount of energy (per atom) as shown in Figure 13.2 and as summarized by the following equation:

$$n + {}^{235}U \rightarrow {}^{236}U \rightarrow f_1 + f_2 + (2 \text{ or } 3)\, n + \gamma + 200 \text{ MeV} \qquad (13\text{-}1)$$

There is much important information contained in equation (13-1). The energy released per fission is about 200 million electron volts (Mev),* most of which is contained in the kinetic energy of the two fission fragments. When the small amount of energy released per atom is multiplied by the many atoms which are undergoing fission, large amounts of heat are produced from just a small amount of uranium. The heat is then used in the boiler of a steam-electric plant, the details of which will be outlined in the next section. The complete fissioning of 1 g of uranium would release about 2.3×10^4 kWh of heat which is equivalent to about 3 tons of coal or 14 barrels of oil. The complete fissioning of about 7 pounds of pure uranium would create enough heat to run a 33% efficient, 1000 MWe nuclear plant at full capacity for a day.

*Small amounts of energy are often expressed in terms of electron volts. One electron volt is the energy acquired by an electron as it passes through a potential difference of one volt, and is equal to 1.602×10^{-12} erg. 200 MeV is therefore about 3.2×10^{-4} erg, or about 8.9×10^{-18} kWh.

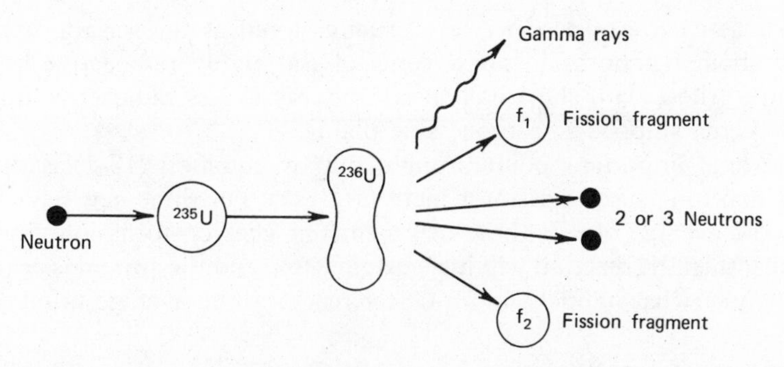

FIGURE 13.2 The fissioning of uranium-235 creates two radioactive fission fragments, f_1 and f_2. Most of the 200 MeV of liberated energy is contained in the kinetic energy of the fission fragments.

The second important point contained in equation (13-1) relates to the fission fragments, f_1 and f_2, which are always radioactive. They are the source of most of the controversy regarding nuclear-fission reactors. The particular fission fragments that result from the fissioning of a given uranium atom, are not predictable except on a statistical basis. The sum of the atomic numbers of the two fragments must equal 92 (the number of protons in the original uranium nucleus) and the sum of their mass numbers must be equal to 236 minus the two or three neutrons which are released. One typical fission reaction might be the following:

$$n + {}^{235}_{92}U \rightarrow {}^{236}_{92}U \rightarrow {}^{95}_{38}Sr + {}^{139}_{54}Xe + 2n \tag{13-2}$$

The fission fragments would then decay according to the following reactions:

$${}^{95}_{38}Sr \underset{\beta}{\overset{0.8m}{\rightarrow}} {}^{95}_{39}Y \underset{\beta,\gamma}{\overset{11m}{\rightarrow}} {}^{95}_{40}Zr \underset{\beta,\gamma}{\overset{65d}{\rightarrow}} {}^{95}_{41}Nb \underset{\beta,\gamma}{\overset{35d}{\rightarrow}} {}^{95}_{42}Mo \text{ (stable)}$$

and

$${}^{139}_{54}Xe \underset{\beta,\gamma}{\overset{41s}{\rightarrow}} {}^{139}_{55}Cs \underset{\beta,\gamma}{\overset{9.5m}{\rightarrow}} {}^{139}_{56}Ba \underset{\beta,\gamma}{\overset{83m}{\rightarrow}} {}^{139}_{57}La \text{ (stable)}$$

The numbers above the arrows in the reactions are the half-lives (s = seconds, m = minutes, d = days, y = years) and the symbols below the arrows refer to the forms of radiation released during the transition. Notice that the mass number of each element in a given decay chain always remains constant (because the only particle emissions will be betas), and the atomic number increases by one during each transition. Each decay chain ends with a stable isotope.

Radioactive waste products are usually stored at the reactor site for up to about 6 months to allow some of the highly radioactive fission products (those with short half-lives) to decay to less radioactive forms, before being shipped to reprocessing plants.

The final important point brought out by equation (13-1) is that a single neutron causes a fission which releases two or three new neutrons. If on the average one of these new neutrons goes on to fission another nucleus, then the reaction will just sustain itself, and the system is *critical*. This is just what happens in a nuclear reactor. The average number of

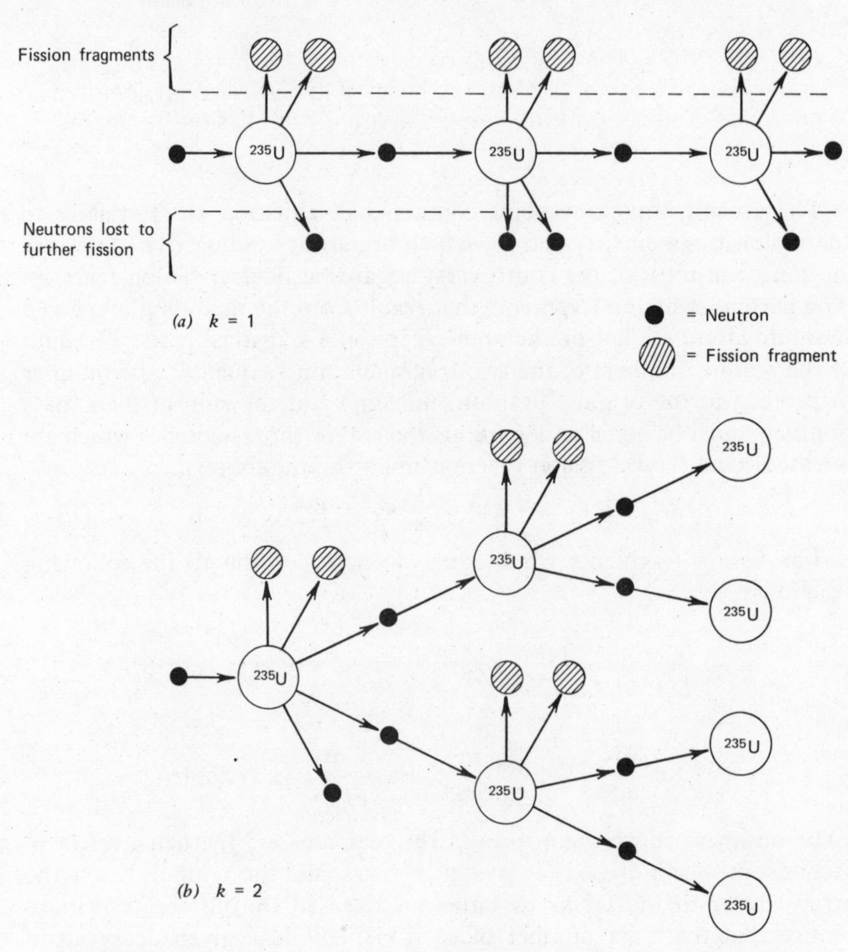

FIGURE 13.3 Nuclear chain reactions: (*a*) Critical, self-sustaining reaction, $k = 1$; (*b*) supercritical, with each fission causing two more fissions to occur, $k = 2$.

neutrons that go on to cause new fissions is called the *multiplication factor* (k). For $k = 1$, the system is critical and the reaction will release a constant amount of power. For $k > 1$, the system is *supercritical,* and the energy released will grow exponentially as in an atomic bomb. Figure 13.3 diagramatically represents the chain reactions which correspond to $k = 1$ and $k = 2$. For $k < 1$, the reaction will die out.

What happens to the neutrons that do not participate in fissions? Some are lost by escaping from the surface of the fuel, some are captured by the ^{238}U atoms in the uranium (converting it to plutonium-239), and some are captured by other elements in the system and are effectively lost.

For a reactor to supply a steady power output requires that just the right number of neutrons be moving around in the fissile ^{235}U. To ensure that there are enough neutrons requires that the concentration of the isotope ^{235}U be greater than its natural concentration of only 0.7 % (fuel enrichment). To keep the system from going supercritical, control rods of cadmium or boron are used. These materials are good neutron absorbers. Moving the control rods deeper into the fuel slows the reaction, and withdrawing them allows it to speed up.

13.6 *Nuclear Reactors*

We have seen how the fission reaction releases thermal energy which can be used in the boiler of a steam-electric power plant. However, as Figure 13.4 illustrates, the fissioning of uranium at the reactor site is only one of many important processes which are essential to the complete nuclear power system.

The first step is of course to mine the uranium. Uranium ore is very dilute in uranium, so it must be processed to increase the concentration. The concentrated U_3O_8 is then converted to uranium hexafluoride (UF_6) which is sent to an enrichment plant where the concentration of the scarce isotope ^{235}U is increased from about 0.7% to perhaps 2 or 3 %. The enrichment plant uses a gaseous diffusion process that is very expensive and requires, itself, great amounts of electricity. The first such plant was built during World War II as part of the Manhattan Project since highly enriched uranium can be used in atom bombs. In fact, the technical sophistication and expense of such diffusion plants has been an important obstacle to the spread of nuclear weapons capability.

The enriched uranium is then converted into a powder and fabricated into small cylindrical pellets of uranium dioxide, UO_2. These pellets are perhaps 1/2 inch long and 3/8 inch in diameter and are not very dangerous since they are only mildly radioactive and emit alpha particles

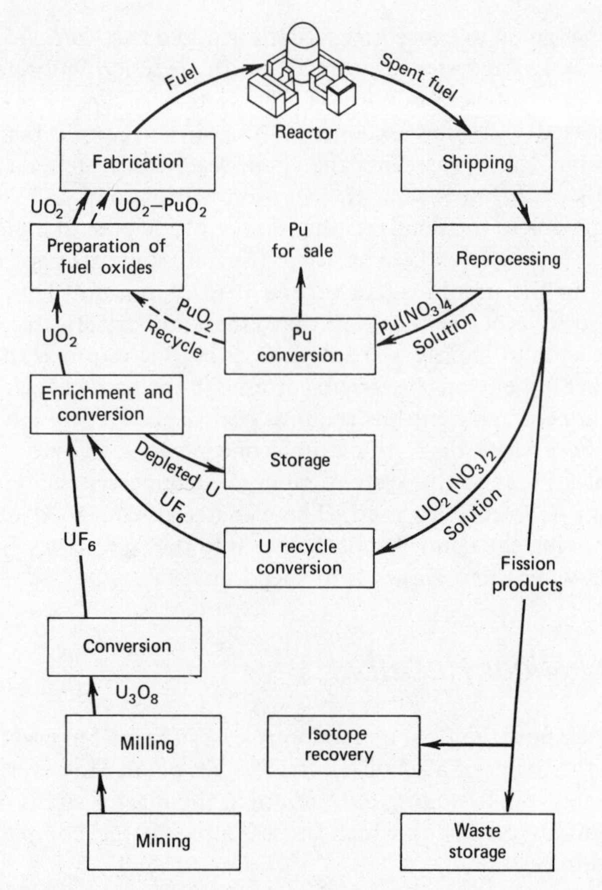

FIGURE 13.4 Nuclear fuel cycle for light water reactor (Federal Power Commission 1971).

which have essentially no pentrating power. These pellets are loaded into long tubes, called fuel rods, which are usually made of an alloy of zirconium called Zircoloy. A square array of perhaps 220 of these fuel rods, along with a number of control rods, is structurally bound together in a unit called the fuel assembly. A number of fuel assemblies are then loaded into the reactor core which is immersed in water in the reactor vessel as shown in Figure 13.5.

The water serves two functions in a nuclear reactor. First, it acts as a coolant, carrying the heat away from the core to the rest of the steam plant where it can be used. Without the coolant, the fuel rods would quickly become so hot they would melt. The second function of the water

The 429 MW nuclear generating station at San Onofre, California. (Photograph by G. S. Masters.)

is to slow down the "fast" (high-energy) neutrons that are liberated during fission. Through multiple collisions with the water *moderator,* the neutrons gradually lose energy until they become "slow" (low-energy) neutrons which are required by ^{235}U for fissioning. Ordinary water is usually used as the moderator and such reactors are called *light-water reactors* (LWR).*

There are three different power plant designs currently in use for capturing the energy from the fission reaction and converting it into electricity. They are the pressurized-water reactor (PWR), the boiling-water reactor (BWR), and the high-temperature gas-cooled reactor (HTGR), as shown in Figure 13.6.

Pressurized-water reactors operate with two coolant loops. In the first loop, the water is kept under high pressure to prevent it from boiling. Its heat is transferred to the secondary loop which operates the turbine. This design helps prevent any radioactive wastes which may get into the

*A few reactors use "heavy" water, D_2O, where D is deuterium (heavy hydrogen) which is hydrogen with a neutron in the neucleus.

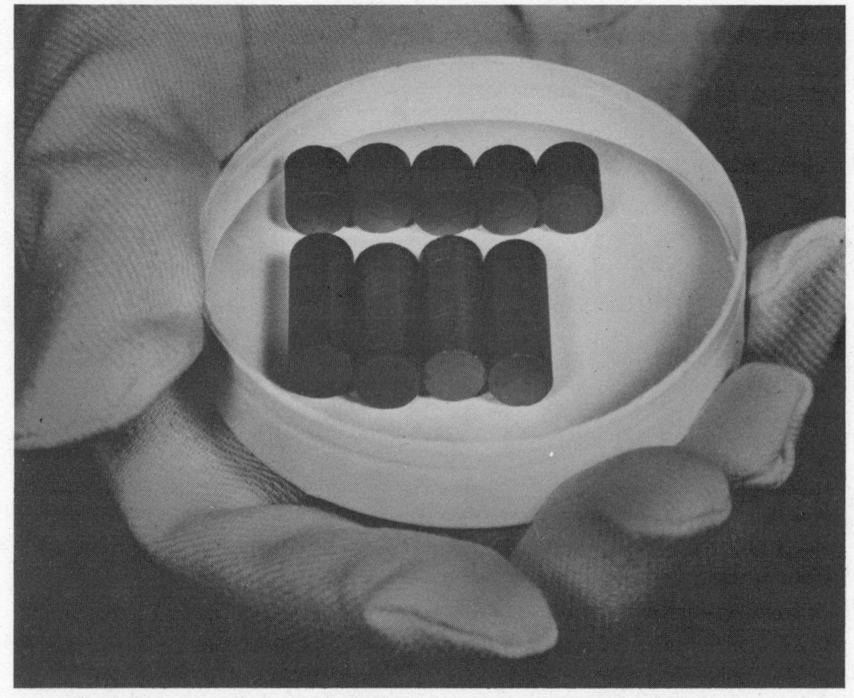

To fuel commercial nuclear power plants, uranium dioxide is fabricated into dense, black cylindrical pellets of about the diameter of a little finger and less than 1 inch long. (Courtesy Pacific Gas and Electric Company.)

primary loop from getting into the condenser cooling water, which of course is released into the environment. Boiling-water reactors operate with a single loop. Since the steam becomes radioactive it is necessary to shield the entire steam system, including the turbine, from the environment. High-temperature gas-cooled reactors are not used much in this country. They do have an advantage of operating at a higher temperature than the BWR or PWR types, so they have a higher efficiency and hence require less fuel and less cooling water.

13.7 *Radioactive Wastes*

We have seen that the production of radioactive fission products is part of the fission process. Approximately once every year or so, a nuclear reactor must be shut down for refueling and for removal of these radioactive waste products. In addition to fission products, some materials become radioactive due to the intense neutron fields around the reactor core; they are called neutron activation products. It is important to distinguish between the small amount of radioactivity which is routinely released at a reactor site, and the large quantities which are periodically removed for reprocessing and disposal.

During operation of a reactor, the primary coolant tends to pick up some radioactivity (fission products and activation products) which are removed, and, on a controlled basis, released to the local environment. There are a number of such radionuclides in reactor effluents including ^{3}H (tritium), ^{58}Co, ^{60}Co, ^{85}Kr, ^{89}Sr, ^{90}Sr, ^{131}I, ^{131}Xe, ^{133}Xe, ^{134}Cs, ^{137}Cs,

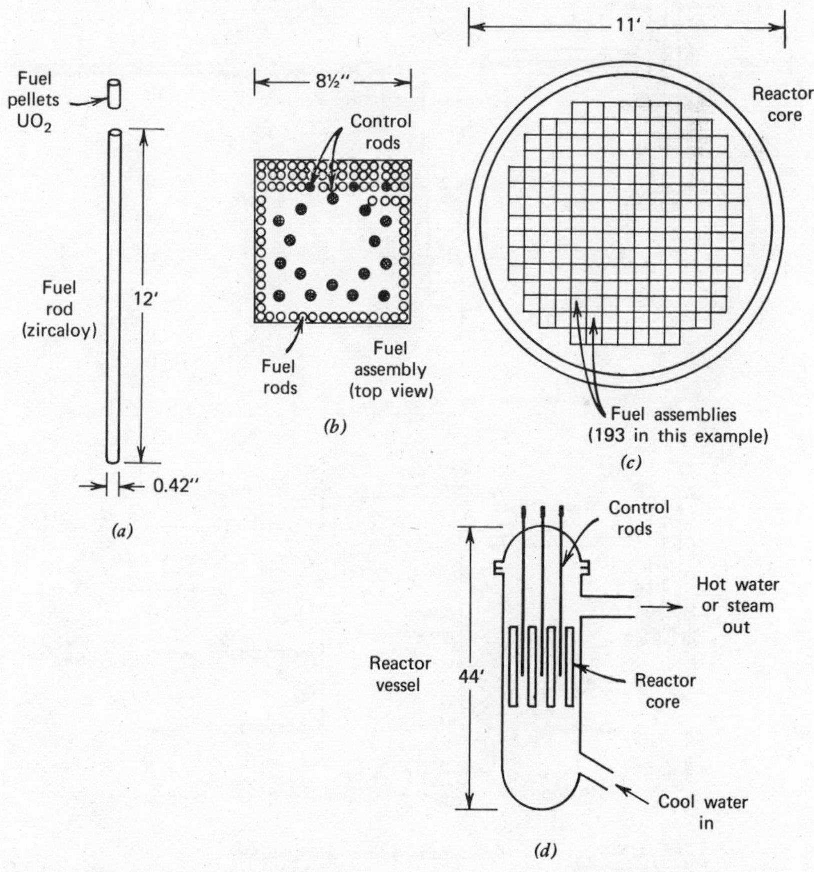

FIGURE 13.5 Components of nuclear reactor: (*a*) Fuel pellets and fuel rods, (*b*) fuel assembly, (*c*) reactor core and (*d*) pressure vessel. Dimensions are typical.

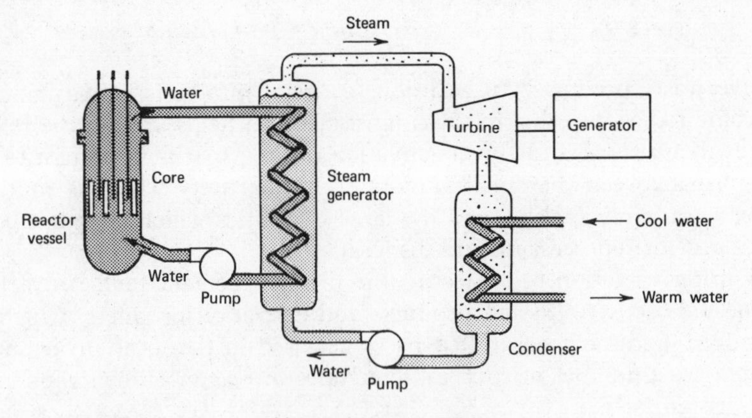

(a) Pressurized–water reactor

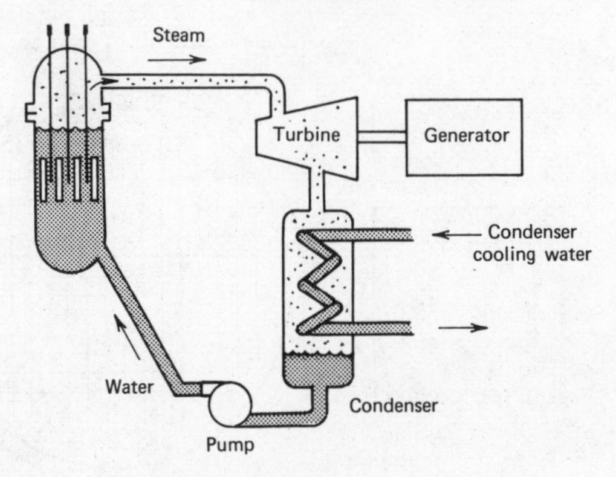

(b) Boiling–water reactor

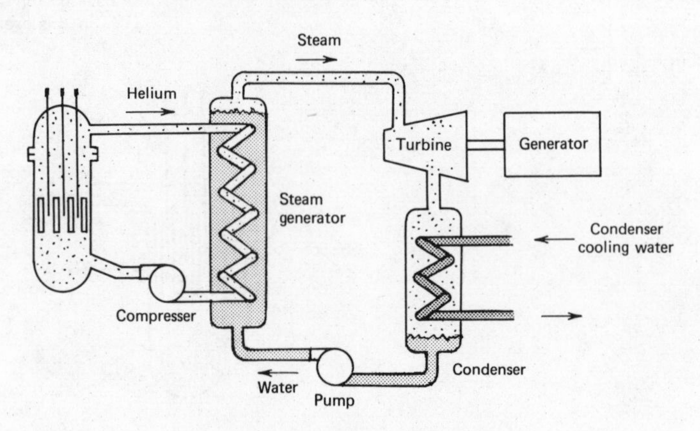

(c) High–temperature gas–cooled reactor

FIGURE 13.6 Three nuclear reactor designs: (*a*) Pressurized-water reactor (PWR), (*b*) boiling-water reactor (BWR), (*c*) high-temperature gas-cooled reactor (HTGR).

This is a pressurized-water reactor vessel, without its top, being prepared for installation at Diablo Canyon. (Courtesy Pacific Gas and Electric Company.)

and ^{140}Ba. The two that have probably received the most attention are tritium, with a half-life of 12.3 years, and krypton-85 with a half-life of about 9 years. The long half-lives of these two makes it possible for them to accumulate in the environment. However, these low-level releases are of relatively minor consequence in the evaluation of environmental effects of nuclear power (see Section 13.4).

Much more important are the highly radioactive spent fuel assemblies which are periodically removed from the site for reprocessing. The spent fuel assemblies are usually stored at the site for about 4 months to allow some of the radioactivity to decay. They are then loaded into shipping casks which must be able to satisfactorily withstand the following sequence of accident conditions:

1. A 30-foot drop in its most vulnerable position, onto a hard flat surface.
2. A 40 inch drop onto a 6 inch diameter steel "spike."
3. A fire at 1475°F for 30 minutes followed by
4. Immersion in water

The number of such casks loaded with spent fuel assemblies that must be transported to fuel reprocessing plants each year, is going to increase rapidly. As shown in Figure 13.7 there were approximately 30 such casks shipped in 1970 and there may be nearly 9500 shipped in 2000. In fact by 2000 there may be an average of 85 such casks in transit at any given time. It is to be hoped that there will be no accidents involving the equivalent of a 31-foot fall.

At the fuel reprocessing plant the fuel assemblies are chopped into pieces and dissolved in acid, and the uranium which had not been burned

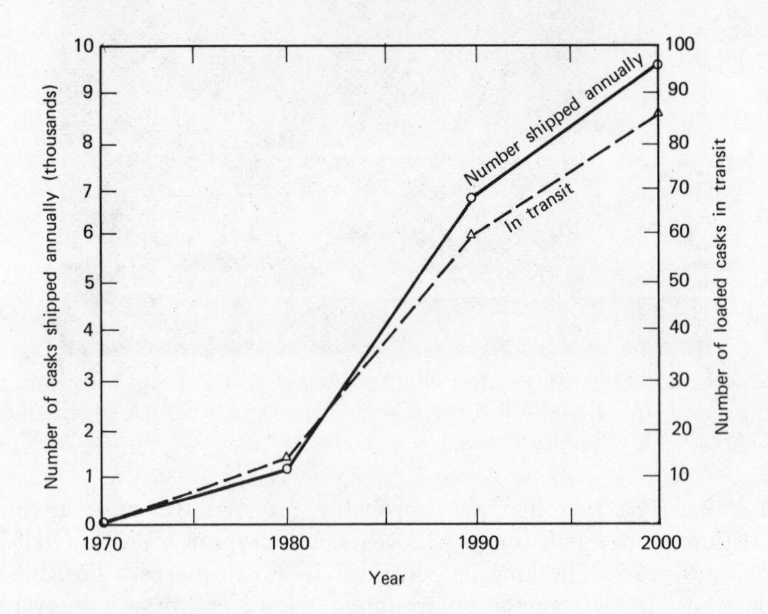

FIGURE 13.7 Predicted increase in spent fuel casks shipped annually and number in transit on any given day (National Academy of Engineering 1972).

at the reactor is removed for reprocessing into new fuel elements. Plutonium, which is produced in the reactor, is also recovered possibly for use in nuclear weapons. Some radioisotopes are saved for medical use. But by far the majority of the radioactive remains are concentrated in liquid or sometimes solid form and removed from the reprocessing plant for ultimate storage. Two constituents are of particular interest—strontium-90 and cesium-137, with half-lives of 28.8 and 30 years, respectively. When first stored, these two comprise only about 2% of the radioactivity of the mixture, but after 30 years they account for about 98% of it. To get some idea of the length of time required for these wastes to decay to innocuous levels, consider that it takes about 20 half-lives for a given amount of radioactive material to be reduced by a factor of 1 million. The concentration of strontium-90 in high-level liquid wastes is about 100 Ci per gallon, so after 20 half-lives or almost 600 years, its concentration would be reduced to 1×10^{-4} Ci per gallon which is still 100,000 times the allowable level in drinking water (Hogerton 1971).

It is estimated that by the year 2000 the accumulated beta activity in the nation's radioactive wastes will be about 209 billion Ci (National Academy of Engineering 1972). Drinking water standards allow only one billionth of a curie of beta activity per liter of water. Thus by 2000 we will have accumulated enough radioactivity to contaminate about 2×10^{20} liters of water, or about 50×10^{18} gallons! This is double the total amount of fresh water in the world (including groundwater and glaciers) and is about one-eighth of the total amount of water in the world (see Table 4.1), including the oceans!

The big question is just where can this extremely dangerous waste be stored for the hundreds or thousands of years required, so that it cannot find its way back into the environment? Much of it at present is stored in huge 600,000 gallon steel tanks at sites near Richland, Washington; Idaho Falls, Idaho; and on the Savannah River near Aiken, South Carolina. The wastes generate so much heat that they are continuously boiling in the tanks and must be cooled and stirred constantly.

The tanks are quite unsatisfactory representing at best a very short-term solution. They have already begun to leak. According to the General Accounting Office, by 1970 some 227,400 gallons of high-level, radioactive liquid waste had already leaked into the ground at the Richland, Washington site (Lewis 1971). In June 1973, another 115,000 gallons were discovered to have leaked into the ground at the same site. Should these wastes seep down to the water table they could contaminate the area's water supply for hundreds of years. In fact, low-level wastes which are routinely released into the soil have already contaminated the groundwater under the Richland site and some has made its way into the Columbia River (*Los Angeles Times*, July 23, 1973).

The ultimate storage scheme is still (at the end of 1973) unknown. In June, 1970, the Atomic Energy Commission (AEC) announced it had selected a site near Lyons, Kansas as its first Federal repository for permanent storage of these radioactive wastes. The plan called for the burial of solidified wastes in salt beds at a depth of from about 500 to 2000 feet underground. Testifying before the Joint Economic Committee on February 23, 1972, Milton Shaw of the AEC gave assurance in the strongest terms that the Lyons salt mine waste storage would be completely safe and proof against all intrusions of water. The Director of the Kansas Geological Survey, however, said "the Lyons site is a bit like a piece of swiss cheese, and the possibility for the entrance and circulation of fluids is great" (Hambleton 1972). The Geological Survey's position prevailed and the AEC has since dropped its proposal to use the Lyons site.

This raises one of the frequently heard criticisms of the AEC regarding the impossibility of its acting effectively as both regulator and promoter of nuclear power. Far too often they have been less than candid with the public on sensitive issues.

There have been other proposals for the disposal of wastes including rocketing them to the sun and storing them under the Antarctic ice cap (Zeller 1973). However, the present situation is that we are beginning to rapidly accumulate more and more of these very dangerous wastes without knowing how we will ultimately store them. This is our legacy to future generations.

13.8 *Reactor Safety*

The second major area of the nuclear power controversy is in regard to their safety. It can be argued that the probability of an accident is extremely small and that the safety record so far is excellent. On the other hand, critics point out that should an accident occur, it could result in the single largest catastrophe ever to strike the United States. So we are left with the necessity of evaluating the product of a very low probability event times a very high danger should the event occur.

It should first be emphasized that the concentration of ^{235}U in light-water reactors is too low to cause the reactor to explode like an atomic bomb. However, it would be possible to incur a small explosion from expanding steam should the reactor core melt down. It would be hoped that the containment shell around the reactor would be able to withstand such an explosion without releasing too much radiation to the environment.

What is a meltdown and how could it occur? Basically, the temperature of the reactor core is controlled by the rate of flow of cooling water and the reaction rate in the reactor. If these factors are not balanced, the reactor can overheat and melt the core. The molten core would melt through the pressure vessel and anything else that might be in its way, eventually making its way into the earth below in what has come to be known as the "China syndrome." It is likely that a good portion of the contained radioactivity would be released to the environment.

Notice that this syndrome could occur even if the control rods are immediately inserted into the core to stop the fission reaction. The heat generated by the radioactive wastes, which may account for as much as 10% of the total heat developed in an operating reactor, is sufficient to cause a meltdown. Thus it is the constant flow of cooling water which is essential for reactor safety.

Should equipment failure, human error, sabotage, or a severe earthquake result in a loss of coolant, perhaps by rupturing one of the main coolant lines, the core is supposed to be immediately flooded with water from an emergency-core cooling system (ECCS) as suggested in Figure 13.8. The ECCS has become a focal point of nuclear reactor safety criticism because the systems have never been adequately tested. Tests which were carried out at the National Reactor Testing Station in Idaho in 1970 and 1971 on semiscale models indicated that the cooling water

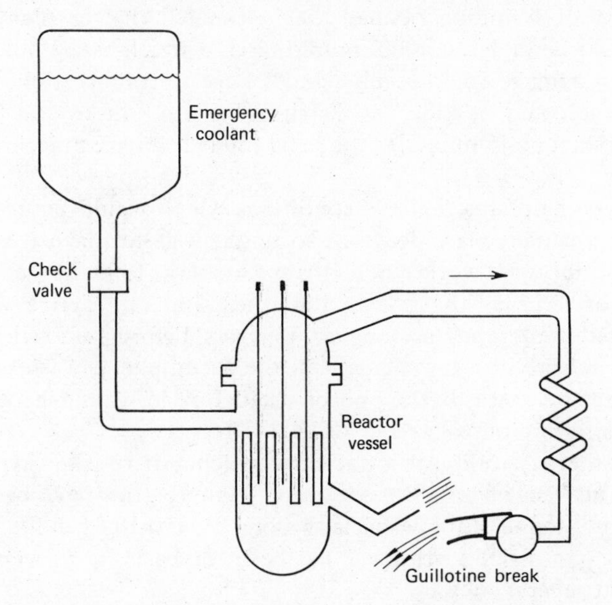

FIGURE 13.8 Schematic version of a loss of coolant accident. The emergency coolant must flood the reactor almost immediately to prevent meltdown.

would probably not reach the core, and full scale tests are not scheduled until 1976 (Forbes, Ford, Kendall, Mackenzie 1972). In the meantime, we must rely on present, untested ECCS systems and the low probability of an accident occurring.

Until recently the AEC has maintained that the chances of a serious loss of cooling accident were so small as to defy meaningful estimation. However, they now estimate that the chances of an accident which could result in a release of radioactivity may be as high as one in a thousand for a given reactor for a given year (Gillette 1973). By the year 2000 when 1000 reactors may be operational, this estimate suggests that one accident could occur every year. While the amount of radiation which would be released in this high probability accident would be minimal, serious damage would be done to the reactor itself.

Just how serious is a major reactor accident in which a large amount of radioactive waste is released into the environment? In 1957, the AEC published a report entitled *Theoretical Possibilities and Consequences of Major Accidents in Large Nuclear Power Plants*, usually referred to by its document code designation, WASH-740. The worst accident that they could postulate involved the release of 50% of the radioactive wastes from a reactor of 500 MW thermal power (approximately 165 MWe) after 180 days of operation. Assuming the reactor was located 30 miles from a city of 1 million people, they estimated that as many as 3400 people could be killed, 43,000 injured, and as much as $7 billion worth of property damage could result. Under adverse combinations of conditions, people could be killed at distances up to 15 miles and injured at distances of about 45 miles. As much as 150,000 square miles might have to be evacuated.

The AEC maintains that the conditions which would be necessary to cause this postulated accident are so unlikely as to make it essentially an impossibility. However, note that a modern 1000 MWe reactor is six times as large as the one in the study and radioactive wastes are accumulated sometimes as long as $1\frac{1}{2}$ years before removal. In other words the release of a few percent of the accumulated wastes in such a modern reactor, towards the end of its fuel cycle, would be equivalent to the 50% release postulated in WASH-740.

Besides the possibility of a nuclear accident, there is always the possibility of an "act of God" or sabotage. Many battles have been fought over the question of siting, especially along the coast of California, where the danger of a major earthquake has been the basis for preventing construction at several sites.

Another interesting controversy regarding nuclear power is over the Price-Anderson Act which limits liability in the event of an accident, to

$560 million, which is about 8% of the $7 billion potential estimated above. And of that $560 million, only about $95 million is carried by private insurance companies, the rest is government liability. Critics claim that if nuclear power plants were required to carry adequate insurance, the cost would be so great that nuclear power would be uneconomical.

The fact that the safety record for nuclear reactors has been excellent, is a point which is always advanced in defense of the overall safety of these systems. However, obviously the 100 or so reactor-years of accumulated experience is insignificant compared to projected nuclear usage. The 1000 reactors in the year 2000 would accumulate 100 or so reactor-years in just a few months.

Although there have been no serious accidents associated with commercial light-water reactors in the United States, there have been a number of incidents in various noncommercial facilities. There have been a total of 26 occasions in the history of the AEC when the power level of fissile systems became uncontrollable because of unplanned or unexpected changes in system reactivity, with a total of six deaths attributable to criticality accidents (AEC 1971).

In 1961, during a shutdown period, the Army Low-Power Reactor, SL-1, located at the National Reactor Testing Station in Idaho, had a meltdown due to the accidental withdrawal of a control rod by a technician. Three technicians were killed. Some gaseous fission products escaped to the atmosphere and were carried downwind of the reactor.

In October 1967, a fire occurred in a production reactor at the Windscale Works in England, which resulted in significant quantities of radioactive material being released into the air, contaminating a section of land downwind of the reactor.

In 1952, a severe nuclear excursion occurred in a test reactor at Chalk River, Canada. The excursion caused a portion of the core to melt and some radioactive materials were released to the environment.

In October 1966, the Fermi Experimental breeder reactor, just outside of Detroit, suffered a meltdown when a piece of sheet metal broke loose inside the cooling system plugging some coolant holes. The radiation released was contained in the shell.

These various accidents can be interpreted in several ways. One way is to say that they demonstrate that accidents do occur and that we have been lucky that they have all been rather minor. The other interpretation is to say that these accidents lend credence to the argument that the safety features do work since there has been no major release of radioactivity in any of them. For more details on the accidents themselves, see Eisenbud (1963) or AEC (1971).

13.9 Conclusions

There is little question that the major sources of energy for the generation of electricity in the near future will be coal and uranium. Coal resources are adequate and most of the environmentally destructive aspects associated with its use can be significantly improved with reasonable changes in technology and corporate attitude.

There is, however, great pressure to find an acceptable alternative to fossil fuels and at present the most developed option is controlled nuclear fission. The environmental advantages of nuclear reactors are significant: no air pollution, no strip mining, no oil spills, no depletion of fossil fuel reserves. Further, the increase in thermal pollution associated with nuclear power may only be a temporary disadvantage as present light-water reactors are replaced by future, more efficient breeder reactors, as will be described in the next chapter.

Against these environmental advantages must be weighed the great *potential* for environmental harm if something goes wrong. The release of only a small percentage of the accumulated radioactive waste products from a single reactor could kill or injure thousands and result in incalculable delayed somatic and genetic effects.

Moreover, the decisions we make now on the methods for storing radioactive wastes are going to affect future generations for perhaps thousands of years. We must be absolutely sure that these wastes cannot possibly come into contact with the biosphere over a period of time that dwarfs the lifetime of any of man's fragile institutions.

Nuclear power may very well be the answer to our electric energy needs for a long time into the future. However, to rush them into use before the safety of all aspects of the complete system can be ensured would be foolhardy to say the least.

Bibliography

Arena, V. (1971). *Ionizing radiation and life*. St. Louis: C. V. Mosby Co.

Eisenbud, M. (1963). *Environmental radioactivity*. San Francisco: McGraw-Hill.

Federal Power Commission (1971). *The 1970 National Power Survey*, Washington, D.C. Dec.

Forbes, I. A., Ford, D. F., Kendall, H. W., and Mackenzie, J. J. (1972). Cooling water. *Environment* 14 (1).

Foreman, H. ed. (1970). *Nuclear power and the public*. Minneapolis: Univ. Minnesota Press.

Gillette, R. (1973). Nuclear safety: AEC report makes the best of it. *Science* 179 Jan.

Gofman, J. W., and Tamplin, A. R. (1971). *Poisoned power*. Emmaus, Penn.: Rodale.

Hambleton, W. W. (1972). The unsolved problem of nuclear wastes. *Technol. Rev.* March/April.

Hogerton, J. F., Kline, J. G., Kupp, R. W., and Yulish, C. B. (1971). *Nuclear power waste management*. Public Affairs and Information Program, Atomic Industrial Forum. New York: March.

Klement, A. W., Jr., Miller, C. R., Minx, R. P. Shleien, B. (1972). *Estimates of ionizing radiation doses in the United States, 1960–2000*. U.S. Environmental Protection Agency, Office of Radiation Programs. Rockville, Md. Aug.

Lewis, R. L. (1971). The radioactive salt mine. *Bull. Atom. Sci.* June.

Lindop, P. and Rotblat, J. (1971). Radiation pollution of the environment. *Bull. Atom. Sci.* Sept.

National Academy of Engineering (1972). *Engineering for resolution of the energy-environment dilemma*. Committee on Power Plant Siting. Washington, D.C.

Tamplin, A. R., and Gofman, J. W. (1970). *"Population Control"! Through Nuclear Pollution,* Chicago: Nelson-Hall.

U.S. Atomic Energy Commission (AEC) (1957). *Theoretical Possibilities of Major Accidents in Large Nuclear Power Plants*. WASH-740, March.

U.S. AEC (1971). *Operational Accidents and Radiation Exposure Experience Within the AEC, 1943–1970*. Division of Operational Safety, WASH-1192. Fall.

U.S. AEC (1973). *The Safety of Nuclear Power Reactors (Light Water-Cooled) and Related Facilities*. WASH-1250, July.

Zeller, E. J., Saunders, D. F., and Angino, E. E. (1973). Putting radioactive wastes on ice. *Bull. Atom. Sci.* Jan.

Questions

1. If all of the empty space between atomic nuclei in a 200 pound man could be reduced to zero, what would be the approximate volume of the remainder?
   ans. 10^{-9} cc

2. What values of a and b would complete the following:

$$^{266}_{88}\text{X} \rightarrow \alpha + ^{a}_{b}\text{Y}$$

$$^{a}_{15}\text{X} \rightarrow \beta + ^{32}_{b}\text{Y}$$

   ans. 262, 86; 32, 16

3. The half-life of iodine-125 is about 60 days. If we were to start with 64 g of it, about how much would remain after half a year? After 1 year?
   ans. 8 g; 1 g

4. Why don't random slow neutrons in nature, such as might be caused

by cosmic radiation, cause self-sustaining chain reactions in the ^{235}U that exists in nature?

5. Why are uranium dioxide fuel pellets not radiologically dangerous?

6. Spent-fuel assemblies are shipped in casks which must be able to withstand a 30 foot fall. What would be the speed of an object after falling 30 feet? Is it comparable to the speed of a train carrying such a cask?
   ans. 44 ft/sec (30 mph)

7. Explain the fundamental difference between the measurement units, curie and rad.

8. Suppose a reactor core is made up of 193 fuel assemblies and each assembly consists of 205, 12 foot long, fuel rods. Estimate the number of 1/2 inch fuel pellets in the core.
   ans. 11.4 million

9. Find out whether there exists a program for permanent storage of high-level radioactive wastes.

Chapter 14 New and Future Energy Sources

There are a number of new approaches to the generation of electricity which could potentially greatly alleviate the drain on our fossil fuel and uranium reserves. In most cases these new techniques would also be environmentally less damaging than present power generation schemes.

In this chapter emphasis will be on the following four approaches: breeder reactors, geothermal power, solar power, and thermonuclear fusion. All four have the potential to become important sources of energy in the near future.

In addition there are other energy sources which will be mentioned including wind power, tidal power, ocean temperature differences, and garbage power. It appears, however, that these will be of lesser importance on a global scale, though they may have local application.

14.1 *Fast-Breeder Reactors*

It was pointed out in Section 11.6 that the reserves of uranium-235, which present nuclear reactors consume, are quite limited. Recall that the isotope ^{235}U comprises only about 0.711% of naturally occurring uranium, while ^{238}U comprises about 99.283%. The purpose of breeder reactors is to make use of the much more abundantly available isotope ^{238}U in the fission reaction, thus multiplying the amount of usable uranium by a factor of about 140. Further, since the cost of the electricity from a breeder could be nearly independent of the fuel cost, it would be economically possible to use fuel which is considerably more expensive than the present $8 a pound uranium, which multiplies the potential fuel reserves by even more than the original factor of 140. If breeder reactors are not developed, then due to a lack of fuel, the whole fission reactor program will essentially end before it ever really gets off the ground.

Besides the advantage of extending uranium reserves by a factor of hundreds, breeder reactors would also be designed to operate at higher temperatures than light-water reactors. As was pointed out in Section 12.5, this higher temperature will result in greater efficiency (around 40 %) so there would be less thermal pollution.

How does a breeder reactor work? The claim is often made that such a reactor would create more fuel than it consumes, which sounds incredible, but of course this is somewhat of a distortion of the truth. In actuality, a breeder converts ^{238}U, which itself cannot be fissioned, into plutonium-239, which can. The ^{239}Pu is then "burned." In the breeding process somewhat more ^{239}Pu is created from the uranium than is burned to maintain the reaction, hence the origin of the above statement.

Materials such as ^{239}Pu or ^{235}U which can be fissioned are called *fissile* materials, while isotopes such as ^{238}U which can be converted into fissile materials are called *fertile* materials. There are two fertile isotopes which could be used in breeders: ^{238}U or thorium-232, but is the ^{238}U which is presently receiving most of the attention. The conversion of these fertile isotopes, by neutron bombardment, into fissile isotopes is represented by the following pair of reactions:

$$^{238}_{92}U + n \rightarrow {}^{239}_{92}U \xrightarrow[\beta]{23.5m} {}^{239}_{93}Np \xrightarrow[\beta]{2.35d} {}^{239}_{94}Pu$$

and

$$^{232}_{90}Th + n \rightarrow {}^{233}_{90}Th \xrightarrow[\beta]{22.2m} {}^{233}_{91}Pa \xrightarrow[\beta]{27d} {}^{233}_{92}U$$

In the first reaction, ^{238}U absorbs a neutron and is converted into ^{239}U which then spontaneously transforms into neptunium-239, which in turn transforms into ^{239}Pu. The ^{239}Pu, which does not occur in nature, is relatively stable, having a half-life of 24,390 years. Another neutron can cause the plutonium to fission which releases energy for the reactor.

We can see then, that one neutron is required to convert ^{238}U to ^{239}Pu, and then another is required to cause the plutonium to fission—a total of two neutrons required per fission. If only two neutrons were given off by the fission, and if they each were to take part in a new reaction, then the process could just sustain itself. These two neutrons would go on to convert one more atom of ^{238}U into ^{239}Pu and then cause the ^{239}Pu to fission, giving two new neutrons, and so on.

By definition, a breeder reactor produces more fissile material (plutonium) than is consumed to maintain the reaction. Thus, for breeding, more than two neutrons are required per fission, so that the excess neutrons can create extra atoms of plutonium from the ^{238}U. Figure 14.1 shows how a breeder which releases $2\frac{1}{2}$ neutrons per fission on the average, would need to burn only two out of every three plutonium atoms created. The ratio of the amount of fissile material produced to the amount consumed, called the *conversion ratio* (or breeding ratio), would be 3/2 in this example.

In any discussion of the desirability of breeder reactors, it should be borne in mind that the most serious, negative aspects of conventional light-water reactors still apply to breeders. Since the energy source is fission, the problems of handling and disposal of dangerous radioactive wastes still must be solved, and there still exists the potential for an accidental release of radiation at the site, such as could occur following a loss-of-coolant accident. There are, in addition, several special properties of breeders which compound the problems.

Breeder reactors using plutonium fuel must use a different coolant around the reactor core than conventional light-water reactors. Recall that ^{235}U requires slow (low-energy) neutrons to instigate fission. The neutrons released during fission are fast neutrons which must be slowed by the water moderator in a conventional light-water reactor.

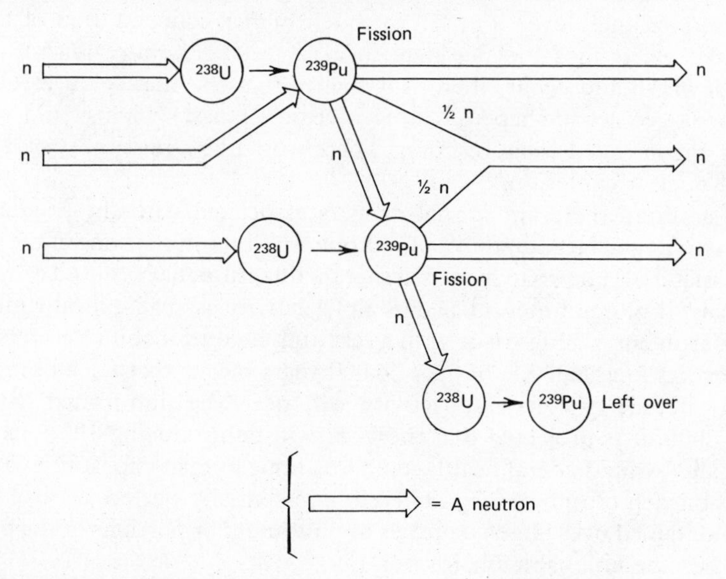

FIGURE 14.1 Ideally, if each fission of Pu-239 releases 2½ neutrons, then for every three atoms of Pu-239 created, only two will be burned leaving one left over; thus more fissile material (Plutonium-239) is created than is consumed.

In the breeder, however, the reactions require high-energy, fast neutrons (which is why they are referred to as *fast* breeder reactors). Therefore, the neutrons emitted during fission must not be slowed down or else the reaction cannot proceed, and so it is not possible to use water as a coolant. Instead there are designs which would use either helium as the coolant or liquid sodium. The AEC has chosen to pursue development of the liquid metal fast-breeder reactor (LMFBR) which uses liquid sodium as the coolant. It hopes to have the "first" (not counting the Fermi breeder reactor that melted down in 1966) demonstration model operating in the early 1980s. To that end, the United States is allocating far more money to the LMFBR than for any other energy program.

There are some difficulties encountered with the coolants in breeder reactors. The sodium in the LMFBR becomes highly radioactive and special precautions must be taken to ensure its containment. Liquid sodium reacts vigorously with water and burns intensely if exposed to air.

Both the gas-cooled and sodium-cooled breeders would have special problems if there should be a loss of coolant accident. In a LWR if the water coolant is lost, then the chain reaction tends to stop by itself because criticality depends on the moderator. In a breeder, however, the reaction could remain critical after the loss of coolant which increases the chance of a sudden meltdown. Further, a much higher concentration of fissile material exists in a breeder than in a typical light-water reactor. If a meltdown should occur, there is a chance that the fuel could assume a different geometric shape, creating a critical mass, which would result in an uncontrolled chain reaction. There would be a possibility of a very small nuclear explosion.

In addition there are special dangers associated with the production and handling of large amounts of plutonium. Plutonium is one of the most toxic materials known to man. Microgram quantities have caused cancer in animals. If plutonium is inhaled its alpha activity can cause locally intense bombardment of lung tissue with a resulting high probability of inducing cancer. Its long half-life of over 24,000 years means that for all practical practical purposes, its radioactivity will never be diminished. By the year 2000, it is projected that the U.S. will be producing 100 tons of it annually, while Federal health standards limit human exposure to a total body burden of only 0.6 μg. Therefore, essentially perfect control must be maintained over large quantities of plutonium or portions of the planet may become uninhabitable forever.

The production of plutonium is the most difficult step in the production of atomic bombs. Somewhere around 7–10 kg is enough to make a bomb, and a year's operation of even a conventional reactor would produce sufficient plutonium to make dozens of bombs. There is the very real danger, with the large quantities of plutonium that would be in transit at any given time, that some of it could be diverted and used to make clandestine atomic bombs. Thus it would become just that much easier for nonnuclear nations to acquire nuclear capabilities, or worse, for hijackers to acquire and use plutonium for atomic blackmail. It sounds like science fiction but it is a danger that must be recognized and dealt with.

14.2 *Geothermal Power*

Geothermal power, the second of the "new and future" energy sources to be considered here, is not actually particularly new but its potential has as yet hardly been touched. In fact, electrical power has been generated from geothermal steam at a site near Larderello, Italy since 1904.

Basically, electric power is generated by extracting steam from beneath the surface of the earth and using it to run a turbine-generator, producing electricity in much the same manner as in any steam-electric power plant. The steam is created when certain geological conditions allow heat from the earth's interior to act on groundwater.

As Figure 14.2 indicates, the heat source is magma, or molten rock, that in some places in the world pushes its way up into the earth's crust. The magma heats a layer of impermeable crystalline rock which is acting as the floor to the groundwater which has percolated down from the earth's surface. The groundwater, which exists in a region of permeable rock, is heated perhaps to 500°F, although because it is under high pressure (being capped by a layer of relatively impermeable rock) it may remain liquid until it is vented at the surface by the well.

Geothermal energy may come in the form of dry steam, wet steam, or just hot water. Dry steam is the least abundant but most desirable form since it can be utilized directly for the production of electric power. The Geysers field in California, the Larderello field in Italy, the Valle Caldera field in New Mexico, and two fields in Japan are the most important dry steam fields known. Wet steam, which is perhaps 20 times more abundant than dry, must have the hot water removed before the steam can be used, and so the technological requirements are more difficult. However, there may be a side benefit associated with the hot water which is removed since (energywise) it would be easy to desalinate it and thus provide a source of large amounts of fresh water.

As of 1971, worldwide capacity of geothermal power plants was only about 800 MWe—something less than the capacity of a single large fossil-fueled plant. Italy had 390 MW capacity installed; the United States, 193 MW; New Zealand, 170 MW; Japan, 33 MW; the Soviet Union, at least 6 MW; and Iceland, 3 MW (Koenig 1972). The only operating installation in the United States is located about 75 miles north of San Francisco, at The Geysers, California. As of late 1973, The Geysers

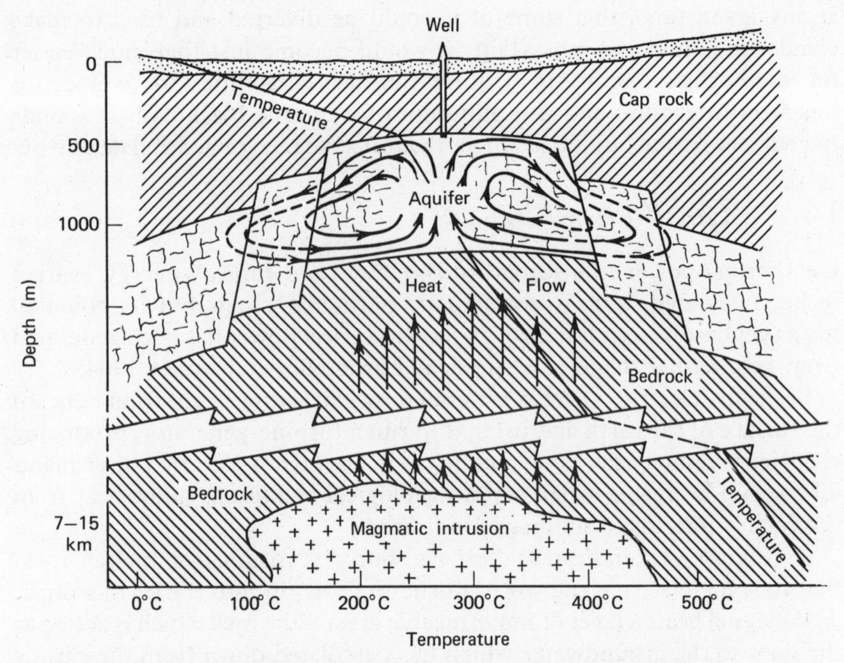

FIGURE 14.2 Origin of the steam beneath the earth's surface, (Facca 1973).

had increased its production to 396 MW and its capacity is expected to exceed 900 MW by 1977. The ultimate capacity is estimated to be around 1000 or 2000 MW (Barton 1972).

There is a vast, wet steam field in the Imperial Valley in California, estimated to have the potential to produce around 20,000–30,000 MW of electric power (Rex 1971). This field could, in addition, provide 5–6 million acre-feet of fresh water per year which could be used in part to help dilute the excessively salty Colorado River on its way into Mexico (see Section 4.5).

At this early stage it is difficult to estimate the total geothermal potential in the U.S., but, some place it at from 100 thousand-MW to 10,000 thousand-MW (Rex 1971). Table 14.1 shows an estimate (Peck 1972) of the potential installed capacity in the U.S. if an extensive research and development program is begun immediately. By 2000, there could be 395,000 MW on line, which would be about 21% of total U.S. generating capacity. The equivalent of 4.6 billion barrels of oil per year would be saved, which is about 50% of projected imports. If these estimates are anywhere close to being accurate, geothermal power could be of great importance to the nation.

The Geysers geothermal power plant. Units 3 and 4 are in the lower left portion of the photo while units 7 and 8, which brought the capacity up to 290,000 kW in 1972, are on top of the hill. Steam at the right and middle is from producing wells. (Courtesy Pacific Gas and Electric Company.)

TABLE 14.1 Potential U.S. Geothermal Electric Power Capacity

	1975	1985	2000
Installed geothermal capacity[a] (thousands of MWe)	0.75	132	395
Percent of total U.S. capacity[b]	0.15	14	21
Electrical energy ($\times 10^9$ kWh/yr)[c]	5.9	1041	3114
Oil equivalent[d] (billion barrels per year)	0.008	1.5	4.6

[a]Peck 1972.
[b]Dupree and West 1972.
[c]90% load factor.
[d]3412 Btu/kWh and 5.8×10^6 Btu/bbl of oil used at 40% conversion efficiency.

Let us consider some of the environmental effects associated with geothermal power. First, we must realize that the steam does not have nearly the temperature and pressure that normal steam-electric plants do. Hence the overall efficiency of the plant is considerably lower (recall Section 12.5), being only about 14.3% on the older units at The Geysers. The greater amount of cooling water required, compared to a conventional plant, could lead to increased thermal pollution.

At The Geysers, this is avoided by using cooling towers which release the extra heat to the atmosphere. The makeup water for the cooling towers comes from the steam itself as shown in Figure 14.3, and thus no outside source of water is required. As the figure shows, the geothermal steam is first passed through a centrifugal separator to remove rock particles and dust before passing through the turbine. The steam is then condensed and sent through the cooling tower where about 75–80% of it is evaporated. Part of the cooled water is used in the condenser and the rest, which contains many undesired contaminants such as boron and ammonia, is returned to the ground.

While reinjection of the polluted steam condensate back into the ground, rather than into a body of water, solves the pollution problem, there is some concern that such reinsertion could trigger earthquakes. It is also possible that the opposite may be true, that is, the reinjected water could allow rock fractures to slide a little at a time to slowly relieve stresses rather than having a sudden major movement. This is an area of concern that will be closely monitored.

Air pollution could be a significant problem since geothermal steam may contain noxious gases such as hydrogen sulfide which are released from the cooling tower.

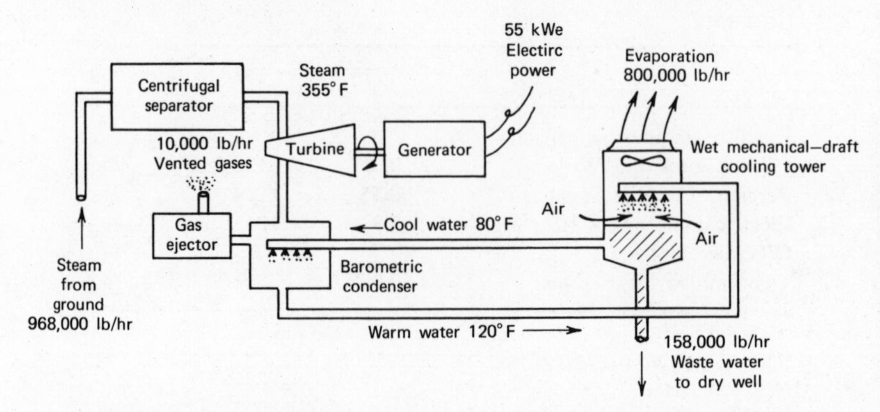

FIGURE 14.3 Simplified schematic of the 55 kW geothermal power units at The Geysers (data from Barton 1972).

Finally, it must be realized that geothermal power plants are depleting a resource that is only very slowly renewable, so that a steam field can be expected to last for only a limited length of time. Too little is known to make very good estimates but the range of expectations varies all the way from 30 to 3000 years (Rex 1971). Also since more water is removed than is replaced, there could be the danger of land subsidence.

To conclude, it seems that as much as one-fifth of total U.S. generating capacity in the year 2000 could be geothermal power. Since most of the geothermal resources are located in the Western states, this power could supply the majority of their demands, and do so with much less environmental damage than coal-fired plants and much less danger than nuclear reactors. It is unfortunate that such a development of geothermal power is quite unlikely due to our committment to other energy sources.

14.3 *Solar Power*

One of the most promising sources of energy for the future is the sun. With relatively minor advances in technology, it would be possible to capture and utilize sufficient quantities of solar energy to meet a good portion of our future energy demands. The sun's energy could be used not only for the generation of electricity but also for the very important sector of the energy picture—heating and cooling.

Referring back to Table 12.3, it can be seen that nearly 85% of the energy consumed in residences is for the two simple requirements of space heating and water heating. Moreover, Table 11.3 indicates that about 25% of *total* U.S. energy consumption is for various types of heating and cooling. What these figures imply, of course, is that if solar power can be used for a fair number of these applications there is the possibility of eliminating a sizable portion of the nation's energy problem.

Solar water heaters are already quite extensively used in such places as Israel, Japan, Australia, and even Florida. Such systems are extremely simple, requiring hardly more than a flat-plate solar collector and an insulated storage tank. Water flows through blackened tubing connected to the plate and carries heat to the storage tank. If the tank is placed slightly higher than the collector, the water will circulate naturally, without pumps, due to the density differences of hot and cold water (thermo-siphon circulation) as shown in Figure 14.4. A typical household version might use a collector area of about 40 square feet with a 75 gallon storage tank.

Solar space heating is also quite feasible. In fact, a number of experimental solar homes have been built in this country which use relatively

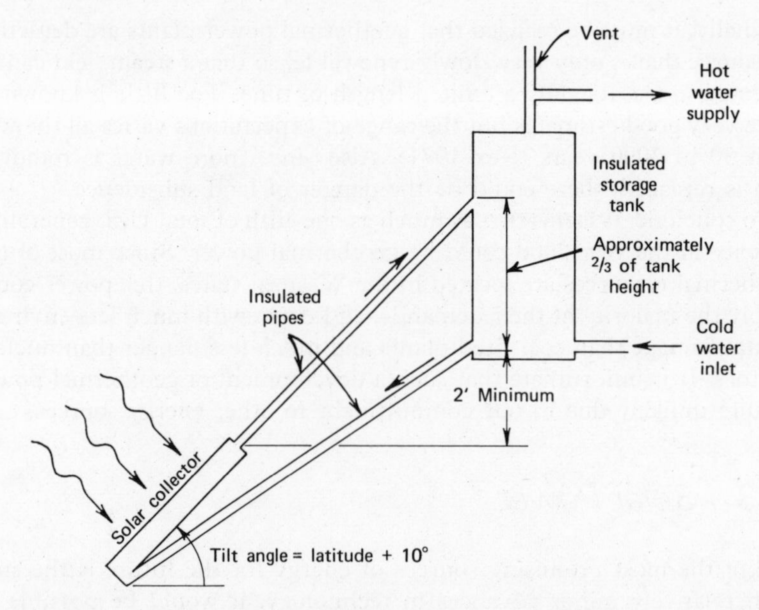

FIGURE 14.4 Simple thermo-siphon water heating unit (after Malik 1969).

simple systems comprised of a solar collector, a heat storage unit, and appropriate heat distribution and control systems (for example, Thomason and Thomason 1973). Solar cooling, by the absorption refrigeration process, is also possible though some development work needs to be done to make these systems economically practical. For most areas of the country, a reasonably sized rooftop solar collector would be able to supply most, if not all, the heating and cooling needs of a typical house.

How much power does the sun deliver to us? The terrestrial solar flux with the sun directly overhead is about 100 W/ft^2 but the daily average in the U.S. is more like 60 W/ft^2. The actual amount depends on such variables as the location, season, time of day, and local weather conditions. A useful measure of the available energy is the daily *insolation* which is the total radiant energy, including the direct solar beam and diffuse sky radiation from the scattered beam, incident on a unit area in the horizontal plane. In Tucson, Arizona, for example, half the days in June have a daily insolation that exceeds 0.81 kWh/ft^2.* In December, however, the median insolation there drops to 0.34 kWh/ft^2 (Bennett 1967). In contrast, the median values in Madison, Wisconsin, in June and

*Since much of the discussion will have to do with the generation of electricity, we will use the electrical units of energy.

December are 0.68 kWh/ft^2 and 0.14 kWh/ft^2. These data for various American cities are presented in Table 14.2.

To get a rough feel for these numbers, recall (Section 12.3) that the average house in the U.S. daily consumes something like 120 kWh of energy for all purposes—heating, cooling, cooking, electricity, etc. In Los Angeles, in December, that amount of energy falls on a horizontal area of 425 ft^2. Tilting the collector can offset some of the losses and a total area of about 1000 ft^2 would be about right.

Let us now turn to a consideration of the exciting possibility of utilizing the sun as an endless, relatively nonpolluting source of energy for the generation of electricity. First it must be asked whether the magnitude is sufficient. If we use 0.45 kWh/ft^2 (thermal) as an approximation for the average daily insolation across the U.S., and 3.1 million square miles as the area (48 states), we can calculate that the U.S. yearly receives about 15×10^{15} kWh of solar energy. This is 1700 times the estimated electric energy consumption for the year 2000. If we could capture solar energy and convert it to electrical energy at an efficiency of 20%, we could meet the total electrical demand in the U.S. in 2000 with about 0.33% of the nation's land area. Figure 14.5 shows the collector area required for various conversion efficiencies.

There are several approaches to capturing the sun's energy and converting it into electricity. The conceptually most straightforward approach is by means of solar cells. Solar cells are used extensively in space vehicles where they achieve efficiences on the order of 14%, and higher efficiencies are theoretically possible. While at the present time, this approach may have some application to small-scale power generation, the cost is prohibitive for large installations. Also realize that solar cells only produce

TABLE 14.2 Median Daily Insolation in Various U.S. Cities in June and December[a]

Location	June, ($kWh/ft^2/day$)	December, ($kWh/ft^2/day$)
Atlanta	0.63	0.26
Boston	0.60	0.15
Lincoln, Nebraska	0.67	0.20
Los Angeles	0.68	0.28
Madison	0.68	0.14
Tucson	0.81	0.34
Washington, D.C.	0.65	0.18

[a]Source: After Bennett 1967.

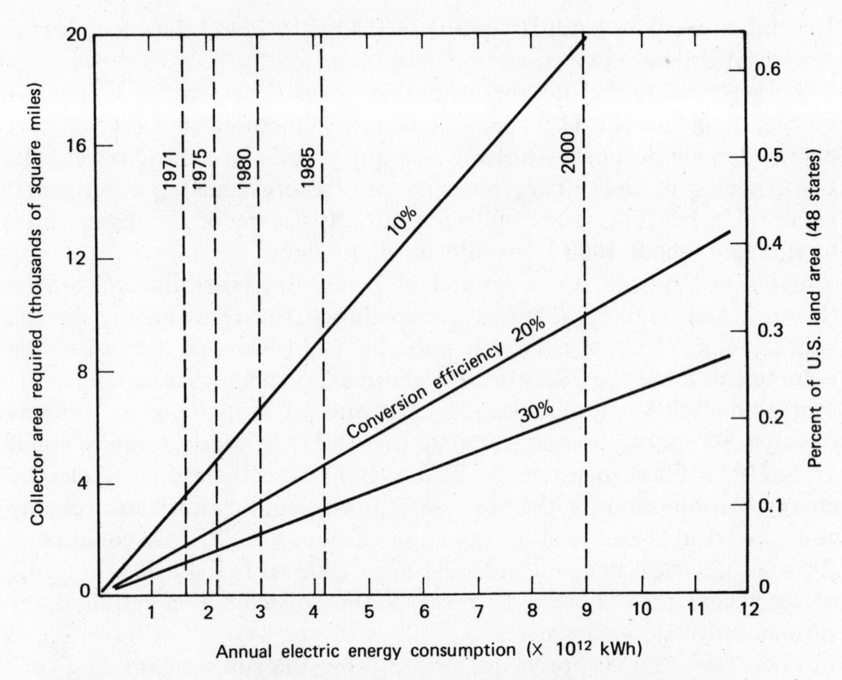

FIGURE 14.5 Solar collector area required to generate electric power for various conversion efficiencies from solar energy to electric energy. Based on a U.S. annual average insolation of about 4.6×10^9 kWh per square mile per year. (Projected electric energy consumptions based on Dupree and West 1972).

electricity during the day so some kind of energy storage system must be included.

The most exotic approach to solar power has been suggested by Glaser (1968) and involves placing into synchronous orbit 22,300 miles above the earth, a huge solar collector which would capture the sun's energy in solar cells and then beam it back to earth by means of a microwave transmitter, to be collected by a large receiving antenna on earth. The size of the solar panel would depend on the conversion efficiency of the solar cells, but as an example, at 18% efficiency it would require a 5 mile by 5 mile panel just to supply the needs of New York City! While this scheme is imaginative, it is unlikely that it could be economically or technologically feasible in the near future.

The solar systems most likely to succeed, if they receive sufficient funding, are land-based photo-thermal schemes. That is, the sun is used to heat some liquid, which in turn runs a turbine-generator pair. The husband-and-wife team of Aden and Marjorie Meinel (1971) at the University of Arizona have suggested that a significant portion of the nation's power needs in the year 2000 could be met with the simple system shown in Figure 14.6.

As can be seen, the right half of the system is a conventional steam-electric power plant. In the left loop, energy is collected by an array of solar panels that focus the sun's rays on to tubes containing flowing liquid metal, perhaps sodium. The heat-collecting loop transfers its energy to a large thermal-storage tank and heat exchanger which must be designed to store enough heat to carry the system through periods of perhaps several days duration, during which there is minimal or no sunshine. In the heat exchanger, energy is transferred to the steam loop which runs the turbine.

To make this system compatible with existing turbine-generator equipment, it is necessary to achieve steam temperatures of around 540°C (1000°F). To achieve such temperatures in the heat-collecting loop, a combination of two approaches is proposed: (a) The suns rays would be *focused* onto the tubes of liquid metal, and (b) the tubes would be coated with special *selective coatings* which readily allow energy in the visible portion of the spectrum to pass into the tubes but which do not allow the infrared (heat) energy from the hot liquid to pass back out again (just like the greenhouse effect). It is hoped that advances in the area of selective coatings will eliminate the need for focusing which would make the system operable even in conditions of diffuse light.

The Meinels propose a 1 million MW (electrical) National Solar Power Facility (operating at full capacity this would just about supply the demand for the entire nation in 2000) to be located on 13,700 square miles

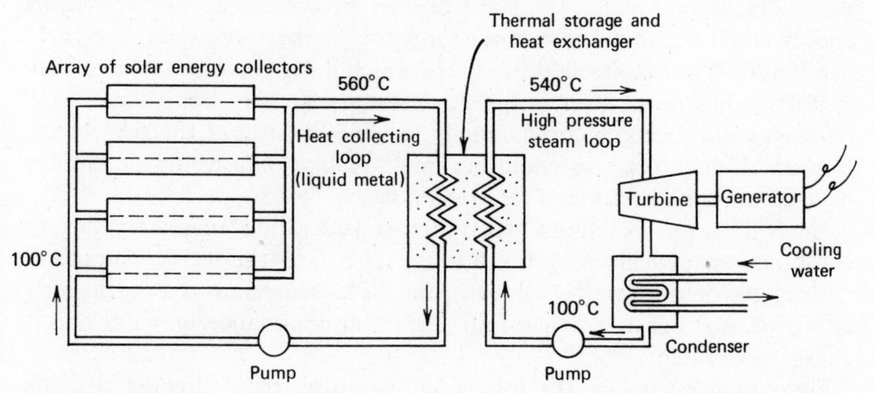

FIGURE 14.6 The Meinel solar power system (Meinel and Meinel 1971).

of desert land (Meinel and Meinel 1971). This is a huge area, about 9% of the size of California. For perspective, though, consider that we have about 500,000 square miles of farms in the U.S., which are also in the business of collecting solar energy. In fact, proponents of solar power would like to call their installations "solar farms" to suggest the analogy with biological farms.

Perhaps a more striking comparison can be made with the amount of land which is going to be strip mined for coal. By 1965, the U.S. had already disturbed over 2000 square miles for strip mining of coal and that number will at least double by 1980. In the long run, more land would be affected by strip mining than would be affected by solar collection.

The environmental effects of solar energy are not negligible even though the "fuel" is pollution-free. Besides the large area whose ecology would be rather completely changed there is the problem of thermal pollution. The Meinels hope to achieve an overall plant efficiency of around 31% so the amount of waste heat to be disposed of would be about the same as for nuclear fission. Since the best sites would be in the desert, large quantities of water would have to be brought in perhaps from the Gulf of California. This water could be desalinated as a side benefit of the solar plant, but care would have to be exercised in the disposal of the residual salty brine.

14.4 *Fusion Power*

While nuclear *fission* involves splitting atoms which have very heavy nuclei, *fusion* is somewhat the opposite, being the fusing together of two very light atomic nuclei. It is the process by which the sun derives its energy and it is also the source of energy for the thermonuclear, or hydrogen bomb. While controlled fusion, in which more power is released than is put in, has never been achieved, most researchers hope that fusion reactors could become commercially feasible by around the turn of the century. If such reactors could become a reality, man would potentially have an unlimited source of electrical power.

Several fusion reactions are under consideration involving various combinations of helium ($^4_2$He), lithium ($^6_3$Li), and the heavy isotopes of hydrogen—deuterium ($^2_1$D) and tritium ($^3_1$T). Deuterium is a particularly attractive fuel because it naturally occurs in abundance in water and is cheaply extracted.

Because it occurs at the lowest temperature, the following reaction involving deuterium and tritium is considered one of the most promising:

$$^2_1\mathrm{D} + {}^3_1\mathrm{T} \rightarrow {}^4_2\mathrm{He} + \mathrm{n} + 17.6\ \mathrm{MeV} \qquad (14\text{-}1)$$

For each fusion of a deuterium atom and a tritium atom, 17.6 MeV (million electron volts) is released. Deuterium is available in ordinary water in the ratio of 1 deuterium atom for each 6500 hydrogen atoms, and it costs only about 8¢ a gram to extract it.

While the supply of deuterium from ordinary sea water is virtually inexhaustible, the tritium does not occur in nature and hence must be manufactured. One source of tritium is the neutron bombardment of Lithium-6, as follows:

$$^{6}_{3}\mathrm{Li} + \mathrm{n} \rightarrow ^{4}_{2}\mathrm{He} + ^{3}_{1}\mathrm{T} + 4.8\ \mathrm{MeV} \qquad (14\text{-}2)$$

Relative to the abundance of deuterium, lithium is scarce and hence it is the component which would limit the total energy potentially available from this reaction. Because of the limited reserves of lithium, the energy thus available would be approximately the same as was available from our initial supply of fossil fuels, that is a several hundred year supply (Hubbert 1969). Thus while it appears the deuterium-tritium (D-T) reaction may be the first to be achieved, it does not seem to have the potential to be the ultimate solution to the energy problem.

There are however, reactions that depend only on the plentifully available isotope deuterium, as the initial fuel, and it is likely that eventually they will be technologically feasible. When two deuterium atoms are fused, the following two reactions are about equally likely to occur:

$$^{2}_{1}\mathrm{D} + ^{2}_{1}\mathrm{D} \rightarrow ^{3}_{2}\mathrm{He} + \mathrm{n} + 3.2\ \mathrm{MeV} \qquad (14\text{-}3)$$

$$^{2}_{1}\mathrm{D} + ^{2}_{1}\mathrm{D} \rightarrow ^{3}_{1}\mathrm{T} + \mathrm{p} + 4.0\ \mathrm{MeV} \qquad (14\text{-}4)$$

When reaction (14-4) occurs, the tritium that is formed reacts with a deuterium as given in (14-1). Moreover, the $^{3}_{2}\mathrm{He}$ also can react with a deuterium as shown in (14-5)

$$^{2}_{1}\mathrm{D} + ^{3}_{2}\mathrm{He} \rightarrow ^{4}_{2}\mathrm{He} + \mathrm{p} + 18.3\ \mathrm{MeV} \qquad (14\text{-}5)$$

The net effect of reactions (14-1), (14-3), (14-4), and (14-5) is the conversion of six deuterium nuclei into two helium nuclei, two hydrogen nuclei (two protons), and two neutrons, with a release of 43.1 MeV of energy.

$$6\ ^{2}_{1}\mathrm{D} \rightarrow 2\ ^{4}_{2}\mathrm{He} + 2\mathrm{p} + 2\mathrm{n} + 43.1\ \mathrm{MeV} \qquad (14\text{-}6)$$

Thus each deuterium nucleus can produce about 7.2 MeV. At this rate 1 gallon of sea water could supply the energy equivalent of about 330 gallons of oil (see Question 8). A cube of water just 550 feet on a side could supply as much energy as the entire 10 billion barrels of Alaskan oil. The oceans could therefore represent an essentially endless source of energy for man if the deuterium-deuterium (D-D) reaction could be realized.

How can fusion be obtained? It is necessary to bring enough particles together, for a long enough time, and at a high enough temperature for fusion to occur and, of course, more energy must be released than is required to cause the reaction. The temperature must exceed what is known as the "ignition temperature," which, for the D-T reaction is around 40 million degrees Kelvin (compare this to the sun's temperature which is about 20 million degrees). At very high temperatures such as this, the electrons and atoms are separated and the gas exists in an ionized form called a *plasma*.

In addition to meeting the temperature requirement, the product of the plasma density times the containment time must exceed a certain value called the Lawson criterion. As yet, no system has been able to meet all three of the requirements—temperature, plasma density, and confinement time—but researchers are close enough to feel that, with sufficient funding, they may be able to do so by the early 1980s.

In view of the high temperatures involved, it would be impossible to find a material which could contain the plasma without itself melting. However, since the plasma consists of charged particles, it is possible to contain it using magnetic fields, and finding the best configuration for such "magnetic bottles" has been the focus of most of the fusion work. The details of these magnetic containment schemes are beyond the scope of this short section, and the interested reader is referred to the references at the end of the chapter.

Let us turn to a consideration of some of the environmental consequences of fusion. First realize that since there are no radioactive fission fragments, the problem of radioactivity is inherently less severe than that for fission reactors. If the D-T fuel cycle is used, there would be a problem containing and handling the radioactive tritium (half-life of 12 years, beta emitter). However, this problem is greatly simplified since the tritium is returned as fuel and need not be permanently stored. Also, there would be induced radioactivity in the reactor's structural materials from neutron bombardment, so they would periodically have to be removed and disposed of somehow.

With regard to safety, fusion reactors are inherently much more desirable than fission reactors since there is no such thing as a nuclear excursion. The danger from sabotage or natural disaster is thus eliminated and there is no problem with diversion of weapons-grade materials (plutonium).

With regard to thermal pollution, D-T fusion reactors should operate in the 50–60% efficiency range so thermal pollution would be less than for any of the other electric-power generation schemes. Also there is the possibility for direct conversion of the energy into electricity without going through the steam cycle. In a design suggested by Post (1971) of the Lawrence Radiation Laboratory, the charged particles from the plasma would be collected on electrodes creating a direct-current (rather than A-C). Such a scheme might yield thermal efficiencies on the order of 80–90% thus greatly reducing thermal pollution.

Thus it is very likely that nuclear fusion will be the ultimate energy source for the future, at least for the generation of electricity. The supply of fuel is essentially unlimited and the environmental impact would be relatively mild. It will be many years, however, before we could hope to generate significant quantities of power.

It is interesting to speculate on how the United States might meet its future demands for electric energy. Figure 14.7 has been drawn to indicate the mix of energy sources which could carry us well into the next century, while placing limited emphasis on nuclear fission. The projection is optimistic in that it is based on a reordering of priorities in the government's funding of energy research, but it is technologically not unrealistic.

14.5 *Other New Sources of Energy*

As the energy crisis intensifies, attention is increasingly being directed toward unusual sources of energy that in the past have been overlooked because fossil fuels were so cheap and convenient. In this section we

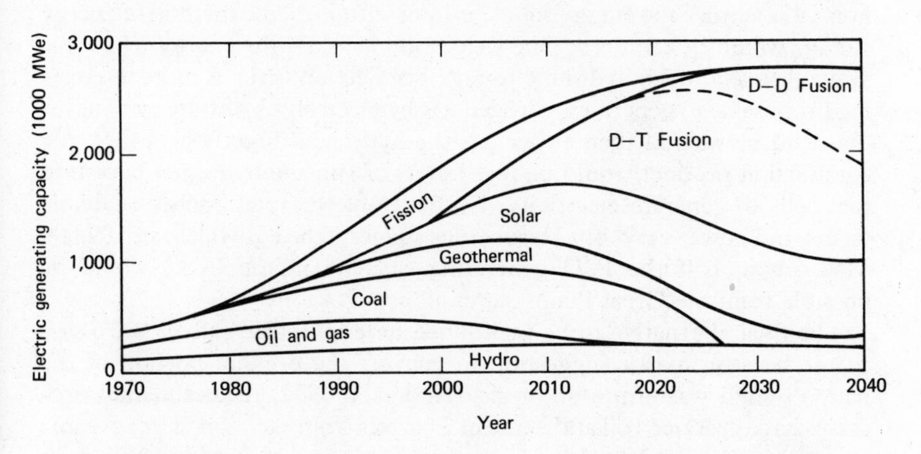

FIGURE 14.7 Speculated sources of energy for the generation of electricity in the U.S. minimizing the necessity for fission. A radical revision of research priorities would be required for this figure to be accurate.

shall mention four alternate energy sources that could be locally important but which, right now, do not appear to be competitive on a large scale with the other approaches already mentioned in the chapter.

There are scattered throughout the world, a number of locations where it would be possible to generate sizable amounts of electric energy by taking advantage of the tides. By damming a coastal basin, it is possible to generate electricity with a hydraulic turbine as the basin periodically fills and empties. The first such installation began operation in 1966 at La Rance estuary in France, with a generating capacity of 240 MW. There is a site on the U.S.-Canadian border at Passamaquoddy Bay that has long been considered for a large tidal plant but as yet no action has been taken. While individual sites could generate significant amounts of electric power, the global contribution that could be made by the tides is insignificant.

Another proposal has been made to take advantage of the temperature difference that exists between the surface waters and waters at a depth of about 2000 feet in the Gulf Stream off Florida (Anderson and Anderson 1966). Temperature differences of 30 or 40°F are typical so the upper limit of efficiency would be somewhere around 8% (see Question 6). Even though the efficiency of an actual power plant might be more like 2%, the amount of energy available is sufficient to supply the entire electrical needs of the United States far into the future.

Another inexhaustible source of energy is the wind, and proposals have been made to establish a network of huge windmills to capture it (Heronemus 1972). Since the energy source is intermittent, some method of energy storage would be required. One suggestion is to use the energy when it is available for pumped hydraulic storage (see Section 12.1). A more interesting proposal is to generate hydrogen gas by electrolysis of water whenever the wind blows and then either use the hydrogen directly as a fuel (the combustion product would be just water) or run the hydrogen back into fuel cells to generate electricity. Estimates of the total energy available from wind power vary, but Heronemus suggests the equivalent of 189,000 MW (about half the 1973 generating capacity in the U.S.) would be possible from the Great Plains states alone.

The final alternative to be mentioned here might be called "garbage" power because it is a suggestion to convert the organic portion of the nation's solid waste into oil or gas (Anderson 1972). Experimental processes have produced oil at the rate of 2 barrels from each ton of dry organic material and gas at the rate of 5 standard cubic feet of methane from each pound of metal- and glass-free urban refuse. Of the total potential of about 1.1 billion barrels of oil which could be so produced per year, Anderson estimates that about 170 million barrels per year would be obtainable from readily available sources—that is, sources such as feedlots and municipal dumps where the wastes are available in concentrated form.

14.6 Conclusions

This chapter has outlined a number of rather untested but promising approaches toward meeting the energy demands of the future. While most of the techniques described are directly applicable to the generation of electricity, some would also be able to meet the demands in other sectors of the energy picture. Moreover, electric power could be used to generate hydrogen gas by electrolysis of water, and the hydrogen could subsequently be used as a pollution-free fuel in applications where the electric form of energy is not practical.

In spite of the difficult technology and uncertain safety associated with nuclear fission, there is little question that the dominant trend in the near future shall be toward fission reactors. What's more, since conventional light-water reactors have such limited fuel resources, we must realize that the real impact of fission reactors will be made by breeders. When weighing the advantages and disadvantages of fission, it is therefore necessary to deal with the environmental implications of breeder technology and the associated production of large amounts of plutonium.

What should be apparent from this chapter is that there are a number of energy sources that could become important in the future if given the proper funding. The proposed 1973 Federal research and development budget for energy is as follows (in millions of dollars): nuclear fission, \$356; fossil fuels, \$136; nuclear fusion, \$65; solar energy, \$4; geothermal energy, \$3; and, related technologies, \$55 (Hammond 1972). Unless these priorities change, it is unlikely that some of the alternate energy sources will ever get off the drawing boards.

Bibliography

Anderson, J. H., and Anderson, J. H., Jr. (1966). Thermal power from sea water. *Mech. Eng.* April.

Anderson, L. L. (1972). *Energy potential from organic wastes: A review of the quantities and sources.* Dept. of Interior, Info. Circular 8549, Bureau of Mines. Washington, D.C.

Barnea, J. (1972). Geothermal power. *Sci. Am.* Jan.

Barton, D. B. (1972). The geysers power plant—a dry steam geothermal facility. Compendium of First Day Papers presented at the First Conference of the Geothermal Resources Council, El Centro, California, Feb.

Benedict, M. (1971). Electric power from nuclear fission. *Technol. Rev.* Oct./Nov.

Bennett, I. (1967). Frequency of daily insolation in Anglo North America during June and December. *Solar Energy* 11 (1).

Bowen, R. G., and Groh, E. A. (1971). Geothermal—earth's primordial energy. *Technol. Rev.* Oct./Nov.

Coppi, B. and Rem, J. (1972). The Tokamak approach in fusion research. *Sci. Am.* July.

Cowen, R. C. (1972). The promised land of fusion. *Technol. Rev.* Jan.

Dupree, W. G., Jr., and West, J. A. (1972). *United States energy through the year 2000.* U.S. Department of the Interior. Washington, D.C. Dec.

Facca, G. (1973). The structure and behavior of geothermal fields. *Geothermal Energy*, UNESCO, Earth Sciences 12, Paris.

Fenner, D., and Klarmann, J. (1971). Power from the earth. *Environment* Dec.

Ford, N. C., and Kane, J. W. (1971). Solar power. *Bull Atom. Sci.* Oct.

Geesaman, D. P. (1971). Plutonium and the energy decision. *Bull. Atomic Sci.* Sept.

Glaser, P. E. (1968). Power from the sun: Its future. *Science* 162: Nov.

Gough, W. C., and Eastlund, B. J. (1971). The prospect of fusion power. *Sci. Am.* Feb.

Hammond, A. L. (1972). Energy options: Challenge for the future. *Science* Sept.

Heronemus, W. E. (1972). The U.S. energy crisis: Some proposed gentle solutions. Joint Meeting A.S.M.E. and I.E.E.E., West Springfield, Mass., January 12, 1972 (also in *Congressional Record-Senate,* Vol. 118, pp. E.1043-9, February 9, 1972).

Hickel, W. J. (1972). *Geothermal energy, a national proposal for geothermal research.* Geothermal Resources Research Conference, Seattle, Wash., Sept. 18–20.

Hubbert, M. K. (1969). Energy resources. In National Academy of Sciences, *Resources and man.* San Francisco: W. H. Freeman.

Koenig, J. B. (1972). The worldwide status of geothermal exploration and development. First Conference of the Geothermal Resources Council, El Centro, Cal. Feb.

Lidsky, L. M. (1972). The quest for fusion power. *Technol. Rev.* Jan.

Lubin, M. J., and Fras, A. P. (1971). Fusion by laser. *Sci. Am.* June.

Malik, M. A. S. (1969). Solar water heating in South Africa. *Solar Energy* 12: (3).

Meinel, A. B. (1971). A proposal for a joint industry-university-utility task group on thermal conversion of solar energy for electrical power production. Arizona Power Authority, Phoenix, April 27.

Meinel, A. B., and Meinel, M. P. (1971). Is it time for a new look at solar energy? *Bull. Atom. Sci.* Oct.

Mills, R. G. (1971). The promise of controlled fusion. *IEEE Spectrum,* Nov.

National Academy of Sciences (1972). "Solar Energy in the Developing Countries: Perspectives and Prospects," Report of Ad Hoc Panel of Board on Sci. & Tech. for Intern. Dev. Washington, D.C. March.

Peck, D. L. (1972). *Assessment of Geothermal Energy Resources.* USDI. Panel on Geothermal Energy Resources, Sept. 25.

Post, R. F. (1971). Fusion power, the uncertain certainty. *Bull. Atom. Sci.* Oct.

Rex, R. W. (1971). Geothermal energy—the neglected energy option. *Bull. Atom. Sci.* Oct.

Thomason, H. E., and Thomason, H. J. L., Jr. (1973). Solar houses/heating and cooling progress report. *Solar Energy* 15: (1).

U.S. Senate (1972). Committee on Interior and Insular Affairs. *Energy Research Policy Alternatives.* Ser. No. 92-30. Washington, D.C. June 7.

Questions

1. In an ordinary light-water reactor, what minimum number of neutrons must be released per fission in order to maintain the chain reaction? In a breeder how many are required? Can you relate the difference to the need for a higher concentration of fissile material in a breeder?

   ans. 1; 2

2. In a loss of coolant accident why would the reaction in a breeder be more likely to remain critical than in a light-water reactor?

3. Suppose a solar water heater consists of a 40 ft^2 collector and a 75 gallon storage tank (625 pounds of water in the system). What is the maximum possible rise in water temperature (assuming 100% heat transfer efficiencies) on a day when the panel is exposed to 1500 Btu/ft^2?

   ans. 96°F

4. The average daily residential requirement for electric energy is 19 kWh. If the efficiency of conversion from solar to electrical energy is 10 % what minimum horizontal collector area would be required for a home in Tucson, Arizona, in December?

   ans. 550 ft^2

5. A portable radio requires about 0.15 W of power. With 10% efficient solar cells and an average solar power density of 60 W/ft^2, what minimum solar cell area would be required?

   ans. 3.6 in.^{2}

6. What minimum collector area would be required for a 30% efficient (solar to electrical) 1000 MWe solar power plant in Tucson in December? Assume the power plant operates at 80% of capacity, 24 hours a day.

   ans. 6.7 mi^2

7. Combine reactions (14-1), (14-3), (14-4), and (14-5) to obtain reaction (14-6).

8. It is relatively simple to calculate that 330 times as much energy can be derived from a volume of water by D-D fusion, as can be derived from the combustion of the same volume of oil. Since there are 6.023×10^{23} H_2O molecules in 18 cc of water, how many deuterium atoms (at 1 deuteron per 6500 hydrogen atoms) would be contained in 1 cc of water? At 7.2 MeV per deuterium atom how much energy could be released from 1 cc of water compared to 0.23 MeV/cc of oil?

9. The surface temperature in the Gulf of Mexico in February is about 76.5°F while in August, it is 86°F. At a depth of 2000 feet the February temperature is 48°F and the August temperature is 43°F. Calculate the Carnot efficiency of a heat engine operating between these temperatures in February and August.

   ans. 5.1%; 7.9%

10. Which of the eight alternative energy sources described in this chapter can only be used for the generation of electricity and which can be applied in other areas?

Chapter 15

Mineral Resource Depletion

In Chapter 11 we considered the drain on the world's energy resources caused by industrialization and we began to see how the depletion of those resources raises questions of international importance that are very difficult to deal with. In this chapter we shall expand our consideration to include all of the mineral resources that are required by a technological society.

The industrialized nations of the world developed during an era of abundant resources, and, indeed, their continued advancement is tied directly to a constant supply of cheap minerals. As domestic high grade ores are depleted, the developed nations are turning increasingly to foreign sources of minerals—a shift which promises to dominate international relations in the future.

It is of the utmost importance to evaluate the extent of the earth's remaining resources. If it can be determined that they are ample, then it would not be unreasonable for all countries to pursue resource-consuming activities if the result is an improvement in the quality of life. However, as we shall see, there is considerable evidence to indicate that the opposite may be true with the result that the already overdeveloped countries are scrambling for the remaining reserves while the vast majority of mankind sinks further into abject and permanent poverty.

15.1 *Minerals Consumption in the United States*

Mineral resources may be conveniently categorized as being either energy resources, metallic minerals, or industrial minerals. The energy resources were discussed in Chapter 11, where it was noted that world reserves of oil and natural gas are quite limited; coal is rather more abundant; and reserves of uranium can be considered small or large depending on whether breeder reactors replace present generation light-water reactors.

The nonmetallic, or industrial, minerals includes a wide variety of substances. By tonnage, the largest group comprise the building materials such as rock, sand, gravel, cement, plaster, and clay. Though these are of obvious importance, they are not in short supply and they play almost no role in the international marketplace and so are of no interest here.

Also included in the nonmetallic minerals are the fertilizer materials which supply the nitrogen, phosphorus, and potassium which are so essential to increased agricultural outputs around the world. Large amounts of nitrogen, in the form of nitrates, used to be available only in the natural deposits which occurred in Chile, but it is now possible to synthetically fix nitrogen from the air so there is no lack of resources. Phosphorus is obtain from phosphate rock, and although world reserves are abundant, it is not very evenly distributed around the world. Northern Africa, the U.S.S.R. and the U.S.A. have most of the reserves. Potassium, as potash, is also quite abundant in the world but the U.S. is not so well endowed, and we presently must import almost half of our consumption (mostly from Canada which has about 40% of the world's reserves).

It is, however, the metals which need to be focused upon in this chapter. Metals can be categorized as being: (a) Iron or ferro-alloy elements, (b) nonferrous metals, or (c) precious metals.

Iron and the ferro-alloy elements are of vital importance in every industrial economy. Iron is, of course, the principal component of steel, and steel is essential in many applications, most especially, heavy construction. The elements which are added to plain carbon steel to improve such properties as strength, ductility, and resistance to corrosion, are called ferro-alloy elements—the principal ones being chromium, cobalt, columbium (niobium), manganese, molybdenum, nickel, tantalum, titanium, tungsten, vanadium, and zirconium.

Some of the major functions of the most important ferro-alloy metals are listed in Table 15.1 along with some indication of whether or not there exist satisfactory substitutes. It should be noted, that manganese, cobalt, and tungsten must be considered essential. Moreover, as Figure 15.1 indicates the U.S. supplies a negligible portion of its manganese and cobalt consumption and will soon be relying heavily on imports for tungsten also.

As Figure 15.1 indicates, the United States has become a nation critically dependent on foreign sources for all of the ferrous metals except molybdenum, of which we are a net exporter.

TABLE 15.1 Principal Ferro-Alloy Metals—Their Functions and Possible Substitutes

Metal	Functions	Substitutes
Manganese	Chiefly used as a desulfurizing agent in steel making. As an alloy it improves rolling and forging and adds to the strength, toughness, wear resistance, hardness, and hardenability. Also used in dry-cell batteries and chemicals.	There are no satisfactory substitutes.
Chromium	Used in stainless, tool, and alloy steels; prevents rust. Also used as a refractory and as a chemical.	No substitutes in stainless steel, but titanium can substitute in some corrosion applications. Aluminum can substitute in some decorative applications.
Nickel	Used in various alloys, mostly stainless steel. Increases toughness, stiffness, strength and ductility. Also used in nickel-cadmium batteries.	Substitutes are available for most uses of nickel. Cobalt can be substituted in many instances.
Tungsten	Gives hardness and toughness to steel. Used for high-speed cutting tools, also as filaments in light bulbs.	One of the most essential metals, though substitutes for most uses are available, such as molybdenum and titanium carbide, but they are seldom as good.
Molybdenum	Alloy for stainless and tool steels, imparting excellent corrosion resistance with increased strength at high temperatures.	Substitutes are available for most uses but none are competitive.
Cobalt	Principal uses in alloys, especially cobalt steel for permanent magnets, also high-temperature, high-strength tool steels.	There are no substitutes for the most important uses, e.g., magnets. Vanadium, tantalum, and nickel are substitutes for other applications.

Iron itself, which is the second most abundant metal in the earth's crust, is in only a little better shape than the alloy metals. In the U.S. the easily accessible high-grade ore is almost gone though there is still a fair amount of low-grade taconite ore. As Figure 15.1 indicates, the total

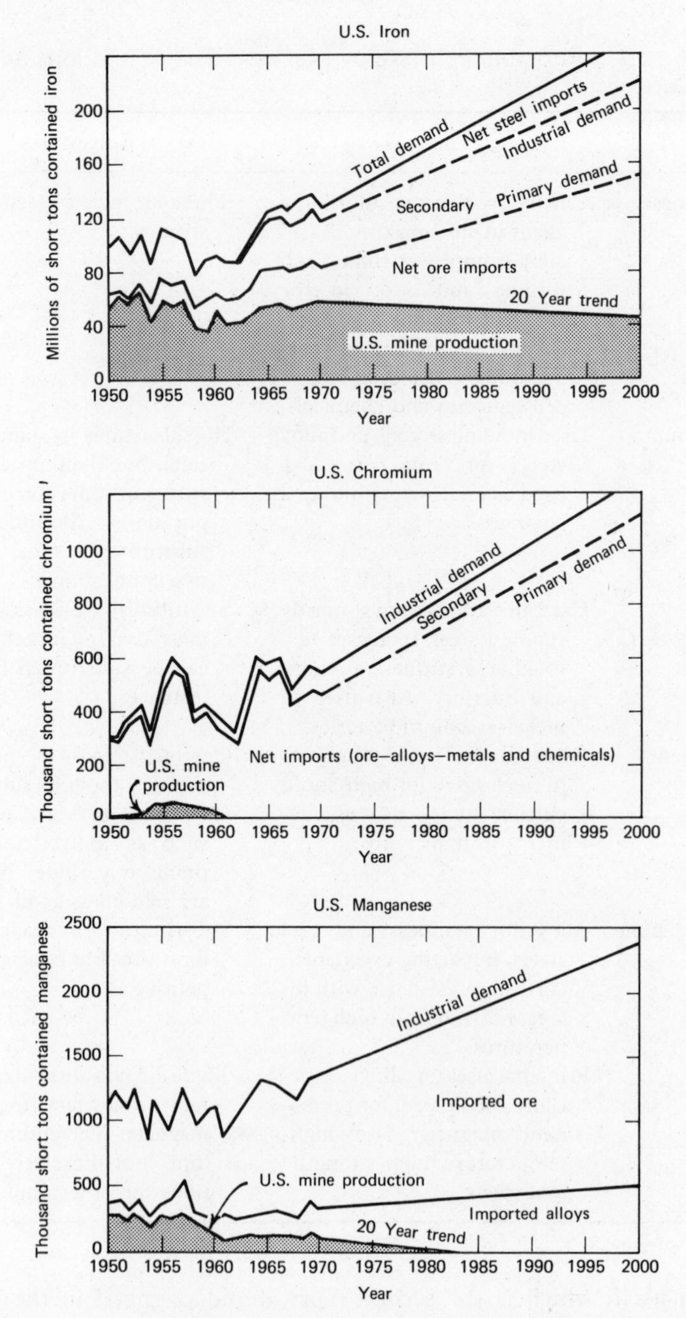

FIGURE 15.1 Iron and ferro-alloy metals, domestic production and demand (National Commission on Materials Policy 1972).

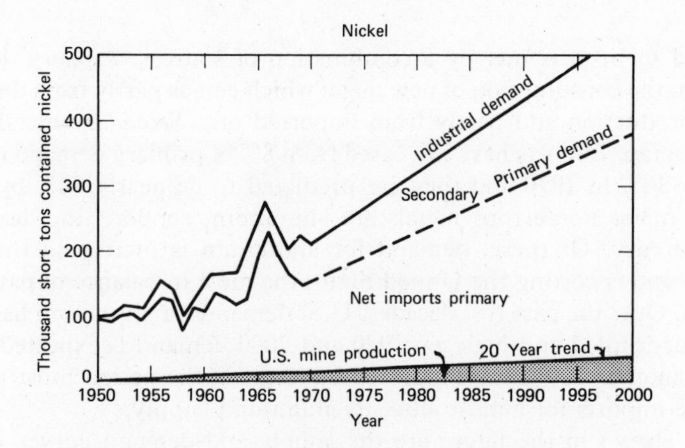
Nickel
Thousand short tons contained nickel
500
400
300
200
100
0
Industrial demand
Primary demand
Secondary
Net imports primary
U.S. mine production
20 Year trend
1950
1955
1960
1965
1970
1975
1980
1985
1990
1995
2000
Year

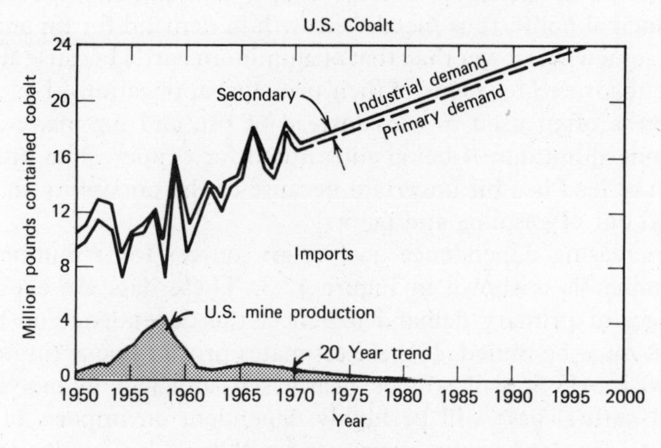
U.S. Cobalt
Million pounds contained cobalt
24
20
16
12
8
4
0
Secondary
Industrial demand
Primary demand
Imports
U.S. mine production
20 Year trend
1950
1955
1960
1965
1970
1975
1980
1985
1990
1995
2000
Year

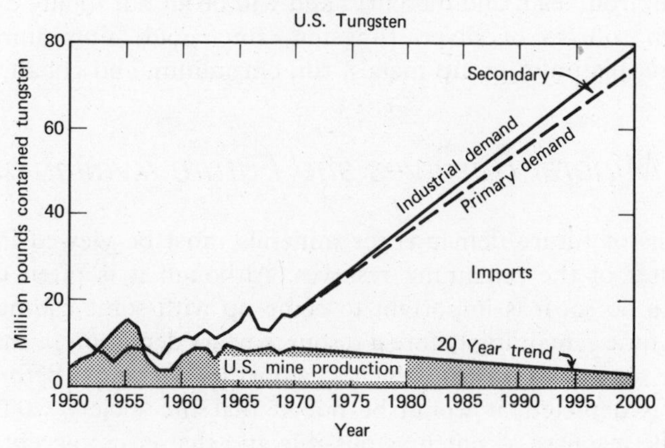
U.S. Tungsten
Million pounds contained tungsten
80
60
40
20
0
Secondary
Industrial demand
Primary demand
Imports
20 Year trend
U.S. mine production
1950
1955
1960
1965
1970
1975
1980
1985
1990
1995
2000
Year

demand for iron is met by a combination of sources. *Primary* demand refers to the consumption of new metal which comes partly from domestic mine production and partly from imported ore. *Secondary* metal is recycled scrap. Imports have increased from 8% of primary demand in 1950 to over 30% in 1970 and they are predicted to be nearly 70% by 2000.

The major nonferrous metals are aluminum, copper, zinc, lead, tin, and mercury. Of these, demand for aluminum is increasing the most rapidly and is costing the United States the most in balance of payments deficits. Over the past two decades, U.S. demand for aluminum has more than quadrupled and between 1970 and 2000 demand is expected to increase another 700%. As Figure 15.2 indicates, the nation must rely on massive imports for almost all of its aluminum supply.

Also shown in the figure are the supply-and-demand curves for the other principal nonferrous metals. Growth in demand for tin and copper has been somewhat slower than that of aluminum partly because aluminum is being substituted for many of their principal applications. For example, aluminum is often used in cans instead of tin, and for many electrical applications aluminum is being substituted for copper. The future consumption of lead is a bit uncertain because of the possibility that it may be phased out of gasoline and paints.

The increasing dependence on foreign sources for a number of important minerals is shown in Figure 15.3. These data are expressed as percentages of primary demand to reflect the dependence on new material that must be mined. If these estimates prove to be accurate, by the year 2000 the U.S. will still be nearly self-sufficient in molybdenum, coal, and natural gas; will be highly dependent on imports of copper, petroleum, iron, lead, and mercury; and will be all but totally dependent on foreign sources of silver, tungsten, zinc, gold, aluminum, nickel, manganese, platinum group metals, tin, chromium, and cobalt.

15.2 *Mineral Reserves and Future Availability*

Projections of future demands for minerals must be viewed in relation to estimates of the remaining reserves. Although it is often extremely difficult to do so, it is important to come up with some estimate of the length of time remaining before a resource nears depletion so that society may have some lead time in which to adjust its demands. Before a given resource is depleted, it would be hoped that the society would be recycling the resource as much as possible and that either acceptable substitutes would be found or that the demand would cease. A sudden shortage could be crippling to an industrialized nation and might prompt any number of undesirable responses which might range from an abandonment of concern over the environmental effects associated with acquiring the mineral to an agressive military action designed to acquire or maintain access to the remaining reserves.

There is a growing controversy over the sufficiency of the mineral resources to sustain continuing domestic and world economic development. On the optimistic side, in 1972, the Director of the U.S. Geological Survey, V. E. McKelvey stated that:

> Personally I am confident that for millennia to come we can continue to develop the mineral supplies needed to maintain a high level of living for those who now enjoy it and raise it for the impoverished people of our own country and the world (USDI 1973).

Somewhat the opposite opinion was expressed by the Committee on Resources and Man (1969) when they concluded that:

> . . . true shortages exist or threaten for many substances that are considered essential for current industrial society: mercury, tin, tungsten, and helium, for example. Known and now-prospective reserves of these substances will be nearly exhausted by the end of this century or early in the next, and new sources or substitutes to satisfy even these relatively near-term needs will have to be found.

Before proceeding further, recall the meaning of the terms reserves and resources as presented in Chapter 11. *Reserves* are quantities of materials that can be reasonably assumed to exist and which are producible with existing technology and under present economic conditions, and *resources* include in addition to reserves, all potential reserves which might become available under future economic conditions and technology.

Thus, while an estimate of the reserves of a given mineral gives an indication of how much can be produced under current conditions, it should not be interpreted as the total which will ultimately be producible. Changes in economic conditions, advances in technology, and new discoveries are constantly adding new quantities to the reserves. Thus there is a dynamic nature to these estimates; as reserves are consumed, new reserves are added. However, since total resources are finite, there is no way for consumption to continue indefinitely, so the question is not, *are* we going to run out, but *when* are we going to run out.

A rough but simple indicator of the future availability of a mineral is the *static reserve index,* or static index, which is obtained by dividing the estimate of reserves by the present annual rate of consumption. If no additions were to be made to the reserves and if present rates of consumption were not to increase in the future, this index would give us the number of years remaining until the mineral would be depleted. Obviously the estimate is crude because additions will likely be made to the reserves

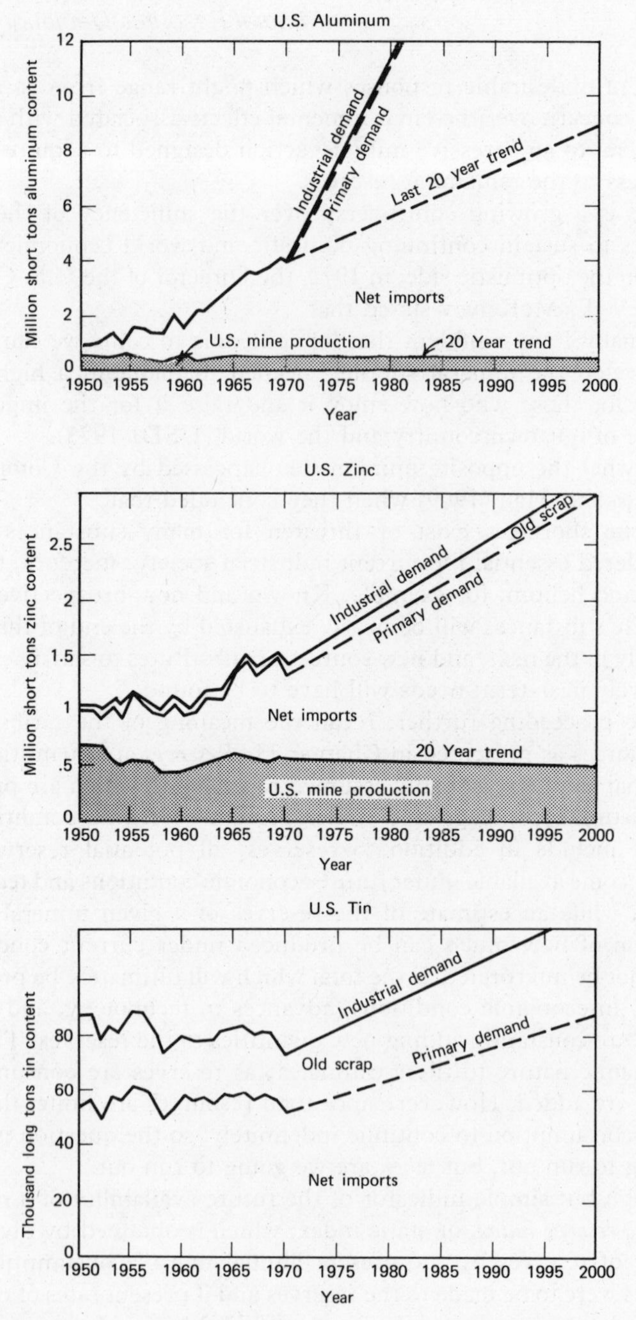

FIGURE 15.2 Domestic production and demand for principal nonferrous metals (National Commission on Materials Policy 1972).

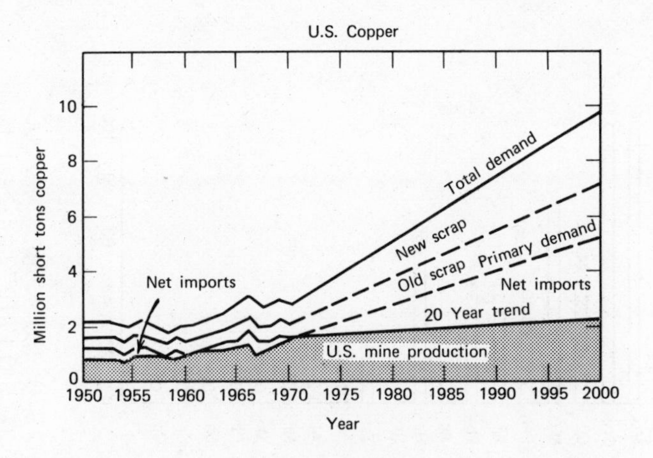
U.S. Copper
Million short tons copper
0
2
4
6
8
10
Total demand
New scrap
Old scrap
Primary demand
Net imports
Net imports
20 Year trend
U.S. mine production
1950 1955 1960 1965 1970 1975 1980 1985 1990 1995 2000
Year

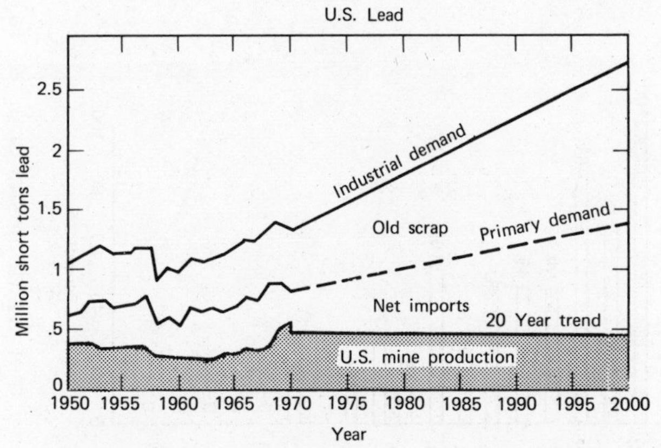
U.S. Lead
Million short tons lead
0
.5
1
1.5
2
2.5
Industrial demand
Old scrap
Primary demand
Net imports
20 Year trend
U.S. mine production
1950 1955 1960 1965 1970 1975 1980 1985 1990 1995 2000
Year

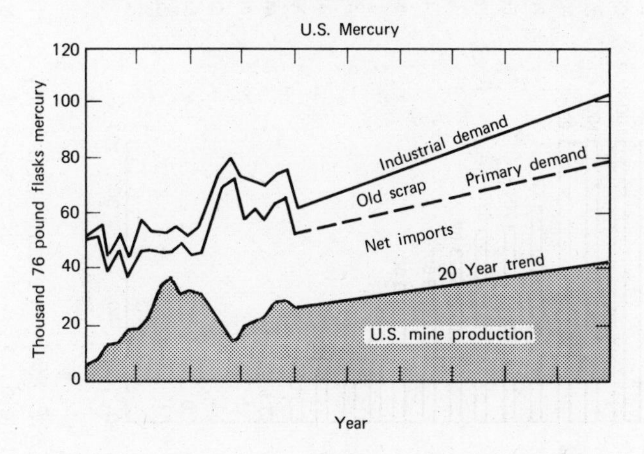
U.S. Mercury
Thousand 76 pound flasks mercury
0
20
40
60
80
100
120
Industrial demand
Old scrap
Primary demand
Net imports
20 Year trend
U.S. mine production
Year

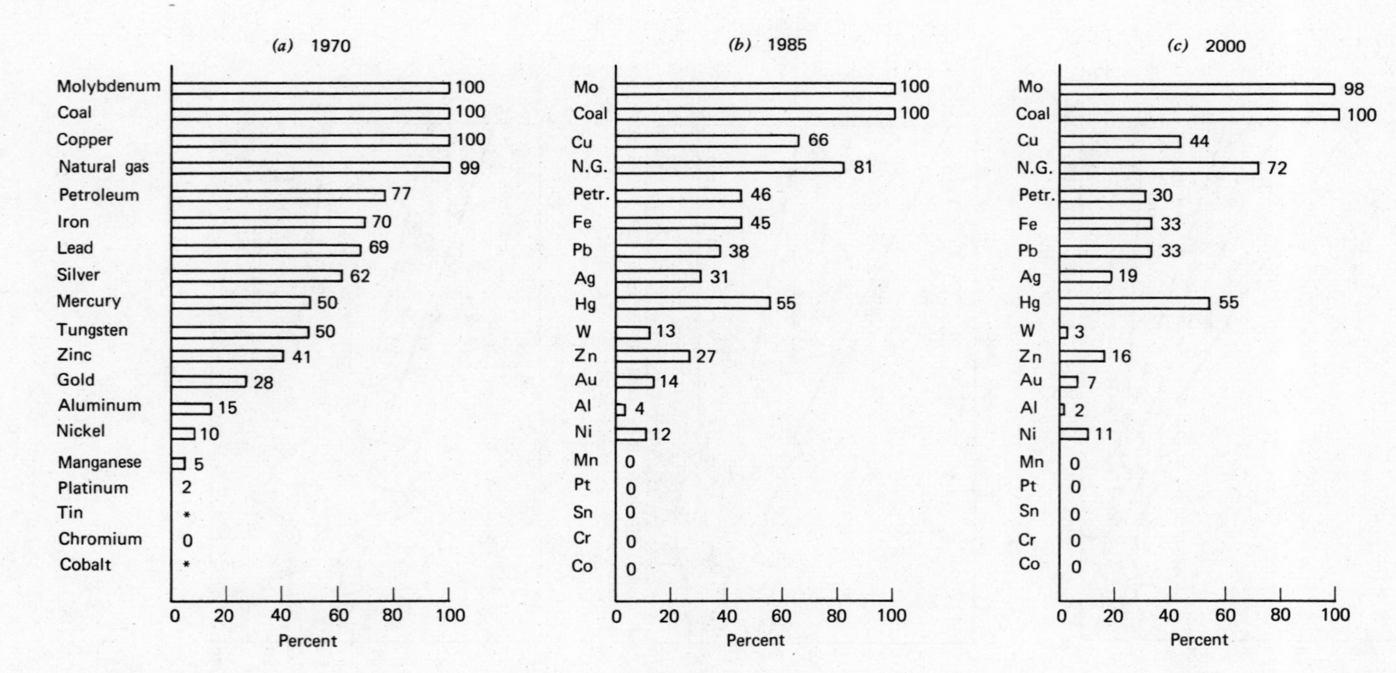

FIGURE 15.3 Percent of primary demand met by domestic sources (*a*) 1970, (*b*) 1985, (*c*) 2000. *Withheld to avoid disclosure of company confidential data—but small. (U.S.D.I. 1972, except for petroleum, natural gas, and coal, Dupree and West 1972.)

and consumption will undoubtedly increase, however these two factors to some degree offset each other.

Using zinc as a not unusual example we can see in Figure 15.4 that rising reserves and production have tended to keep its static reserve index nearly constant for the 10 year period 1962–1971. Thus, while in 1962 it may have appeared that zinc reserves would only last another 22 years, 10 years later there were still 22 years worth of reserves remaining. On the other hand, tin, for example, had a static reserve index of 26 years in 1962 and in 1971 it had dropped to 18 years.

Table 15.2 summarizes the latest data on global reserves, production, and static index for a number of minerals.

In an effort to determine which of these minerals are going to be the most critical a comparison is made in the table between the predicted cumulative demand for each mineral from 1971 to 2000 and the 1971 reserves. If the ratio of these numbers exceeds 1 then the mineral will be

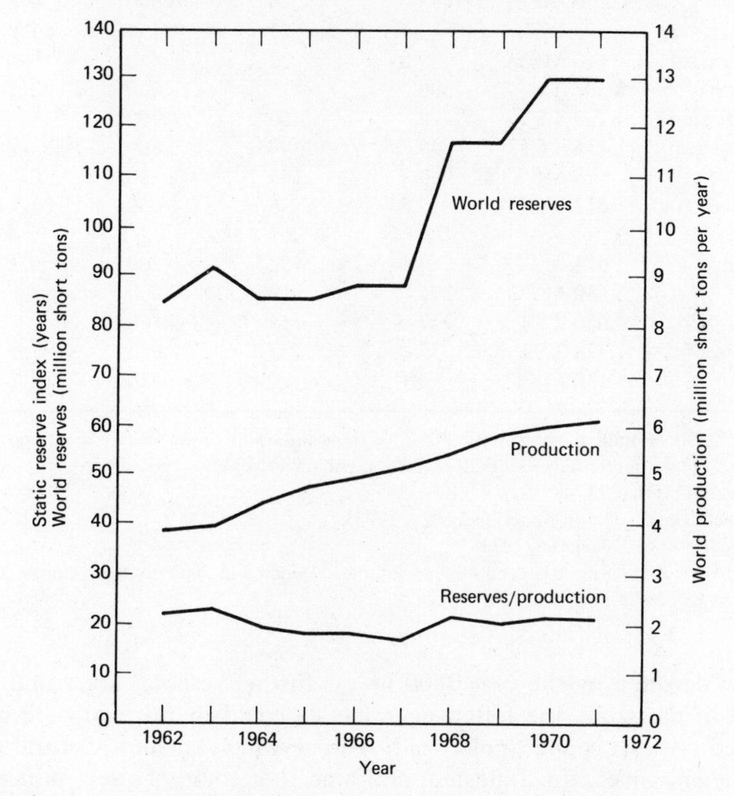

FIGURE 15.4 World reserves, production, and static reserve index for zinc, 1962–1971 (USDI 1973a).

TABLE 15.2 World Mineral Reserves and Production Data, With A Comparison between Cumulative Demand to 2000 and Present Reserves

Mineral	Global[c] reserves, 1971	Production,[c] 1971	Static index 1971, years	Cumulative[d] demand 1971–2000	Cumulative demand through 2000/1971 reserves
Aluminum[a]	3,280 MST	12.5	252	1,400	0.4
Chromium[a]	132 MST	2.0	66	100	0.8
Coal[a]	9.6×10^6 MST	3,300	2,900	0.1×10^6	0.01
Cobalt	2,730 TST	26	105	900	0.3
Copper	340 MST	6.7	51	450	1.4
Gold	1,000 MTO	46.5	21	1,600	1.6
Iron	96,700 MST	455[b]	210	18,000	0.2
Lead	103 MST	3.7	27	150	1.5
Manganese[a]	646 MST	9.0	72	440	0.7
Mercury	5,300 TF	306	17	13,000	2.5
Molybdenum	4,700 TST	78	60	4,900	1.04
Natural gas	1,755 TCF	40	44	2,600	1.5
Nickel	75 MST	0.7	106	30	0.4
Petroleum	632 Bbbl	17.6	36	1,050	1.7
Platinum group	624 MTO	4.1	152	200	0.3
Silver	5,480 MTO	295	19	17,000	3.1
Tin	4,680 TST	257	18	9,800	2.1
Tungsten[a]	1,378 TST	37	37	2,050	1.5
Zinc	130 MST	6.1	21	280	2.1

Notes: MST = million short tons; TST = thousand short tons; MTO = million troy ounces; TF = thousand flasks; TCF = trillion cubic feet; Bbbl = billion barrels.

[a]1970 data USDI 1972.

[b]National Commission on Materials Policy, 1973a.

[c]*Commodity Data Summaries*, 1973.

[d]Straight line calculation based on median estimated demand in 2000 given in *Mineral Facts and Problems*, 1970.

totally depleted by the year 2000 unless further reserves are found. As listed in the table, the following minerals could in this sense be considered critical: copper, gold, lead, mercury, molybdenum, natural gas, petroleum, silver, tin, tungsten, and zinc. For some of these minerals, namely mercury, silver, tin, and zinc, present reserves must be more than doubled by 2000 to keep up with demand.

It should be apparent by now that it is very difficult to make precise estimates of future resource availability. However, certain important conclusions can still be drawn. For a number of vital minerals, the reserves are so marginal that new reserves must be added at least as fast as they are consumed or else the mineral will be exhausted in just a few years time. With consumption increasing exponentially then, new reserves must increase exponentially, and we are familiar with the explosive implications associated with such growth. As time goes on and it becomes increasingly difficult to find new ores, the grade of ore drops and the costs both to the economy and the environment multiply.

15.3 *Environmental Effects*

The amount of environmental damage that we are willing to accept in exchange for these minerals, may become the limiting constraint on their exploitation. The mining and processing of lower grade ores causes the disruption of larger areas of land and results in greater volumes of mine tailings to be disposed of. The already severe problems of surface mining

The Bingham Canyon Mine of Kennecott's Utah Copper Division is the largest producer of copper in the United States. Almost 2½ miles wide and about ½ mile deep, it is the largest excavation in North America. (Courtesy Utah Copper Division, Kennecott Copper Corporation.)

and acid wastes would be compounded. Moreover, the smelting of ores such as zinc and copper is a significant source of air pollution, and the cost of control equipment may restrict the expansion of these facilities in this country (it is not unlikely that these environmentally "dirty" operations may be exported to other countries in the future).

As the grade of ore decreases, the energy required to extract and process the mineral increases rapidly. Already (1968) the primary metals industry consumes 5.3×10^{15} Btu per year, or about 9% of total U.S. energy (Office of Science and Technology 1972). The problems of resource depletion and energy are thus closely related. In fact it is often asserted that if energy were cheap enough we could mine very low-grade ores or even plain rock to acquire all the minerals we would ever need. Cheap energy, however, is only one aspect of the increased cost of mining marginal ores. The capital and labor costs and the need to dispose of large volumes of unusable waste, and the costs of the chemical operations involved would all tend to maintain the expense even if energy were cheap.

Note too that environmental damage is not restricted to the land. Already around 17% of the world's petroleum is produced offshore (McKelvey and Wang, 1969) and that percentage is going to increase rapidly. The danger of oil well leaks and blowouts as well as the routine oil spills that are a regular part of the transporting and processing of petroleum are bound to increase. Moreover, other minerals may soon be mined from the seabed, most especially manganese, with as yet unknown environmental effects.

15.4 *Imports and Balance of Payments Problems*

The United States has been a net importer of minerals only since about 1930, and by 1950 the value of imports exceeded exports by only $2 billion. By 1970, that deficit was close to $8 billion and there is reason to believe it will soar to about $73 billion annually by the year 2000. These estimates assume prices will not increase over 1970 levels, so they are quite likely to be conservative.

Figure 15.5 shows the developing deficits for energy, metals, and nonfuel-nonmetallic minerals for the United States. With regard to the fuels, there is already considerable concern over the huge outflow of dollars that will be required to purchase Middle East oil, but what is not so often appreciated is that even more dollars may be spent buying the metals that will be needed if our economy continues to grow as expected.

The major contributors to the growing deficit for minerals are petroleum, aluminum, iron, and copper, which together accounted for 85% of the deficit in 1970 and will account for about 74% in the year 2000 (Figure

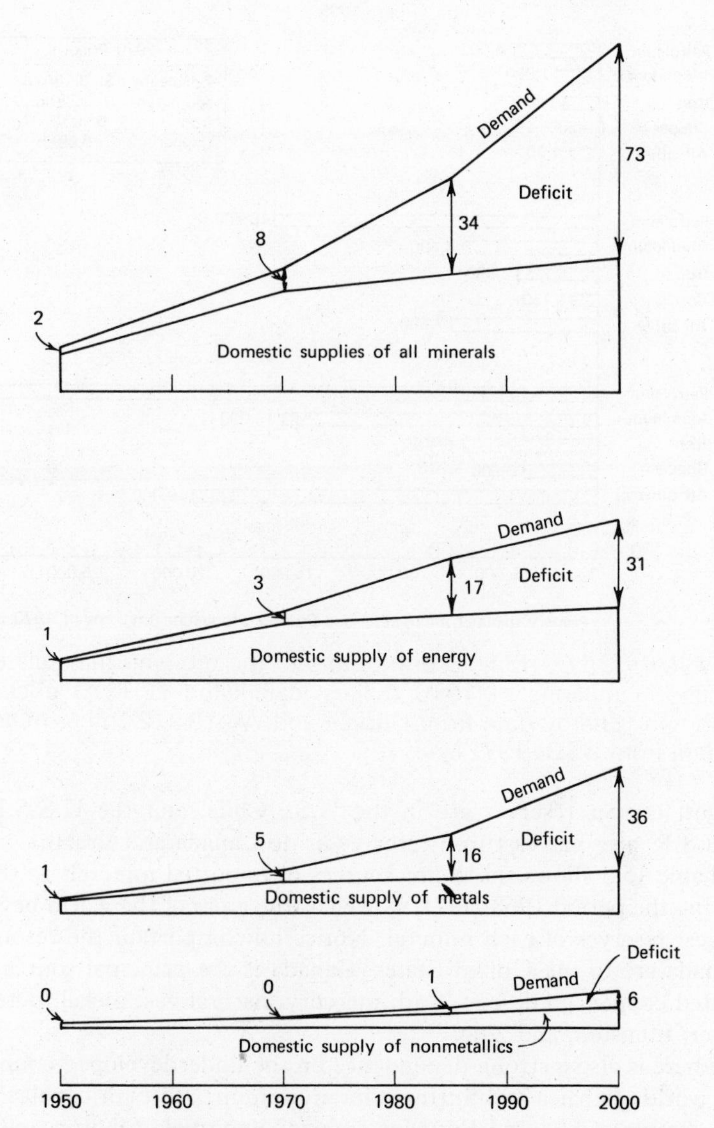

FIGURE 15.5 Developing deficits for primary minerals (U.S.) in billions of dollars (with 1985 and 2000 at 1970 prices) (energy data from Dupree and West 1972; other minerals from USDI 1972).

15.6). Major reserves of petroleum are in the Middle East—most especially Saudi Arabia, Kuwait, and Iran. The U.S.S.R. also has large petroleum reserves. The majority of the world's aluminum deposits are in Guinea (Africa), Australia, and Surinam (South America). The

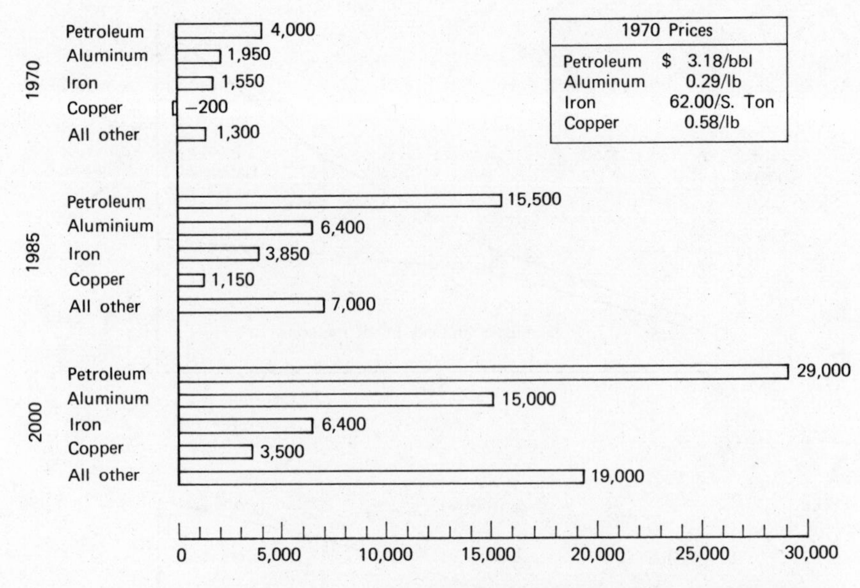

FIGURE 15.6 Major contributors to the growing minerals deficits (in millions of 1970 dollars, calculated at 1970 prices as shown (Energy data from Dupree and West 1972; other minerals data from USDI 1972).

largest copper reserves are in the U.S., Chile, and the U.S.S.R. The U.S.S.R. also has vast iron reserves as do Canada and Brazil.

Table 15.3 shows the major sources of imported minerals to the U.S. during the period 1968–1971, as well as the areas of the world having the largest reserves of each mineral. Notice how important the resources of Canada are to the United States. Canada is the principal source of imported copper, gold, iron, lead, mercury, natural gas, nickel, petroleum, silver, tungsten, and zinc.

There is also a strong dependence on the underdeveloped countries of the world: Jamaica and Surinam for aluminum; Zaire for cobalt; Turkey for chrome; Chile and Peru for copper; Venezuela for iron and petroleum; Brazil and Gabon for manganese; Malaysia and Thailand for tin; and Peru for tungsten.

It can be argued that the flow of dollars into these underdeveloped nations will provide the capital needed for their development and indeed to some extent this is true. Unfortunately, all too frequently the main effect is simply to increase the income of the already rich elite of the country, further separating them from the great majority of the people.

Nearly all the aluminum demand in the U.S. is met by imports. This is a bauxite refinery in Surinam. (Courtesy UNFAO.)

Moreover, in many cases, the profits that are obtained from resource exploitation are kept by the multinational corporations involved and very little remains in the country.

It is however, the vast sums of money that are predicted to flow into the oil-producing countries of the Persian Gulf that is the major concern of the developed countries of the world. Akins (1973) for example, estimates the 1975 annual revenue to these countries will be over $16 billion and by 1980 it could be $58 billion. The cumulative income to the Arab countries from 1973 through 1980 will probably exceed $210 billion. What they do with this money will be a matter of crucial importance to the world economic system.

15.5 *Control of Resources*

It is becoming apparent to many of the resource-producing, less developed countries that they must act to maintain control over their own resources. One approach has been to form international alliances among exporting nations such as the Organization of Petroleum Exporting Countries

TABLE 15.3 Sources of U.S. Imports and Major Reserves

Mineral and percent of primary demand from imports, 1970	Source of imports 1968–1971 and (%)[a]	Countries having[b] highest reserves in 1970, %
Aluminum[c] (85%)	Jamaica (38) Surinam (21) Australia (13)	Guinea (34) Australia (34) Surinam (11)
Chromium (100%)	South Africa (34) U.S.S.R. (29) Turkey (19) Philippines (15)	South Africa (76) Rhodesia (15) U.S.S.R. (6)
Coal (0%)	Export	U.S.S.R. (63) U.S.A. (16) China (12)
Cobalt (~100%)	Zaire (42) Belgium-Luxembourg (33) Norway (9)	New Caledonia-Australia (27) Zaire (27) Zambia (14)
Copper (~0%)	Chile (25) Canada (25) Peru (23)	U.S.A. (24) Chile (17) U.S.S.R. (10)
Gold (72%)	Canada (53) Switzerland (18) Burma (9)	South Africa (60) U.S.A. (8) Communist Countries (20)
Iron (30%)	Canada (53) Venezuela (29) Liberia (6)	U.S.S.R. (43) Brazil (11) Canada (14)
Lead (31%)	Canada (26) Peru (21) Australia (20)	U.S.A. (35) Canada (14) Communist Countries (16)
Manganese (95%)	Brazil (34) Gabon (32) South Africa (7)	Gabon (16) South Africa (8) Communist Countries (39)
Mercury (50%)	Canada (54) Spain (20) Mexico (15)	Spain (50) Yugoslavia (9) U.S.A. (7)
Molybdenum (0%)	Export	U.S.A. (51) U.S.S.R.-China (24) Chile (14)

[a]USDI 1973a.
[b]USDI 1972.
[c]Includes aluminum metal, bauxite, and alumina (from Source *b*).

TABLE 15.3 Continued

Mineral and percent of primary demand from imports, 1970	Source of imports 1968–1971 and (%)[a]	Countries having[b] highest reserves in 1970, %
Natural gas (1%)	Canada (95)	U.S.S.R. (29)
	Mexico (5)	U.S.A. (20)
		Algeria (10)
Nickel (90%)	Canada (89)	Cuba (24)
		New Caledonia (23)
		Canada (13)
Petroleum (23%)	Canada (44)	Saudi Arabia (21)
	Venezuela (20)	U.S.S.R. (13)
	Indonesia (6)	Iran (12)
Platinum (98%)	U.K. (47)	South Africa (47)
	U.S.S.R. (30)	U.S.S.R. (47)
	Canada (5)	Canada (4)
Silver (38%)	Canada (60)	U.S.A. (24)
	Peru (16)	Mexico (13)
	Mexico (9)	Communist Countries (36)
Tin (~100%)	Malaysia (66)	Thailand (34)
	Thailand (26)	Malaysia (14)
	Bolivia (7)	Indonesia (13)
Tungsten (50%)	Canada (71)	China (73)
	Peru (18)	U.S.A. (7)
Zinc (59%)	Canada (55)	Canada (26)
	Mexico (15)	U.S.A. (23)
		Communist Countries (16)

(OPEC). Formed in 1960, OPEC currently comprises 11 nations* which produce more than half of the Free World's oil needs, and control the major portion of its reserves. Other groups have been formed in copper and tin.

The importance and power of such cartels can only increase as the rest of the world competes for the resources. However, as prices are driven upward by competition among the industrialized nations, the probability that the resource-poor underdeveloped countries will ever be able to acquire the resources necessary for their development diminishes.

*Kuwait, Saudi Arabia, Iraq, Abu Dhabi, Qatar, Libya, Algeria, Iran, Venezuela, Indonesia, and Nigeria.

Another step which some exporting nations have found necessary to keep profits from leaving their countries along with the resources, is nationalization of the resource-producing industries. The nationalization of the copper mines in Chile is an example of great importance to the United States. Moreover, exporting nations are wisely recognizing the benefits to be derived by processing their minerals into more valuable forms before export. By shipping steel instead of iron ore, alumina instead of bauxite, fuel oil instead of crude, the value of the resource increases greatly. Of course, while this justly benefits the exporting country, it increases the cost to the importing nations.

Competition for resources is not necessarily restricted to the economic sphere, especially when the resources are of vital importance. When highly industrialized nations are dependent on foreign resources, there is always the danger that military action may be taken to ensure access to those resources. Fortunately, the risk of a global confrontation in the immediate future is somewhat lessened by the high degree of mineral self-sufficiency that presently characterizes the Soviet Union and China (see Choucri, 1972, for example).

It is rather more likely that conflicts will arise between the major powers and the exporting countries themselves. It has been speculated, for example, that U.S. involvement in Southeast Asia may have been, at least in part, based on the desire to control the resources of the area (Weisberg 1971). As Table 15.3 indicates, the U.S. is totally dependent on foreign sources of tin, and the majority of world tin reserves are located in Thailand, Malaysia, and Indonesia. Moreover, there is evidence that the continental shelf off of Southeast Asia which already produces over a million barrels a day, may have extensive petroleum deposits (Offshore 1970).

These resources did not escape the attention of Henry Cabot Lodge, then U.S. Ambassador to Vietnam, when he stated in 1965:

> He who holds or has the influence in Vietnam can effect the future of . . . Thailand and Burma with their huge rice supplies to the west, and Malaysia and Indonesia with their rubber, oil, tin, to the south (Weisberg 1971).

15.6 *Unequal Consumption of Resources*

It is difficult to comprehend the incredible differences in per capita consumption of resources that exist around the world. For example, the per capita consumption of steel in the United States is 37 times that of the average underdeveloped country; for aluminum it is 85 times as great;

and for total energy the factor is over 30. Consumption in the U.S. even exceeds that of the average developed country in the world by roughly a factor of 2, as shown in Table 15.4. Approximately 30% of the world's population accounts for about 90% of the annual consumption of mineral resources.

There are several ways to accentuate the gross disparity in resource consumption between the U.S. and the rest of the world. From one point of view, we can ask how long the world's reserves would last if everyone consumed at the same rate as Americans. Or, turning it around a bit, suppose the present quantity of resources consumed each year were to be equally distributed among all the peoples of the world. By what percentage would the per capita American consumption have to decrease? The first approach is illustrated in Figure 15.7, where the number of years required to deplete the world's mineral reserves has been calculated assuming the U.S. per capita demand for primary material were to apply to everyone in the world.* Of the 19 important minerals, 9 would be exhausted within 10 years, and all but three—coal, iron, and aluminum—would be gone by the year 2000. While it is true that reserves are dynamic,

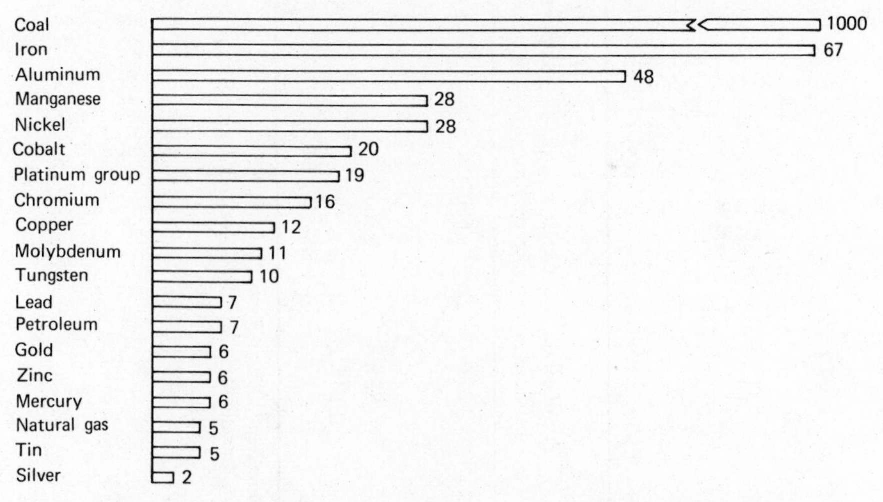

FIGURE 15.7 Number of years to deplete world mineral reserves if world population consumed at U.S. rate (based on 1970 data—see footnote p. 379).

*Calculated from: years = 0.058 × (World Reserves)/U.S. Primary Demand. [World Reserves are taken from Table 15.2 and U.S. Primary demand is taken from National Commission on Materials Policy (1972) for all but silver, gold, platinum group, and molybdenum which are from USDI (1972)].

TABLE 15.4 Annual Per Capita Consumptions of Major Minerals, 1966–1969[a]

Region	Population, millions	Crude steel, kg/person	Refined copper, kg/person	Primary aluminum, kg/person	Zinc, kg/person	Total energy, kg/person coal equivalent
U.S.A.	203	660	9.3	16.9	6.0	10,000
All other developed[b] countries	885	366	4.8	7.3	3.3	3,500
Underdeveloped countries	2,445	18	0.15	0.2	0.17	330

[a]Source: Data after Malenbaum (1972).
[b]Europe, U.S.S.R., Japan, Canada, Australia, New Zealand, Israel, South Africa.

it must be realized that to avoid depletion, the entire quantity of world reserves would perpetually have to be replaced every 2 years for silver; every 5 years for tin and natural gas; every 6 years for zinc, gold, and mercury; every 7 years for petroleum and lead, etc. This evidence suggests the startling conclusion that it will never be possible for the whole world to enjoy the same standard of living that we in the U.S. do today!

The second question asks what the effect would be on the American rate of consumption if it were suddenly decreed that present world consumption were to be evenly distributed among all people. As Table 15.5 indicates, U.S. consumption would have to decrease by anywhere from 61 to 90%, depending on the mineral, with an average decrease of almost 80%. At the same time, the 70% of the world's population that live in

TABLE 15.5 Decrease in U.S. Consumption if World Consumption Were Uniformly Distributed

Resource	U.S. consumption as % of world total, 1970[a]	Decrease in U.S. consumption if resource distributed equally, %[b]
Aluminum	35	83
Chromium	26	78
Coal	16	64
Cobalt	28	80
Copper	27	79
Gold	20	71
Iron	18	68
Lead	37	85
Manganese	15	61
Mercury	22	74
Molybdenum	33	83
Natural gas	58	90
Nickel	27	79
Petroleum	32	82
Platinum group	36	84
Silver[c]	33	73
Tin	33	83
Tungsten	20	71
Zinc	24	76

[a]USDI (1972), includes secondary consumption.
[b]Calculated from: 1 − (5.8%/U.S.%) where U.S.% is column 2.
[c]Excluding communist countries.

underdeveloped countries would be able to increase their consumption by about 700%. There would be no problem convincing over two-thirds of the world that they should be allowed to multiply their consumption but there is little chance of the developed countries voluntarily decreasing theirs.

As Falk states (1972): "A firm conclusion takes shape here: the affluence of the advanced countries is interdependent with the poverty of the poor ones." The highly unstable nature of the present world system is largely based on the uneven distribution of resource consumption, and, as Falk warns:

> It should be realized that domestic revolutions have arisen out of the perception that human misery is not a necessary aspect of human life, but merely a consequence of the fact that economic and political power is being wielded for the sake of the few, at the expense of the many.

15.7 *Reclamation and Recycling*

Since resource constraints preclude the possibility of raising everyone to anything close to the same level of consumption that is already practiced in the most advanced countries of the world, and since present disparities are concomitant with the suffering of the majority of mankind, it is imperative for the developed countries to reverse their present pattern of ever-increasing consumption.

There are at least three independent approaches that need to be taken: Reduce the unnecessary, wasteful portions of demand; design products to last longer and be more easily repaired; and increase the reclamation and recycling of materials.

There are few who would question that a great deal of the consumption in the United States could be eliminated without causing any great reduction in our standard of living. A trip to the local municipal dump provides convincing evidence. The total amount of urban solid waste in the U.S. is huge, having been estimated at about 7 pounds per person per day in 1971, or about 260 million tons per year total (Anderson 1972). Table 15.6 shows the average composition of this waste, and as can be seen, it is largely paper of all kinds, with a relatively small percentage of metals (8%).

Urban solid waste is becoming a difficult problem partly because it is considered in terms of disposal rather than as a resource. As a disposal problem it is serious. Oftentimes in the past it was simply burned in the open at the dump site which of course causes air pollution. Increasingly

TABLE 15.6 Average Composition of Municipal Refuse[a]

Component	Percent by weight
Paper, all kinds	42
Rubbish (wood, grass, rags, plastics, etc.)	22
Food wastes	12
Metals	8
Glass and ceramics	6
Ashes	10

[a]Source: Gough 1970.

it is being buried but sizable areas of land are required—land which is difficult to obtain near the city where the wastes are generated. Incineration of the wastes can yield about 10 million Btu per ton (one-third the heat value of bituminous coal), which can be used to generate electricity. The Union Electric Company in St. Louis, for example, burns 300 tons of municipal refuse per day to generate 12.5 MW of electrical power (EPA 1972). Recall, too, that it is possible to convert organic wastes directly into oil or natural gas (Section 14.5).

While utilizing solid wastes for their energy content would reduce the primary demand for energy resources, it is the possibility of recycling materials that could have the major effect on our consumption of minerals. Consider Figure 15.8 in which the total demand for resources is shown to be met partly by reclamation and partly by virgin materials. By increasing the amount of reclamation, the same total demand can be met

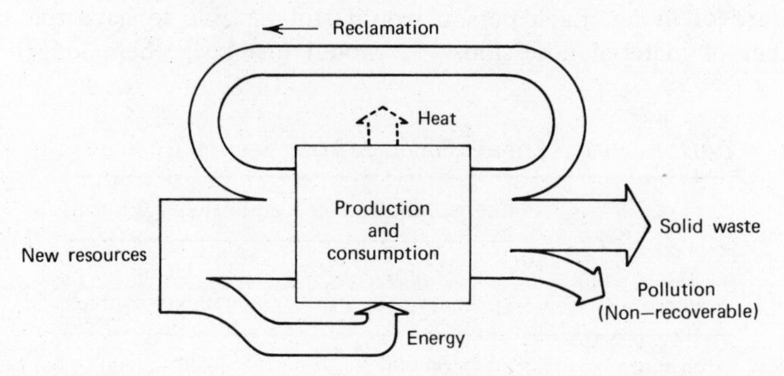

FIGURE 15.8 The production-consumption system. Increasing reclamation and product lifetime decreases the demand for new resources.

with a decrease in the consumption of new materials. To facilitate such reclamation, products would need to be redesigned to simplify the sorting and recycling.

Moreover, by increasing recycling, the total amount of energy required to maintain the system would be decreased since, in general, less energy is required to reprocess scrap than would be consumed producing virgin materials. Table 15.7 indicates the energy savings which result when aluminum, steel, and paper are produced from recycled materials. Through recycling, around 97% of the energy required to produce a ton of aluminum can be saved; for steel the saving is over 50%; and for paper it is nearly 70%. About 2% of the total U.S. energy demand could be saved if all the available aluminum, steel, and paper were to be recycled (National Commission on Materials Policy 1973b).

Figure 15.9 which shows the percentage of 1971 U.S. consumption met by recovered old scrap, indicates that, while sizable reclamation exists for some resources, there are still significant gains to be had by increased recycling. For example, only 4% of the production of aluminum was from recycled scrap, in spite of the enormous reduction in energy that can be gained by recycling. Around 15% of aluminum production is for cans, which means that aluminum reclamation could be increased at least 300% just by returning all throw-away aluminum containers.

The third approach to reducing the primary demand for minerals is to reduce the throughput in the system by designing products which have longer useful lifetimes. Furthermore, if the products are designed to be repaired rather than thrown away, new jobs can be created to ease the economic impact of these proposals. Notice that these two independent mechanisms—increased product lifetime, and increased recycling—would reduce the drain on resources without necessarily reducing the standard of living. Each person would still be able to have the same number of material posessions—he would just keep them longer and

TABLE 15.7 Energy Savings from Recycling[a]

	Energy required (coal equivalent kWh/ton)	
Material	Using primary sources	Using secondary sources
Aluminum	64,000	1,300–2,000
Steel	14,000	6,500
Paper	5,000	1,500

[a]Source: National Commission on Materials Policy 1973b.

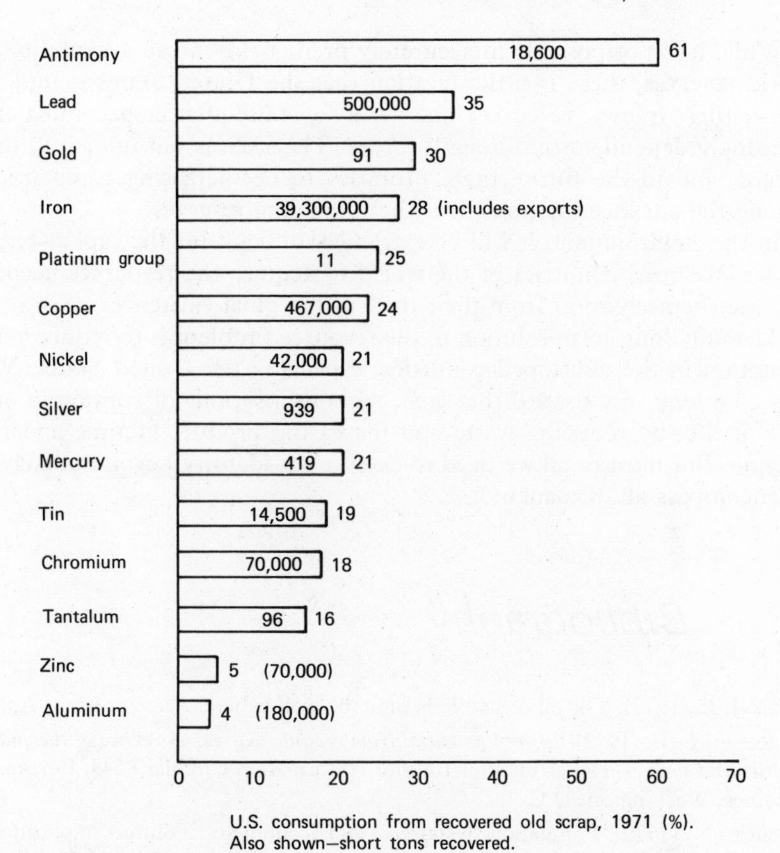

FIGURE 15.9 Old scrap recovery in the U.S. in 1971 as percent of consumption. (data, USDI 1972).

their composition would contain a greater percentage of reclaimed material.

It is not difficult to conceive of a system of economic incentives to encourage a reduction in demand for raw materials. For example, taxes could be imposed on products in inverse proportion to the estimated life of the product, the percent of reclaimed material that it contains, and the percent of the product that could be subsequently recycled.

15.8 *Conclusions*

In spite of the uncertainties and conflicting opinions, the problem of resource depletion is emerging as one of the most complex, far-reaching, and potentially dangerous issues that mankind must face.

While it is not possible to accurately predict the future availability of world reserves, there is little question that the United States is rapidly exhausting its own resources and is, as a consequence, becoming increasingly dependent on foreign sources. The U.S. is not unique in this regard, and in the future there promises to be increasing competition among the advanced nations over the remaining reserves.

In this environment, it will be extremely difficult for the capital-poor, underdeveloped countries of the world to acquire the resources needed to raise themselves up from their present marginal existence.

The only long-term solution to the resource problem is to reduce consumption in the developed countries, especially the United States. We can go a long way toward that goal, without a serious disruption in our way of life, by reducing waste and increasing product lifetime and recycling. But most of all we need to extend our identity beyond ourselves to encompass all of mankind.

Bibliography

Akins, J. E. (1973). The oil crisis: This time the wolf is here. *Foreign Affairs* April.

Anderson, L. L. (1972). *Energy potential from organic wastes: A review of the quantities and sources.* Department of Interior, Information Circular 8549, Bureau of Mines, Washington, D.C.

Choucri, N. (1972). Population, resources, and technology: Political implications of the environmental crisis. In D. A. Kay and E. B. Skolnikoff, eds., *World ecocrisis, international organizations in response.* Madison, Wis.: Univ. Wisconsin Press, 1972.

Committee on Resources and Man (1969). *Resources and Man,* National Academy of Sciences National Research Council. San Francisco: W. H. Freeman.

Dupree, W. G., Jr., and West, J. A. (1972). *Energy through the year 2000.* U.S. Department of the Interior. Washington, D.C. Dec.

Falk, R. A. (1972). *This endangered planet, prospects and proposals for human survival.* New York: Vintage.

Gough, W. C. (1970). *Why fusion?* U.S. Atomic Energy Commission, Div. of Research, Washington, D.C., June.

Hirst, E. and Healy, T. (1973). Electric energy requirements for environmental protection. Presented at Conference on Energy: Demand, Conservation, and Institutional Problems. MIT, Cambridge. Feb. 12–14.

Malenbaum, W. (1973). *Materials Requirements in the United States and Abroad in the Year 2000.* For National Commission on Materials Policy, Washington, D.C., March.

McKelvey, V. E., and Wang, F. F. H. (1969). *World subsea mineral resources* USDI, Geological Survey, Misc. Geol. Investigations, Map I-632. Washington, D.C.

Meadows, D. H., et al. (1972). *The limits to growth.* New York: Universe.

National Commission on Materials Policy (1972). *Towards a National Materials Policy.* Interim Report. Washington, D.C. April.

National Commission on Materials Policy (1973a). *Toward a National Materials Policy World Perspective.* 2d Interim Report. Washington, D.C. Jan.

National Commission on Materials Policy (1973b). *Material Needs and the Environment Today and Tomorrow.* Final Report, Washington, D.C., June.

Office of Science and Technology (1972). *Patterns of Energy Consumption in the United States.* Executive Office of the President. Washington, D.C. Jan.

"Offshore Oceania to be Active Through 1980," *World Oil*, Aug. 15, 1970.

Park, C. F. (1968). *Affluence in jeopardy, minerals and the political economy.* San Francisco: Freeman, Cooper and Co.

U.S. Bureau of the Cnesus (1972). *Raw Materials in the United States Economy: 1900–1969,* Working Paper 35, Washington, D.C., July.

U.S. Department of the Interior (USDI) (1970). Bureau of Mines. *Mineral Facts and Problems, 1970.* Washington, D.C.

USDI (1972). *First Annual Report of the Secretary of the Interior Under the Mining and Minerals Policy Act of 1970.* March.

USDI (1973a). Bureau of Mines. *Commodity Data Summary.* Washington, D.C. Jan.

USDI (1973b). Geological Survey. *United States Mineral Resources.* Professional Paper 820. Washington, D.C.

U.S. Environmental Protection Agency (EPA) (1972). *Energy Recovery from Waste.* Washington, D.C.

Weisberg, B. (1971). Offshore oil boom. *The Nation* March 8.

Questions

1. Discuss the implications of a worsening balance of payments deficit.
2. Write a short paper describing the political situation in any of the underdeveloped countries that the U.S. is dependent upon for raw materials.
3. The following table roughly summarizes the population and resource consumption figures for the U.S., the rest of the developed countries, and the UDCs.

Region	Percent of world's population	Percent of world's resource consumption
U.S.A.	6	30
Other DCs	24	60
UDCs	70	10

(a) Show that the average American consumes roughly 35 times the resources as the average person in a UDC.

(b) If resources are divided evenly, show U.S. per capita consumption would decrease by 80% and UDC per capita consumption would increase 700%.

(c) If everyone consumed at the same rate as the average person in the U.S., show world consumption would increase by a factor of 5.

4. Discuss the relationship between environmental quality and resource depletion.

5. Write a scenario for the future based on a shortage of resources.

6. Make a list of specific proposals for reducing resource consumption in the U.S.

7. Suppose the solid waste from a city of 1 million people is used to fuel a 40% efficient steam-electric power plant. Assuming half of the waste is combustible with a heating value of 10 million Btu/ton, and assuming each person's total solid waste is 7 pounds per day, calculate the size of the power plant.

   ans. 84 MWe

Appendix Population Derivations

To derive equation (2-3) we start by writing the rate of change of population is proportional to the population size

$$\frac{dN}{dt} = rN \qquad \text{where } N = \text{population size}$$
$$r = \text{growth rate}$$

This can be integrated to give

$$\int_{N_0}^{N} \frac{dN}{N} = \int_0^t r\, dt$$

or

$$\ln \frac{N}{N_0} = rt$$

which simplifies to:

$$N = N_0\, e^{rt} \qquad \text{equation (2-3)}$$

where N_0 is the initial population size and N is the size after some time t. The constant e is the base of the natural logarithms and is equal to about 2.718. Usually raising e to some power is most easily done on the slide rule or by using log tables. However, a simple way to make the calculation if no tables or slide rules are available, is to use the exponential expansion

$$e^x = 1 + x + \frac{x^2}{2!} + \frac{x^3}{3!} + \cdots$$

which can be terminated after only a few terms if x is small compared to 1 (for the expansion to work, x must be smaller than 1).

EXAMPLE A1.1 The population of the United States in 1972 was 209 million and growing at the rate, r, of 1.0% per year. If that rate holds constant, what would the population be in 2000?

Solution: Using equation (2-3) with $N_0 = 209 \times 10^6$, $r = 0.01$, and $t = 28$, gives

$$N = 209 \times 10^6 \times e^{0.01 \times 28} = 209 \times 10^6\, e^{0.28}$$

To evaluate the exponential we can either look it up giving $e^{0.28} = 1.323$, or we can try a few terms of the expansion for an approximation

$$e^{0.28} \cong 1 + 0.28 + \frac{(0.28)^2}{2} = 1.319$$

which is not bad. So, the estimate for the population in 2000 becomes

$$N = 209 \times 10^6 \times 1.323 = \textit{276 million}$$

Doubling Time

From equation (2-3)

$$N = N_0\, e^{rt}$$

we can calculate the time required for N to double by setting $t = T_d$, the doubling time, and $N = 2N_0$ for double the initial population:

$$2N_0 = N_0\, e^{rT_d}$$

or simply

$$2 = e^{rT_d}$$

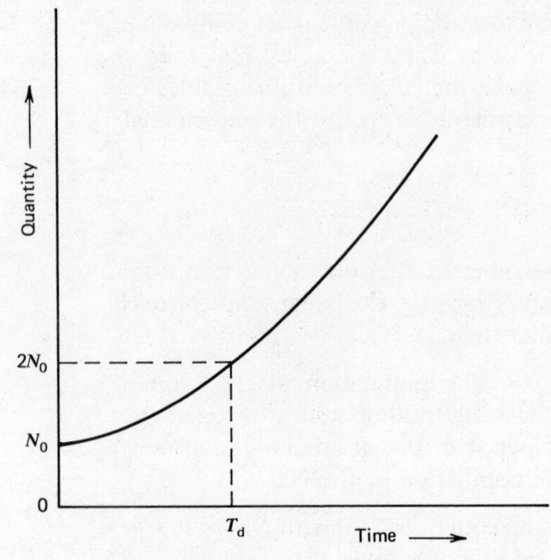

FIGURE A.1 Doubing time.

Taking the log of both sides gives

$$\log_e 2 = rT_d$$

or

$$T_d = \frac{\log_e 2}{r} \cong \frac{0.7}{r}$$

If the growth rate r is expressed as percent per year, then the doubling time in years is given by this handy relation:

$$T_d \cong \frac{70}{r} \text{ years} \tag{2-4}$$

For example if $r = 1\%$, the doubling time is 70 years. If $r = 2\%$, the doubling time is 35 years. If you earn 5% on your bank savings, then it will take 14 years to double your savings.

Index